Discovering Pluto

DALE P. CRUIKSHANK AND
WILLIAM SHEEHAN

Discovering Pluto

Exploration at the Edge of the Solar System

**THE UNIVERSITY OF
ARIZONA PRESS**

TUCSON

The University of Arizona Press
www.uapress.arizona.edu

ISBN-13: 978-0-8165-3431-9 (cloth)
ISBN-13: 978-0-8165-3938-3 (paper)

Cover design by Jennifer Baer
Cover art: photos PIA 20473 and PIA 09113 courtesy of NASA/Johns Hopkins University Applied
Physics Laboratory/Southwest Research Institute

Library of Congress Cataloging-in-Publication Data
Names: Cruikshank, Dale P., author. | Sheehan, William, 1954– author.
Title: Discovering Pluto : exploration at the edge of the solar system / Dale P. Cruikshank and
    William Sheehan.
Description: Tucson : The University of Arizona Press, 2018. | Includes bibliographical references
    and index.
Identifiers: LCCN 2017047625 | ISBN 9780816534319 (cloth : alk. paper)
Subjects: LCSH: Pluto (Dwarf planet)
Classification: LCC QB701 .C78 2018 | DDC 523.49/22—dc23 LC record available at https://lccn.loc
    .gov/2017047625

Printed in the United States of America
♾ This paper meets the requirements of ANSI/NISO Z39.48–1992 (Permanence of Paper).

This book is dedicated to the memory of
Paul S. Cruikshank
(November 9, 1964–July 13, 2015)

Contents

Preface

WITH THE NEW HORIZONS flyby of Pluto in July 2015, the Grand Tour of the Solar System, which began in the 1960s with the Mariner missions and continued with the Voyager flybys of the giant planets, was finally complete. Pluto—originally, when the Grand Tour was first conceived, the ninth major planet of the Solar System—together with its anomalously large moon Charon and the four small satellites of their system, has been surveyed at close range. This book presents the long and fascinating history of searches of trans-Neptunian space by Percival Lowell and his colleagues that led to Clyde Tombaugh's discovery of what seemed at the time to have been Lowell's Planet X, and the subsequent gradual accumulation of facts about this small icy body on the inner edge of the Kuiper Belt. The book is brought up to date with a summary of the scientific results so far about this surprising world and a look forward to the next steps planned in the exploration of the Kuiper Belt.

Pluto, now classified as a dwarf planet but regarded as the ninth major planet of the Solar System from the time of its discovery by Tombaugh at Lowell Observatory in 1930 until it was reclassified by the International Astronomical Union (IAU) in 2006, remains—however classified, and despite its smallish size and distance—one of the most charismatic bodies in the Solar

System. Except perhaps for Mars, no other planet seems of greater interest to the public.

Lowell first identified it as the incarnation of Planet X, a seven-Earth-mass major planet posited because of supposed residuals (unexplained differences between the theory and observed positions) of the planet Uranus. Astronomers hailed its discovery as a triumph of celestial mechanics similar to John Couch Adams and Urbain Le Verrier's calculations leading to the discovery of Neptune in 1846—and it arguably saved the famously broke Lowell Observatory from extinction. From the first, however, there were doubts about its mass and planetary status. By 1976, Dale Cruikshank, Carl Pilcher, and David Morrison discovered methane ice on Pluto. From then on it was clear that Pluto was much too small—only about 2,000 km in diameter, or smaller than Earth's Moon—to have exerted any significant perturbations on the giant planets Uranus or Neptune. The discovery of the largest of its five satellites, Charon, by James Christy at the U.S. Naval Observatory two years later furnished a precise value of the mass.

Since those 1970s discoveries, it was learned that Pluto is a particularly outsized member of the expanse of icy bodies on the edge of the outer Solar System called the Kuiper Belt. Rather than being, as once thought, a major planet in a relatively simple collection of planets, Pluto is representative of a part of the Solar System that contains some two thousand known objects. It is a mini-world or, to use the IAU's nomenclature, a dwarf planet. It and its Kuiper Belt congeners (as well as bodies belonging to the still more distant Oort cloud) point to a Solar System in a state that was once vastly different from what it is now.

This book falls naturally into two parts. Though both authors contributed to all the chapters, the first eight were primarily the work of coauthor Sheehan, while the remaining twelve were chiefly written by coauthor Cruikshank. Cruikshank was a graduate student of Gerard Kuiper and later a planetary scientist specializing in the investigation of the icy outer worlds of the Solar System, and he has had an insider view of the recent science as an investigator on the New Horizons mission.

Acknowledgments

THE AUTHORS WISH to thank the many individuals and institutions that have generously supported their research. In his research of archival materials and for discussions of the events leading to the discovery of Neptune, coauthor Sheehan thanks Brian Sheen and Norma Foster, who helped provide understanding of and guidance to important sites and resources related to John Couch Adams's Cornish roots, and the Truro Public Records Office, which provided some significant Adams documents. There were many discussions and, in some cases, Neptune-related travels with Neptunians Craig Waff, Nick Kollerstrom, Dennis Rawlins, David Dewhirst, Sir Patrick Moore, Richard Baum, Robert W. Smith, Roger Hutchins, and Allan Chapman. Adam Perkins of Cambridge University Library kindly provided assistance in accessing George Biddell Airy's Neptune file and James Challis's papers. The librarian of St. John's College, Cambridge, provided access to the papers of John Adams and his student Ralph Sampson, and Peter Hingley of the Royal Astronomical Society was helpful in more ways than we can name.

James Lequeux, the biographer of Urbain Le Verrier, provided much assistance with the French side of the Neptune discovery saga, while Françoise Launay was always ready at a moment's notice to provide advice and support. Jacques Laskar and Gregory Laughlin provided useful insights and advice into perturbation theory, while Professor Kenneth Young and H. M. Lai are thanked for correspondence related to their 1990 paper, with C. C. Lam, on the perturbations of Uranus by Neptune.

Special Collections at the Howard-Tilton Memorial Library of Tulane University and Howard Plotkin provided guidance and documentation on W. H. Pickering's planet searches. On Percival Lowell and Planet X, Robert Burnham Jr., who spent countless hours on the blink comparator at Lowell Observatory comparing plates taken in the Planet X search with later plates in the search for stars of large proper motions, was an early inspiration; William Graves Hoyt's pioneering researches provided a high standard for those who came afterward, while Arthur A. Hoag generously supported coauthor Sheehan's initial reconnaissance in the Lowell Observatory Archives. The staff of the Houghton Library of Harvard University was a pleasure to work with; Lowell historians, librarians, and preservationists Antoinette Beiser, Lauren Amundsen, Kevin Schindler, and Michael and Karen Kitt were not only professionally helpful but became close friends.

The Lowell Observatory is to be warmly thanked for its generosity in providing photographs from its collection. William Lowell Putnam, the sole trustee of the Lowell Observatory, offered insights into Percival Lowell, his family, and his father, Roger Lowell Putnam, who was sole trustee at the time Pluto was discovered. He also commented in detail on a draft of chapter 5. Clyde W. Tombaugh provided many personal recollections and written comments that have found their way into chapter 6, while Bradford A. Smith, his long-time collaborator at New Mexico State University, furnished many insights into Tombaugh's work at Lowell and after. As principal investigator of the Voyager imaging team, Brad also offered some of his recollections about that epic voyage of outer–Solar System exploration. James W. Christy and his wife, Charlene, were interviewed at length in their Flagstaff, Arizona, home, and gave an interview, full of little-appreciated details, about the discovery of Charon.

Coauthor Sheehan wishes to express special appreciation to his wife, Deborah, sons Brendan and Ryan, animal companions Brady and Ruby, and his many professional colleagues for their patience and understanding during the long periods of reading, traveling for research, immersion in archives, and absorption in writing that a book like this requires.

Both authors thank our copyeditor, Timothy Clifford, for significantly improving the text, and Kevin Kilburn for catching a number of errors in the first printing.

For coauthor Cruikshank, a fair fraction of the story has personal connections, starting in 1958, when he was a summer assistant at Yerkes Observatory. His experience as a darkroom worker making prints for a photographic lunar atlas began an association with G. P. Kuiper that extended for ten years through graduate studies at the University of Arizona and beyond. During that time, planetary astronomy flourished, largely under Kuiper's influence, and began a merger with geology, atmospheric science, and chemistry to become planetary science, a curriculum now taught in many universities.

In the context of the story of Pluto, Cruikshank's personal connections extend through the observational work with telescopes on the ground and chemical apparatus in the laboratory to puzzle out the surface compositions of Pluto and its significant "twin," Triton. Those telescopic studies of Pluto began in 1976 in collaboration with David Morrison and Carl Pilcher, and continued for two decades with Tobias Owen, Thomas Geballe, Catherine de Bergh, Bernard Schmitt, Ted Roush, Robert H. Brown, and Roger Clark. Laboratory work connected with Pluto and many other Solar System bodies benefitted by collaborations with Bishun Khare, Hiroshi Imanaka, Scott Sandford, and Christopher Materese, while Vladimir Krasnopolsky kindly included Cruikshank in his theoretical studies of the atmospheric and surface chemistry of Triton and Pluto.

As a member of the science team on the New Horizons mission to explore the Pluto-Charon system and bodies in the Kuiper Belt beyond, Cruikshank has had the opportunity to work with many other brilliant planetary scientists, including but certainly not limited to Will Grundy, Bernard Schmitt, Alan Stern, Richard Binzel, Leslie Young, Marc Buie, and John Spencer.

In addition to the individuals mentioned above, Cruikshank also acknowledges decades-long friendships and collegial associations with W. K. Hartmann, the late V. I. Moroz, Alan Binder, Cristina Dalle Ore, Max P. Bernstein, and L. J. Allamandola, all of whom have contributed to his continuing education in astronomy, chemistry, and all facets of planetary science. In terms of the story told in this book, G. P. Kuiper's biographer Derek W. G. Sears provided valuable insight into the relationship between Kuiper and

H. C. Urey. Astronomers Seth B. Nicholson and Philip S. Riggs, both of whom have passed on, gave early encouragement to an eager high school student in a small Iowa town.

Cruikshank's wife and life partner, Dr. Yvonne Pendleton, deserves special thanks for sharing her widely recognized expertise on galactic astronomy and organic chemistry of the interstellar medium, and for working side by side at the telescope as we extended our view of the Solar System beyond the major planets and deep into the Kuiper Belt, while pondering together the great questions of human existence and intellect.

Dale P. Cruikshank, Sunnyvale, California
William Sheehan, Flagstaff, Arizona
February 28, 2017

Discovering Pluto

1

Twenty-Seven Years and Three Billion Miles

IT WAS A HOT AND MUGGY early July day in Maryland at the Applied Physics Laboratory in Laurel, less than an hour's drive from downtown Washington, DC. About sixty scientists, spacecraft engineers, celestial navigators, and NASA managers drifted into the main auditorium in Building 200 for the daily briefing. Most were carrying cups of coffee and buzzing about the latest data coming to Earth from New Horizons, the spacecraft closing in on Pluto and its moons at 50,000 kph (31,000 mph). Everyone was eager to get the latest news on the view of Pluto and Charon transmitted back from the spacecraft 3 billion miles away and to get updates on the onboard instruments and other spacecraft systems. The pictures and other information about to be presented had taken nearly four and a half hours, carried by radio waves at the speed of light, to travel across the Solar System to the antennas listening on Earth.

The idea of a spacecraft journey to Pluto began to gain momentum in about 1988, in the wake of the decision to send the Voyager 2 spacecraft, then already some 4 billion km (2.5 billion miles) from Earth and closing in on Neptune on a path that would encounter its largest moon, Triton, but make it impossible to go on to Pluto. Thus, the hugely successful Voyager mission, which had been an offshoot of an even grander concept of a Grand Tour that would have visited all five planets beyond Mars, would veer away from the last planet on the itinerary. The New Horizons mission to Pluto and beyond into yet deeper regions of the Solar System emerged in the early 2000s with a growing awareness of the scientific importance of the smaller planetary bodies beyond Neptune. Launched from Cape Canaveral in January 2006, the piano-size spacecraft had reached its destination nine and a half years later, and on July 14, 2015, just hours away from the flyby, the teams of scientists and engineers—and the public worldwide—were poised to see what was being found.

At 8 a.m. sharp, Hal Weaver, the New Horizons project scientist, called the group to order and issued the daily reminder that everything to be discussed in that meeting was confidential and to be kept entirely within the team—only NASA headquarters could release information about the mission to the public.

By that time the New Horizons mission principal investigator, Alan Stern, was in his chair at the desk adjacent to the podium, and the science team members had assembled in their assigned areas in the front of the room. Stern was from the Southwest Research Institute, while most of the engineers present were affiliated with the Applied Physics Lab, the research organization of Johns Hopkins University that assembled and tested the spacecraft, and now managed its operations. NASA, the National Aeronautics and Space Administration, is the federal agency in which all American planetary missions have been conceived, funded, and managed. NASA has centers of expertise located around the United States. Stern recruited the science team from specialists at the centers, several universities, and private research organizations—they were individuals who had studied the planets for their entire careers, and most had already made important contributions to the understanding of Pluto and its moons as well as the other objects in the outer reaches of the Solar System.

The scientists were loosely organized into four subteams: composition (COMP), particles and fields (PPS), atmospheres (ATM), and geology and geophysics (GGI). All these groups had been assembled at the Applied Physics Lab for several days, analyzing the early data stream trickling in from the spacecraft as it approached the target. The bluish glow of a laptop computer illuminated each person's face as they checked their e-mail and other information that the teams had worked on overnight.

The room was softly buzzing with conversation. It quickly quieted as the first presentation came up on the screens at the front of the room and the appropriate science team presenter began speaking at the podium. Team scientists who had been involved with New Horizons since the earliest study phases more than twenty years ago were there, a bit grizzled by the passage of time and many nail-biting moments along the way. They had brought a few young graduate students and postdoctoral fellows with them to participate and to apply to the incoming data new analytical tools that had passed by a few (perhaps most!) original team members who are now senior citizens. Including students and young scientists in this way initiates them to the complexities and the rewards of exploring planets with spacecraft. Some mid-age scientists had joined the team along the way, particularly in the last two years. The science team was originally kept lean since launch to maintain mission cost levels and because the need for scientists was much reduced during the long flight from Earth to Pluto.

Each day of these plenary sessions, the first up was GGI team leader Jeff Moore from NASA's Ames Research Center or his deputy Bill McKinnon of Washington University. They showed the latest images and reported on the latest attempts to refine the diameters of Pluto and the moons. The most current maps made by Paul Schenk at the Lunar and Planetary Institute from the images obtained thus far were flashed on the screens. Everyone was astonished at the daily improvement in the image detail as the spacecraft raced ahead. The GGI report was always greeted with enthusiasm and antici-pation. The team had viewed dozens of planetary landscapes from across the Solar System from several other space missions over the years. Were Pluto and its moon Charon different, or mostly replays of what had already been seen? The short answer: some of both.

Adding to the excitement and anticipation, Will Grundy, a planetary sci-entist at Lowell Observatory and leader of the roughly dozen scientists in the COMP team, took the podium to show the results of the most recent work on the composition of Pluto and Charon. The COMP team was eager beyond words to extract from the earliest New Horizons data any information they contain about the ices that cover the surfaces of Pluto and Charon. So far, what they knew from telescopic studies from Earth had been confirmed by the spacecraft, but they were on the threshold of far more detail, which would tell them about the chemical makeup of the planetary bodies farther out in the Solar System than had ever been scrutinized so closely. This in turn would inform them about the composition of the solar nebula, the swirling mix of dust, gas, and ice from which the Sun, all the planets and their moons, and myriad small bodies coalesced some 4.6 billion years ago—that singular event that was the origin of the Solar System.

Next was the report from the ATM team by Randy Gladstone of the Southwest Research Institute. The ATM team were avidly waiting for direct information on Pluto's exceedingly thin atmosphere—they knew it was there, but they wanted some very specific information that could only be gleaned from the combination of three critical experiments on the spacecraft. Those observations were yet to come, just after closest approach on July 14, but in the meantime the search for new atmospheric gases using the preliminary data in hand occupied the team.

The space environment around Pluto and Charon involves the interaction of streams of atomic particles from the Sun (the solar wind) with Pluto's thin atmosphere, which was discovered from Earth-based observations. These interactions were anticipated, but largely unknown because Pluto's great distance from the Sun means that solar activity takes several weeks to propagate that far out. By the time a pulse or a stream of solar wind reaches Pluto, it has disconnected from its source 3 billion miles away, thus complicating the interaction and confounding expectations. Ralph McNutt from the Applied Physics Laboratory alternated with Fran Bagenal of the University of Colorado in making the report for the PPS team. Among other things, they were trying to detect molecules or fragments of molecules at exceedingly low concentrations at this great distance from the Sun. No new molecules that day.

Stern and the spacecraft engineers reported that New Horizons was performing perfectly, was exactly on track and on time for its close flybys of Pluto and Charon, and that the data were coming in to the antennas of NASA's Deep Space Network as planned. It could have hardly been more exciting!

The plenary session finished after an hour of presentations and status updates. Each science team recessed to its own work room in Building 200. Showing the plastic, electronically coded security badge issued to each team member, they made their way through the turnstiles, one person at a time. The next morning's session would bring new data, new insights, and surely some new discoveries—and they would be another 1.2 million km (750,000 miles) closer to Pluto by then.

This journey across the Solar System to the July 2015 encounter with Pluto and its moons can be said to have begun not just in 1988, but a full century ago. In 1915, Percival Lowell, well-known for his controversial view on the habitability of Mars as he saw it from his private observatory in Arizona, where Pluto was eventually found in 1930, published his *Memoir on a Trans-Neptunian Planet*. As we discuss in detail in later chapters, Lowell's motivation for the prediction and search for Planet X, as he called it, really began with the discoveries of Uranus in 1781 and Neptune in 1846, the first planets beyond the five (Mercury, Venus, Mars, Jupiter, and Saturn) known since classical times. Thus, while much of this book is about what we can call Pluto's first century, the story begins even earlier, as new planets are found and the stage is set for the epic voyage of the New Horizons spacecraft.

2

A New Planet

Then felt I like some watcher of the skies
When a new planet swims into his ken
JOHN KEATS, "ON FIRST LOOKING INTO
CHAPMAN'S HOMER" (1816)

ANY BOOK CONCERNED with Pluto and the other bodies that inhabit the outer Solar System must inevitably come to terms with the date March 13. It is connected to at least three significant events. First, it was the date in 1781 when William Herschel, a professional musician, amateur astronomer, and telescope builder in Bath, England, discovered a new planet, now called Uranus. It was also the date in 1855 on which Percival Lowell, the individual who provided the impetus to search for another planet, was born. Finally, on that date in 1930, astronomers at Lowell's observatory announced that such a planet did, apparently, exist—its incarnation being the curious object that between then and 2006 was identified as the ninth major planet of the Solar System.

The Greatest Amateur Astronomer of All Time

We begin with Herschel's discovery, which opened the vast frontier of the outer Solar System beyond Saturn, the outermost of the planets known since antiquity.

An immigrant to England from Hanover, where he was born into a musical family in 1738, Friedrich Wilhelm Herschel—who naturalized his name as William Herschel—at first found varied employment as a musician. He was engaged for some years mostly in the north of England as a copyist, performer, and composer. After several years of disappointment and struggle, he achieved a breakthrough as an organist at the Octagon Chapel in the brilliant resort town of Bath. The spa with its healthful water drew the wealthy and boasted a dazzling social season from May to September that included balls and dinners and the largest entertainment market outside London. Herschel's new position assured him a good income and allowed his brother, Alexander, and sister, Caroline, to come over from Hanover and join him in England. Skilled not only in the organ but also in the violin, oboe, and harpsichord, he gave music lessons, mostly to affluent young ladies, since even "well-bred" women of the time were discouraged from wasting their time on anything other than doing useful household chores, such as needlework, or in learning French or music, the only "decorative accomplishments" thought proper for the female sex.[1] He had so many pupils that he had to schedule more than twenty concerts a year to showcase their talents. A composer rather than a performer by inclination, he produced more than a score of symphonies, three oboe concertos, and numerous chamber and voice pieces, many only now being rediscovered. Though sometimes dismissed as sounding like "Mozart gone stale,"[2] at least one musicologist has found his work "arresting, innovative . . . the product of a superb analytic mind driven by an obsession for order and coherence."[3] In any case, Herschel's professional activities— and especially the grind of teaching—began to wear on him. By 1770, he was feeling that music was "an intolerable waste of time." Thereupon he began to devote more of his leisure time to telescope building and astronomy.

Soon after his younger sister, Caroline, joined him, hoping to escape the life of a household drudge slaving for their mother in Hanover by establishing a professional career as a soprano in England, William began immersing himself in reading books on mathematical subjects. As he afterward recalled,

> The great run of business, far from lessening my attachment to study, increased it, so that many times after a fatiguing day of 14 to 16 hrs spent in my vocation, I retired at night with the greatest avidity to unbend the mind (if it may be so called) with a few propositions in Maclaurin's *Fluxions* or other books of that sort.

Figure 2.1 William Herschel, 1785. Engraving by Lemuel Francis Abbott after the famous oil painting now in the National Gallery, London. Courtesy of William Sheehan Collection.

Among other mathematical subjects, Optics and Astronomy came in turn, and when I read of the many charming discoveries that had been made by means of the Telescope, I was so delighted with the subject that I wished to see the heavens and Planets with my eyes thro' one of those instruments.[4]

Though good commercially made telescopes, such as small achromatic refractors, were available, they were very expensive. Herschel decided, as many an amateur has done since, to build his own. He settled on the reflector type and became a diligent grinder of mirrors. He was confident that he could work out how to do this on his own. He had little choice: At the time, only a few other amateur telescope-makers existed in England, and how-to information was hard to find. He taught himself by trial and error. The only abrasives available at the time were sand and emery. The mirrors were made of speculum metal, a rather hard, brittle bronze alloy that required a mirror-maker to be part metallurgist and foundry man as well as optician. Herschel had to select and refine the often-crude raw materials, then melt suitable disks using a small iron furnace and special molds. He eventually discovered that the best molds were those formed from pounded horse dung. The castings had to be slowly cooled so they could be properly annealed. Because speculum metal is much harder than glass, working a mirror was physically demanding and very fatiguing. Moreover, since speculum metal tarnishes in the open air, and is especially susceptible in a damp climate like that of England, any mirror made had to be periodically repolished and refigured. Herschel's mirrors often deteriorated noticeably within a month and became unusable after three or four. To continue to use his telescopes without interruption, Herschel eventually devised the expedient of producing two or three specula for each of his telescopes, deploying them relay-style while he was reworking the tarnished members.

His brother, Alexander, who possessed great mechanical skill, assisted, while Caroline—who continued to hope for more of William's time so she could progress with her singing—looked on in dismay. "To my sorrow," she recalled, "I saw almost every room turned into a workshop; a cabinet-maker making a tube and stands of all descriptions in a handsomely furnished drawing room; Alex putting up a huge turning machine . . . in a bedroom for turning patterns, grinding glasses and turning eye-pieces &c."[5] Before long, Caroline too was pressed into service. She found herself "much hindered in my practice [of music] by my help being continually wanted." By January 1774, Herschel had managed to complete a 4.5-inch mirror, which he mounted in a square wooden tube. On March 1, using a magnifying power of 40×, he recorded his first astronomical observations: of Saturn, which appeared "like two slender arms," and of the "lucid spot in Orion's sword."[6]

Herschel regularly changed his residence as he sought to balance the need for space to pursue his hobbies with the convenience of being near the concert halls where he had professional engagements. Thus he moved from 7 New King Street in Bath (a street he was later to make famous) to a house near the outskirts of town. It offered sheds and stables, which could be turned into workshops, and had a field where he could set up his telescopes. He then moved back to New King Street, into no. 19 (for the first time), then to 27 Rivers Street near the Circus and New Assembly Rooms, where he had his concerts. The Bath actor John Bernard recalled in his *Retrospections of the Stage* that he came to see Herschel at Rivers Street twice a week for singing lessons and found lodgings that "resembled an astronomer's much more than a musician's, being heaped up with globes, maps, telescopes, reflectors, &c., under which his piano was hid, and the violoncello, like a discarded favorite, skulked away in a corner."[7]

By 1778, Herschel's records of his telescope making were becoming lengthy. The most successful instrument produced up to then employed a 6.2-inch diameter, 7-ft focal length mirror, and a mahogany tube conveniently mounted on a folding wooden frame. Two years later, he was using the "seven foot," as he always called it, in a "review of the heavens." By then he had already finished a "first review" using the 4.5-inch telescope described above, in which he had examined all stars magnitude 4 or brighter in the search for new double stars.[8] (On his very first night, he had discovered that Polaris is double.) Now, in a "second review," he extended his survey to all the "Flamsteed stars," stars as dim as magnitude 8 that had been cataloged by the Reverend John Flamsteed of the Royal Greenwich Observatory.

He was thus engaged in March 1781 when he moved back to the house at 19 New King Street. (It is now the William Herschel Museum.) Probably he wished he had never left, since it had the advantage of a good workshop at ground level where he could cast and grind his telescope mirrors and boasted a south-facing garden, "behind the house and beyond its walls all open as far as the River Avon,"[9] where he could set up his telescopes and observe the sky.

In sleuthing for new double stars, Herschel used magnifications that were regarded by his contemporaries as absurdly high. In his survey of the Flamsteed stars, he routinely used a magnification of 227×, and on occasion even higher powers, as much as 460×, 932×, or even more whenever he needed to examine closely a double-star suspect. The superb quality of his optics allowed it. Other astronomers, working with inferior instruments (even those at the Royal Greenwich Observatory were not up to scratch with Herschel's), were skeptical about the use of such high powers. His closest

Figure 2.2 A replica in the William Herschel Museum at 19 New King Street, Bath, of the 7-ft reflector used by Herschel to discover Uranus. Image by William Sheehan, 2009.

friend in Bath, Dr. William Watson Jr., pointed out that opticians generally were happy if they could sell a telescope magnifying 60× or 100×. Watson warned Herschel about using such high magnifications: "Some will think you fit for Bedlam."[10]

The Discovery

On the night never to be forgotten, March 13, 1781, Herschel was alone as Caroline, still tying things up at the Rivers Street residence, had not yet joined him. He set up the 7-ft telescope in the south-facing garden and placed his eye at the ocular. Under the date March 12, he had recorded in his observation log that he looked at Mars at 5:45 a.m. "Mars," he wrote, "seems to be all over bright but the air is so frosty & undulating that it is possible there may be spots without my being able to distinguish them." At 5:53 he added, "I am pretty sure there is no spot on Mars." He next looked at Saturn and noted, "The shadow of Saturn lays [sic] at the left upon the ring."

The next entries were made under the date March 13:

> Pollux is followed by 3 small stars at about 2' and 3'.
> Mars as usual.
> In the quartile near Zeta Tauri the lowest of two is a curious either nebulous
>   star or perhaps a comet.
> A small star follows the Comet at 2/3rds of the field's distance.

These notes show that Herschel observed Pollux and Mars in the morning, then in the evening—presumably after returning from giving concerts—he recommenced his double-star search, with the ocular magnifying 227×. He had reached in his survey the interesting region between M1 (the Crab Nebula) in Taurus and M35, an open cluster in Gemini. Though the telescopes used by astronomers elsewhere would have shown the stars as mere glary "patines of bright gold,"[11] Herschel's showed single stars as "round as a button."[12] He now came across one that was not only "round as a button" but a noticeably larger button than the others bestrewing the field.

Herschel had just observed a planet "swim into his ken," but he did not yet realize it. All he could say for certain was that it was not a fixed star. It had a disk, which became even clearer as he boosted the magnifying power to 460× and 932×. Cautiously, he noted it down simply as a "curious either nebulous star or perhaps a comet."

Measuring its position with his micrometer, the object's motion was apparent to Herschel within a few hours. He continued to track it on successive nights. Whatever it was, his observations showed it to be captured in the Sun's gravitational embrace. It was not a nebula. It was surely a comet, then.

Discovered by an amateur, the "comet" was now turned over to the superintendence of the professional astronomers, including the Reverend Thomas Hornsby at the Radcliffe Observatory, Oxford, and the Reverend Nevil Maskelyne, the Astronomer Royal at Greenwich. Oddly, it refused to develop the usual characteristics of a comet, remaining "perfectly sharp upon the edges, and extremely well defined, without the least appearance of any beard or tail."[13]

It was imperative to compute an orbit. At the time, all methods for doing so were extremely difficult and laborious, including the graphical method that Isaac Newton had introduced in the *Principia* (book 3, proposition 41) and that Edmond Halley had famously adapted in his demonstration that the comets of 1531, 1607, and 1682 had been the single object now bearing Halley's name. A new method, entirely analytical (i.e., based on solving a series of equations) had just been introduced by the French mathematical astronomer Pierre-Simon Laplace. Laplace read his paper to the Paris Academy of Sciences just eight days after Herschel's discovery,[14] and Herschel's "comet" had the distinction of being the first to which Laplace's method—soon to become standard—was applied. However, for some reason Laplace did not communicate the orbital elements he had derived until long afterward (January 1783), so the first published orbits—by Pierre Méchain in Paris; the Ragusan Jesuit Roger Boscovich, then living in Paris; and Anders Lexell, a professor of mathematics in St. Petersburg but then in London where Hornsby may have called his attention to the problem—were computed using older methods.

In the meantime, Maskelyne was beginning to suspect that what Herschel had found was not a comet at all. Writing on April 4, 1781, he noted that the

Pietro Simone Laplace

Figure 2.3 Pierre-Simon Laplace. Color engraving by Luigi Rados after François Joseph Bosio. Courtesy of Wikipedia Commons.

comet's motion was "very different from any comet I ever read any description of or saw" and added on April 23 that it was "as likely to be a regular planet moving in an orbit very nearly circular around the sun as a comet moving in a very eccentric ellipsis."[15]

A more definitive result was obtained on May 8, when Jean-Baptiste Gaspard Bouchart de Saron demonstrated to the Paris Academy of Sciences that the object was very remote—at least 14 astronomical units (AU) from the Sun (1 AU is equals the distance of Earth to the Sun: 150 million km or 93 million miles). If so, then Maskelyne's suspicions seemed to be justified; it did not appear to be a comet at all, but a planet. Boscovich, having juggled several different orbits, concluded that the best approximation was a "circular orbit of large radius." After combining an observation made by Herschel on March 17 with one made by Maskelyne on May 11, he settled on a circular orbit with a radius of 18.93 AU and a rate of motion of only about 4°20' per year.

In late May, the planet slipped away into the glare of the Sun. When it emerged again in July, further positions were obtained by Augustin Darquier. Johann Elert Bode of the Berlin Observatory analyzed them and announced in the authoritative *Berliner Astronomisches Jahrbuch* that Herschel's "comet" was, without a shadow of doubt, a hitherto unknown planet of the Solar System 19 AU from the Sun. The thing was settled. The size of the Solar System had doubled.

Herschel seems to have been one of the last to appreciate the importance of his own discovery. As late as November 1781, when he was awarded the prestigious Copley Medal of the Royal Society, he still seemed reluctant to acknowledge the object's significance. When at last its planetary status could no longer be denied, Herschel was finally in with both feet. He was tasked, as its discoverer, with proposing a name, and he chose "Georgium Sidus," the Star of George, for the then king of Great Britain, George III. Herschel himself loyally continued to use this name until his death in 1822 (as did the *British Nautical Almanac* until 1850). In the end, however, the name that stuck was that proposed by Johann Elert Bode—"Uranus," for the father of Saturn in Greek mythology.

Herschel's discovery had shattered the starry status quo and introduced a new view of the limits of the Solar System. The British historian of astronomy Richard Baum has written,

> Herschel had broken with immemorial tradition. His sighting of a body, identified as a massive world beyond the icy rim of the known, a world so remote as to be barely visible to the unaided eye . . . sent a breath of fresh cosmic air coursing through the dusty corridors of thought.[16]

For Herschel, the planet was the harbinger of a sea change in his life circumstances, earning him a pension from George III, which allowed him to
give up music and devote all his time to astronomy, as he had long wished
to do. He was hardly a man to rest on his laurels. Instead he was, in Shakespeare's words,

Like one that stands upon a promontory,
And spies a far-off shore where he would tread,
Wishing his foot were equal to his eye.
(Henry VI, *part 3, scene 2*)

His promontory was the eyepiece of his splendid telescopes, the far-off
shore the shoals of stars and nebulae among which he would soon be tread-
ing, far beyond the region of the new planet. As the King's "personal astronomer," first at Datchet and then at Slough, near Windsor Castle, his principal
preoccupation was thenceforth to build larger and larger telescopes capable
of greater light grasp and what he called "space-penetrating power."[17] With
these telescopes he would try to make progress on the gigantic problem of
the "construction of the heavens."[18]

Despite the demands of this monumental work, he remained a regular
observer of the planets, especially Saturn, always his favorite, and Uranus,
in which he took a paternal interest. In 1787, using a 20-ft reflector with
an 18.7-inch (47-cm) mirror, he discovered two satellites of Uranus, now
known as Titania and Oberon; two years later he added two to Saturn's
retinue, Mimas and Enceladus. (The names were chosen long afterward
by Herschel's son, Sir John Herschel.) He suspected that he had found two
more satellites of Uranus; one, recorded on April 17, 1801, was almost surely
Umbriel. It was not seen again for fifty years, when the Liverpool amateur
William Lassell reclaimed it. "William Herschel's telescopes were so far
ahead of his time," notes astronomical analyst Dennis Rawlins, "that it took
half a century for Umbriel to be re-discovered—during which time the
satellite had traveled so far that to track it back to April 17, 1801 required
many post-1851 years of Umbriel-orbit-refinement before Herschel's sighting could be confirmed."[19] The discovery of Umbriel is still generally credited to Lassell, for despite Herschel's apparent priority, Lassell's name had

been on the books for so long that he had become (in Rawlins's words) the "dislodgeable discoverer."

A Perturbing Business

At first Uranus's motion seemed to fit a nearly circular orbit around the Sun, with a period calculated as eighty-three or eighty-four years. Soon, however, the planet was observed to be getting ahead of the predicted longitude. This meant that a circular orbit wouldn't do. Instead, the orbit was elliptical, with its eccentricity variously estimated by the mathematicians as follows (where 0 is the eccentricity of a perfect circle):

Oriani	1783	0.0484200
Méchain	1783	0.0430000
Laplace	1784	0.0475870
Fixlmillner	1784	0.0461183

Were Uranus's motion due solely to a force directed constantly toward the center of the Sun and inversely proportional to the squares of its distance to the Sun, its orbit around the Sun would be a perfect ellipse, forever immobile, with neither the plane nor axis ever altered. This perfect elliptical orbit would then suffice to predict its motion for all time. Of course, the actual case is otherwise: the Sun and Uranus do not exist in isolation. All the planets perturb one another, and for the mathematical theory of Uranus (and any other planet) to remain in step with the observed motion, it must account for these perturbing effects (notably, those due to the two largest planets, Jupiter and Saturn).

Recall that the orbit of a planet, comet, or any other body moving around the Sun because of the central force of gravity is a conic section. In the case of a planet, as we noted above, this is an ellipse. To precisely define such an orbit, the following elements are introduced:

i.   the mean distance, or semimajor axis, a
ii. the eccentricity, e

iii. the longitude of the ascending node, Ω

iv. the longitude of the perihelion, *i.e.*, the point of the orbit nearest to the Sun, ω

v. the inclination of the planet in which the orbit lies to some fixed plane of reference, such as the ecliptic, i

vi. the origin of the time, t_0, where the longitude at epoch, ε, is used to specify a known point in the orbit

vii. the mean motion, n

These are depicted in figure 2.4. If the planet's motion were due only to the action of the Sun, the values of these empirically determined elements, e, a, i, Ω, ω, ε, n, would be constants. However, because the planets are gravitationally attracted not only by the Sun but by other bodies, notably the other planets, the elements are quantities that vary over time. The purpose of the branch of celestial mechanics known as perturbation theory is to determine the nature of these variations quantitatively.

To account for the perturbations of one planet on another, the first step is to calculate the forces and accelerations produced by the perturbing planet on the orbit of the planet being perturbed. Small time intervals are considered, from which first-order deviations in the perturbed orbit's elements are derived; these deviations in turn lead to a small secondary change in the forces and accelerations of the second order, which in turn lead to a further change of the third order, and so on. In this way, by a series of approximations, a result is approached ever more closely (asymptotically, in mathematical terms) that defines the changed motion of the planet over time. Because of the smallness of the eccentricities and relative inclinations of the planetary orbits, the perturbations can be developed in a series of sines and cosines of angles, increasing in proportion to time, that converges at least for periods of time that are of practical interest to astronomers. Laplace carried the calculation to the third order in the eccentricities and inclinations. Later investigators carried it still further. For instance, Urbain Le Verrier, a leading figure in later chapters, carried his tables to the seventh order by 1854. (As published in volume 1 of the *Annales de l'Observatoire Impérial de Paris*, the converging series of sines and cosines of angles occupy no fewer than fifty-seven pages. We will not reproduce them here.)

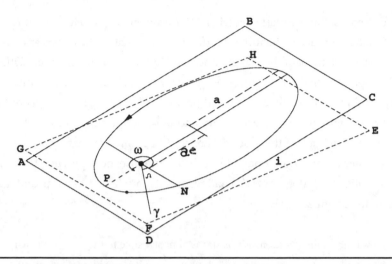

Figure 2.4 Orbital elements. ABCD = orbital plane of a planet or comet around the Sun. EFGH = plane of the ecliptic. i = inclination of the orbital plane to the ecliptic. γ = the First Point of Aries (location of the vernal equinox). ϖ = the direction of the axis with respect to the nodes. N = ascending node. Ω = longitude of ascending node. e = eccentricity, the ratio of the distance between the center of the ellipse and either of its two foci and the semimajor axis a. Diagram by William Sheehan.

In this overview of the general importance of perturbations, we cannot follow the mathematical argument further; suffice it to say, the integration of the above formulas leads to long series of terms in successively higher powers (orders) of the orbital elements multiplied by sines or cosines of angles increasing in proportion to the time.[20] These terms, once calculated, can be conveniently organized in tables that astronomers can consult to avoid repeating all the laborious calculations.

Despite the hideous nature of the bookkeeping involved, Laplace's method had reduced in principle the calculation of planet tables to the application of a straightforward algorithm. Of course, complications arose, the chief of which involved uncertainties in the elements and masses of the perturbed and perturbing planets. The most serious uncertainties involved the masses; these and other uncertainties inevitably led to corresponding uncertainties in the tables.

However, at least in the first decades after Herschel discovered Uranus, there seemed to be no reason to doubt that its motion would be sorted out in due course by the mathematicians. If a problem existed, it seemed to lie not in the province of perturbation theory but in that of the observations—the determinations of the planet's positions in the sky as determined with a telescope designed for the purpose. The very recentness of Uranus's discovery meant that it had been observed only over a very short arc of its orbit. Improved agreement between theory and observation might be expected if only the amplitude of the observed arc could be somehow extended.

Reenter at this point Johann Elert Bode. Suppose, he suggested, that one of the diligent catalogers of stars had recorded it by chance, as a star, and had failed to recognize it. He wrote,

> Now there is the question, as this star is almost visible to the naked eye, why have not astronomers discovered it long ago? . . . It is not altogether unusual for former astronomers, who furnish us with more or less complete observations of the stars of the zodiac, to overlook many 6 and 7 magnitude stars, as I often found by comparison of star catalogues with the sky. . . . It can also well be that this new star . . . was seen and entered in catalogues as a fixed star, especially as the places of the smaller stars were usually based on a single observation, and so the singular motion was not recognizable. I have therefore with this purpose compared the star catalogues of Tycho, Hevel, Flamsteed and Tobias Mayer with the sky and computed where the present star stood in the years of their observations.[21]

The idea was inspired, for in due course Bode came to a star that had been seen in 1756 by Tobias Mayer at the Göttingen Observatory. Mayer had cataloged it as star 964 in Aquarius. Bode found that there was no longer a star in this position, but tracking backward, he found that Uranus had indeed been there at the time. A second prediscovery observation, also found by Bode, turned up in 1785. (This one would prove to be the oldest of what became known as the "ancient" observations—those made prior to 1781. Working through a chilly night on Greenwich Hill on December 23, 1690, the Reverend John Flamsteed had recorded a star of the sixth magnitude. It was given the designation 34 Tau, and lay close to ω Tauri, between the Hyades and Pleiades.) Then, in 1788, three more prediscovery observations were found by the eminent French astronomer Pierre Charles Le Monnier; in this case, they happened to be his own observations. Laplace could not fail to take notice of these developments, and in his book *Exposition du Système du Monde* (1796), he expressed gratification that the new planet indeed seemed to be moving expressly in the manner demanded by perturbation theory:

> The planet Uranus, though lately discovered, offers already incontestable indications of the perturbations which it experiences from the action of Jupiter and Saturn. The laws of elliptic motion do not exactly satisfy its observed positions, and to represent them its perturbations must be considered. This theory, by a very remarkable coincidence, places it in the years 1769, 1756, and 1690, in the same points of the heavens, where Le Monnier, Mayer, and Flamsteed, had determined the positions of three stars, which cannot be found at present: this leaves no doubt of the identity of these stars with the new planet.[22]

In the same work, he added, prophetically,

> There still remain numerous discoveries to be made in our own system. The planet Uranus and its satellites, but lately known to us, leaves room to suspect the existence of other planets, hitherto unobserved.[23]

At the time, even Laplace, the greatest living practitioner of celestial mechanics, could hardly have guessed that the existence of another planet would be first discerned not through the telescope but through its perturbations on Uranus.

3

Gaps

Between Mars and Jupiter: A Planet?

THE DISCOVERY OF URANUS disrupted the eighteenth-century vision of the system of the world represented by the orrery that demonstrated the motions of the six planets known since antiquity: Mercury, Venus, Mars, Jupiter, Saturn and—following Copernicus—Earth. The addition of a new major planet not only doubled the scale of the Solar System but shook the complacency of astronomers who had believed their inventory of its members complete and its architecture fully known. For the first time in history, astronomers would set out purposefully in quest of new worlds.

Their attention was first called not to further bodies on the outer fringes of the Solar System but to a possible planet between Mars and Jupiter, whose existence was posited on the authority of a strange, almost fanciful relationship known as Bode's law or, more properly, the Titius-Bode law: Bode in 1772 having brought to general notice a relationship first noticed a few years earlier by Daniel Titius, a professor of mathematics in Wittenberg. (With this acknowledgment of Titius's contribution, we propose, nevertheless, to refer to it henceforth simply as Bode's law.)

The relationship was this. Take the numerical series 0, 3, 6, 12, 24, 48, and 96, and add 4 to each term; the resulting set of numbers, 4, 7, 10, 16, 28, 52, and 100, closely approximates the relative distances of the planets from the Sun. There is no a priori reason the scheme ought to work. Laplace scoffed at it as no more than a numbers game. Yet work it did, and work extremely well, with one glaring exception: in the position of number 28 was a gap. Expanded to include Uranus, which fit perfectly into the scheme, the table of results is as follows.

Planet	Bode Law Distance (AU)	Actual Distance (AU)
Mercury	0.4	0.39
Venus	0.7	0.72
Earth	1.0	1.00
Mars	1.6	1.52
—	2.8	—
Jupiter	5.2	5.20
Saturn	10.0	9.50
Uranus	19.6	19.20

Bode was convinced that the gap at 2.8 AU must mark the position of a missing planet. As far back as 1772 he had written, "Can one believe that the Founder of the Universe had left this space empty? Certainly not!"[1]

Though Bode himself did not actually look for the "missing" planet, inevitably a man of vision appeared on the scene who proposed to do just that. That man was Baron Franz Xaver von Zach. Born in Pest (now part of Budapest) in 1754, he studied physics there, and after a stint in the Austrian army moved to Paris (1780–83) and London (1783–86) as tutor in the house of the Saxon ambassador, Hans Moritz von Brühl. His acquaintance with men such as famous French astronomer Joseph-Jérôme Lalande, Laplace, and Herschel opened doors and led to his appointment by Ernst II, duke of Saxe-Gotha-Altenburg, to the directorship of the duke's private observatory on Seeberg hill, near his castle at Gotha, which was finished by 1791. Here, von Zach set in motion the grand project of searching among the stars of the zodiac for the trans-Martian planet.

As a first step, von Zach devoted several years to compiling a revision of the (cataloged) stars of the zodiac. By such means, he could compare the stars in the telescope against those in the catalog, and note any interloper. Eventually he realized that even this preliminary to the actual search was too large for any one observer and began to cast around for collaborators. A meeting to discuss the matter was held at Gotha in August 1798. Both Lalande and Bode were in attendance.

Nothing came of it, but von Zach organized another meeting for September 1800, this time to be held at the private observatory of Johann Hieronymus Schröter, a Hanoverian magistrate of the sleepy village of Lilienthal, near Bremen, with a lot of time on his hands and a remarkable enthusiasm for astronomy. Schröter's large reflectors were the most powerful telescopes in the world apart from the giant ones being wielded by Herschel in England. Moreover, in contrast to Herschel, who was by then chiefly mostly absorbed in his great investigations of the stellar universe, Schröter used his telescopes exclusively for studies of the physical features of the Moon and planets. With Europe being torn by the Napoleonic Wars, Lalande and other French astronomers were barred from traveling to Germany. But even without them the meeting proved a great success. Schröter's enthusiasm could be taken for granted. His assistant, Karl Ludwig Harding, and Heinrich Wilhelm Matthäus Olbers, a physician in Bremen by profession and an astronomer by predilection, were also eager to participate.

Olbers was a prodigious talent. He had, while attending to a sick fellow student as a medical student, invented a new method for computing the orbits of comets that proved to be an improvement on Laplace's. Even after establishing a thriving medical practice in Bremen in 1781, specializing in ophthalmology, he remained active in astronomy and set up a private observatory in the upper story of his house. (He was helped by his ability to get by on only four hours of sleep per day.)

Von Zach opened the meeting by presenting his ideas for searching for the missing planet between Mars and Jupiter. He suggested setting up for the purpose the first international association of astronomers, the Astronomische Gesellschaft, nicknamed the Celestial Police. Schröter agreed to serve as president, and von Zach as secretary. Twenty-four astronomers were to be invited to participate in a coordinated worldwide effort in which the constellations of the zodiac would be subdivided into search zones covering 150° of longitude and 7 or 8° of latitude north and south of the ecliptic.

(Note that throughout this book, we use the notation °, ', and " for degrees, minutes, and seconds of arc, respectively, on the sky.) Each participating member was to be responsible for scrutinizing the stars of a zone down to the ninth magnitude. In the astronomer's magnitude scale, magnitude 6 is the faintest star visible to the naked eye, and magnitude 9 is about fifteen times fainter than that.

Like any ambitious plan, it took time to implement, and before they could get underway surprising news arrived. The sought-for planet had apparently already been found—not through a systematic search but, as with Herschel's discovery of Uranus, through the diligent but otherwise-directed labors of the Sicilian monk-astronomer, Giuseppe Piazzi. (Ironically, Olbers had written to Piazzi, inviting his participation, but delayed by the slow mails of a continent at war, the letter had not yet reached him.)

Piazzi oversaw the observatory in the Norman tower of Santa Ninfa, in the royal palace in Palermo. Though his nominal patron was King Ferdinand (the fourth of that name of the Kingdom of Naples and the third of the Kingdom of Sicily), Ferdinand had no interest in astronomy. It was only through

Figure 3.1 Giuseppe Piazzi, director of the Palermo Observatory in Sicily, who with assistance from Niccolò Cacciatore discovered the first asteroid, Ceres, on January 1, 1801. Courtesy of William Sheehan Collection.

the efforts of the viceroy prince of Caramanico that the king was finally persuaded to provide funding, and with it Piazzi, on a visit to England, procured a magnificent 5-ft altazimuthal circle built by the skillful instrument-maker Jesse Ramsden. An altazimuthal circle equipped with a small telescope for pointing is used to obtain accurate positions of stars, and the Ramsden circle was believed to be superior in accuracy to any other instrument at the time. Piazzi, based at what was then the most southerly observatory in Europe, began using it in 1791 in a single-minded effort to revise the great star catalog of the French astronomer Nicolas-Louis de Lacaille. His star positions were typically accurate to within only 2 or 3' of arc. This degree of accuracy allowed Piazzi to establish that stars with proper motions—hitherto known in only a few cases—were more the rule than the exception. This breakthrough led to the discovery of the rapid motion of 61 Cygni, which became known as "Piazzi's flying star" and celebrated as the first star for which a stellar parallax was measured (by Friedrich Wilhelm Bessel in 1838).

Piazzi was pursuing his useful but routine work of revising Lacaille's star catalog on the first night of the new century, January 1, 1801. He had arrived at one of Lacaille's seven-magnitude stars in the constellation Taurus. Another fainter one nearby, of eighth magnitude, was not in the catalog. At first he assumed that Lacaille had simply missed it. Returning to it on the following nights, he found that this star was shifting its position at 4' in right ascension and 3.5' in declination per day. This meant it was not a fixed star at all—it had to have been a body belonging to the Solar System and moving in an orbit around the Sun.

So the story is usually told, with Piazzi alone when he made the discovery. However, as pointed out by asteroid historian Clifford J. Cunningham, he was not alone: Niccolò Cacciatore, his assistant over the past year, was with him.[2] His involvement on the night in question was set down by the British naval officer and traveler Basil Hall, who claimed to have heard it from Cacciatore himself. As Hall wrote in *Patchwork*, published in 1841,

> Cacciatore was Piazzi's assistant in the observatory. . . . As Piazzi was at that time engaged in making the noble catalog of the stars, which has since become so well known, he placed himself at the telescope and observed the stars as

Figure 3.2 The celebrated 5-ft altazimuthal circle built by the English instrument-maker Jesse Ramsden and acquired by Piazzi for the Palermo Observatory. It is the instrument used to discover Ceres. Courtesy of NASA.

they passed the meridian, while Cacciatore wrote down the times, and the polar distances, as they were read off by his chief.

Certain stars passed the [measuring] wires, and were recorded as usual on the 1st of January, 1801. On the next night, when the same part of the heavens came under review, several of the stars observed the evening before were again looked at and their places recorded. Of these, however, there was one which did not fit the position assigned to it on the previous night, either in right ascension or in declination.

"I think," said Piazzi to his companion, "you must, accidentally, have written down the time of that star's passage, and its distance from the pole, incorrectly."

"To this," said Cacciatore, who told me the story, "I made no reply, but took especial pains to set down the next evening's observations with great care. On the third night there again occurred a discordance, and again a remark from Piazzi that an erroneous entry had probably been made by me of the place of the star. I was rather piqued at this," said Cacciatore, "and respectfully suggested that possibly the error lay in the observation, not in the record."

"Under these circumstances, and both parties being now fully awakened as to the importance of the result, we watched for the transit of the disputed star with great anxiety on the fourth night. When lo, and behold! it was again wide of the place it had occupied in the heavens on the preceding and all the other nights on which it had been observed."

"Oh ho!" cried the delighted Piazzi, "we have found a planet while we thought we were observing a fixed star; let us watch it more attentively."[3]

Cacciatore, hardly remembered today, seems to deserve credit with Piazzi for this important discovery.

As with Uranus, the moving object was not surrounded by any cometary nebulosity. Nor for that matter, though Piazzi himself did not say so, did it show a discernible disk. It looked exactly like a "fixed star." Nevertheless, Piazzi did as Herschel had done and announced the discovery of a "comet," though privately he confided to his close friend and fellow astronomer Barnaba Oriani at the Brera Observatory in Milan that he thought it might be "better than a comet."[4] He continued to follow it until February 11, when one of his frequent illnesses (probably brought on by overwork) forced a temporary retreat from the telescope. By the time he was well enough to return to it, the object had been lost in the glare of the Sun.

Because of the slow mails, news of the discovery was delayed, and no one else could observe it. However, Piazzi himself had made enough observations to establish a rough orbit and to work out a distance: 2.8 AU. Bode seized on this to announce the fulfillment of his prediction: the trans-Martian planet evidently had been found! Von Zach agreed, and published an article, titled (in translation) "On a long supposed, now probably discovered, new major planet of our solar system between Mars and Jupiter."[5]

Before being interrupted by his illness, Piazzi had followed the new planet—if planet it was—through an arc of the sky of only 3°. At the time, there were no methods of calculating an accurate orbit from such a small arc. Though diligently searched for, including by Herschel, it proved elusive. In retrospect, searchers likely were misled by their expectation that, like Uranus, it would present a discernible disk. The new find was beginning to look irrecoverably lost, but just in the nick of time Carl Friedrich Gauss, a twenty-four-year-old mathematician of unrivaled genius from Brunswick, in the Duchy of Brunswick-Wolfenbüttel (now Lower Saxony, Germany) came to the rescue. Using his newly discovered method of least squares, Gauss improvised a way to calculate orbits from only a few observations and computed a set of ephemerides for the new planet.[6] Equipped with these, von Zach and Olbers, independently, recovered it among the stars of Leo on New Year's Eve 1801—a day short of a year after Piazzi had first chanced upon it passing through the stars of Taurus. Its actual position in the sky was only half a degree from where Gauss's calculations had put it!

Astronomers, especially in France, took to calling the new object Piazzi, while the discoverer himself proposed the name Ceres-Ferdinandea, after the Roman (and Sicilian) goddess of grain and the Bourbon King Ferdinand. ("Ferdinandea" was soon dropped. Not only had Ferdinand been a reluctant supporter of astronomy, but in France, where many of the leading astronomers lived, Bourbon kings were not exactly in fashion: one of them, Louis XVI, had been guillotined in 1793.)[7]

Astronomers first expected that Ceres—as befitted its Bode's law position between Mars and Jupiter—would prove to be a major planet. However, when Herschel and Schröter tried to measure it, they found it to be a very small body. According to Herschel, it was only 260 km across. (This was an underestimate; the currently accepted value is 950 km.) Moreover, it was not unique. While tracking Ceres from his top-story observatory in Bremen, Olbers, on March 2, 1802, came upon a second body. Once its orbit was computed, its distance was also found to be 2.8 AU; it, too, was moving in the "gap" between Mars and Jupiter. Olbers named it Pallas. It had turned up in the same square degree of the sky in which he had recovered Ceres just over a year before. Could this be a coincidence? Olbers thought not. "Could it be," he wrote to Herschel, "that Ceres and Pallas are just a pair of fragments, or portions of a once greater planet which at one time occupied its proper place between Mars and Jupiter?"[8] If so, the fragments might be left over from an exploded planet and would be expected to periodically return to the site of the explosion.

In that case, other fragments, members of a new class of objects, might be awaiting detection. Though they were generally still known simply as

Figure 3.3 Portrait of a dwarf planet. This image, obtained on March 7, 2015, by NASA's *Dawn* spacecraft from about 13,600 km (8,400 miles) above the surface of Ceres, shows the ancient, heavily cratered surface. Bright spots in a few craters are deposits of salts (evaporites) that have come from below the surface as recently as about 4 million years ago. *Dawn* also discovered patches of frozen water and regions where simple organic chemicals have been deposited. NASA image PIA 19557.

"planets," alternative names were suggested, such as "asteroids," the name Herschel proposed to the Royal Society (though he did not invent the name), and "planetoids," which Piazzi preferred. Whatever the two new objects were called, they were soon joined by two others. Schröter's assistant, Karl Harding, another charter member of the Celestial Police, discovered Juno on September 1, 1804, then Olbers added Vesta on March 29, 1807. (By a remarkable coincidence, first noted by Baltimore astronomical analyst Dennis Rawlins, Vesta turned up in the same "magical" square degree of sky where von Zach and Olbers had recovered Ceres and Olbers had discovered Pallas. The chance of this is about a billion to one!)

After the discovery of Vesta, the Celestial Police disbanded. Olbers did not yet give up, however, and now had the field to himself. Though he discovered a comet in 1815, he found no more asteroids. The following year he, too, gave up. In fact, no more objects would be discovered in the "gap" between Mars and Jupiter until 1845.

In terms of visibility, the Big Four are certainly in a class by themselves: all can be picked up in binoculars, if one knows where to look, while Vesta at times, under dark sky conditions, easily skims into naked-eye range.

Looking Outward

The "gap" between Mars and Jupiter was no longer empty; It was filled by no fewer than four small planets. The expectation of additional planetary discoveries was very much in the air. However, this time astronomers would seek to close not a gap suggested by a dubious succession of numbers but one in the accuracy with which the tables of Uranus accounted for the planet's observed motions in its orbit. The astronomers who attempted to close this gap would rely on the increasingly sophisticated methods of perturbation theory, and they, too, would find that its closure required positing a new planet.

Laplace, in the *Exposition du système du monde*, had affirmed his unshakable belief that "the law of universal gravitation . . . represents all the celestial phenomena even in their minutest details." The initial difficulty in calculating an elliptic orbit for Uranus did nothing to undermine his assurance because, within a few years of Herschel's discovery, the planet offered "already incontestable indications of the perturbations which it experiences from the action of Jupiter and Saturn. The laws of elliptic motion do not exactly satisfy its observed positions, and to represent them, its perturbations must be considered."[9] Piazzi's friend Oriani, in 1789, was the first to attempt to account for these perturbations, but it was Jean Baptiste Joseph Delambre, a pupil of Lalande, who first arrived at a result. By using the perturbation theory of Laplace and the prediscovery observations of Flamsteed and Mayer (one each) and Le Monnier (three), Delambre produced tables, published in 1790, that accounted to an accuracy of 1' or 2' of arc for every observation except those of 1690 and 1790.

The observational record was complete from 1791, but before then, going back to Flamsteed's observation of 1690, it was gappy at best. Astronomers hoped to produce a more continuous record by uncovering additional prediscovery observations from the star catalogs, and they were successful. A trickle of new observations surfaced. Thus in 1810, the German astronomer Friedrich Wilhelm Bessel found one that the Reverend James Bradley had made at the Greenwich Observatory in 1753. A few years later, Johann Karl Burkhardt, who had once worked with von Zach at the Gotha Observatory, added two more by Flamsteed from 1712 and 1715. Then, in 1818, the trickle turned briefly into a torrent because of the successful sleuthing of Alexis Bouvard, a mathematical astronomer at the Paris Observatory and the leading expert at the time on the motions of the giant planets.

Bouvard's was a classic rags-to-riches tale. He was born in 1767 in a hut at Contamines-Montjoie, between the now well-known ski resorts of Chamonix and Megève, in what was then the Duchy of Savoy (the hut still exists), and passed his early life as a shepherd, "without means or prospects." Intending to improve his lot, at eighteen he went to Paris, where he attended free lectures in mathematics, discovered a talent for the subject, and was brought to the attention of Laplace himself, who employed him as a computer. He eventually assisted the great man in performing the detailed and laborious calculations of the monumental *Mécanique Céleste*. In addition, for many years Bouvard supervised the calculation of the ephemerides of the *Annuaire du Bureau des Longitudes*. On the death of Delambre in 1822, Bouvard succeeded him as director of the Paris Observatory, remaining in the role until his death in 1843.

Figure 3.4 Alexis Bouvard. Courtesy of Wikipedia Commons.

Bouvard had published tables of Jupiter and Saturn in 1808, but their accuracy was marred by the considerable errors in the masses of the two planets. In addition, shortly after they appeared, Laplace noticed that Bouvard had made a mathematic sign error in the terms for the "great inequality" of Jupiter and Saturn. This was a problem that Laplace himself had famously resolved in 1784. Since five times the mean motion of Jupiter is a little more than twice that of Saturn, the cumulative effect of the resulting periodic disturbance produces a very large perturbation in the motions of the two planets. With an incorrect expression for the great inequality of Jupiter and Saturn, not only would the tables not accurately represent the long-term motions of Jupiter and Saturn but, because these planets are also the main perturbers of Uranus, they would undermine the accuracy of its motion as well.

In 1820, Bouvard decided to revise his flawed tables for Jupiter and Saturn using better values of the masses and correcting the erroneous expression Laplace had found, and also to calculate new tables of Uranus, of which Delambre's were still the best. As a preliminary, he decided to see if he could find any additional prediscovery observations; he suspected the richest trove might be found in the records of the late Pierre Le Monnier, who had filled no fewer than fifteen folio volumes with his observations made between 1736 and 1780.

Le Monnier's main interest was always the motions of the Moon, and he made many positional measurements of the Moon's path through the zodiac with purpose-built instruments that he set up in his observatory in the former convent of the Capuchins, on the rue St. Honoré. At the time, an exact catalog of the stars along the zodiac did not exist, so in making his measurements, Le Monnier made it his practice to systematically reobserve his stars—both the bright ones and what he called the "little" ones. Bouvard guessed that he might find positions of Uranus among the "little" stars.

A measured star position included, at a minimum, the right ascension and the declination. In Le Monnier's time, each of these was measured with entirely separate instruments. The declinations were taken with a transit circle mounted between two stone piers; the right ascensions were taken with a mural quadrant. The latter was made as stiff as possible and mounted on the face of a great stone pier or wall. A lightweight telescope attached to it could turn about the center, but because it was supported on one side only, a flexure or bending outward of either the telescope or circle or both was always a potential source of errors. (Later astronomers would design improvements to eliminate some of these problems.) In Le Monnier's time, in addition to the two basic coordinates, additional data was collected: the effects of refraction by the air (which made a star observed near the horizon appear half a degree out from its position as measured overhead) and effects due to various motions of Earth, including precession, the aberration of starlight, and nutation. An observation taking only a few minutes to make at the telescope thus required a great deal of mind-numbing calculation afterward to correct it for these effects. The corrections were referred to as "reductions," and the corrected observations were said to have been "reduced." For obvious reasons, the reductions were often put off until long after the observations themselves were entered, and sometimes they were not carried out at all—deferred in the hope that eventually some future drudge might be found to undertake them, perhaps decades later.

Obviously, an astronomer engaged in this kind of work would be expected to be, by temperament and practice, meticulous in the extreme. Yet, according to Bouvard, when he opened Le Monnier's observing books he was surprised to find them *"dans le plus grand désordre."*[10] Le Monnier's writing, Bouvard claimed, was so imperfect that it was sometimes impossible to read the figures; his clock was so irregular that it was difficult to determine its rate; his meridian transits were so carelessly recorded that errors of several seconds of arc or more were frequent. Nevertheless, Bouvard somehow managed to identify from these haphazard records nine new prediscovery observations of Uranus in addition to the three Le Monnier himself had published. One, according to a comment Bouvard is reported to have made to the later director of the Paris Observatory François Arago, was recorded on a paper bag that had contained hair powder![11]

The most important of these were six observations, recorded over a nine-day period in January 1769, that led Bouvard to comment that if only Le Monnier had compared the observations from day to day, he would without doubt have discovered Uranus. Failing to do so, he "was deprived of the honor of a beautiful discovery."[12]

Bouvard's "sloppy Pierre" accusation remained unchallenged for a century and a half. Then, in 1970, historian and analyst Dennis Rawlins looked up Le Monnier's actual records during a visit to the Paris Observatory. He found that all twelve of the prediscovery observations had been entered properly—recorded in ink in a continuous prebound register. Moreover, Rawlins found, the six observations over nine days in January 1769, which had seemed to put beyond question the carelessness of his methods, had taken place when Uranus was near a stationary point in its annual retrograde loop. Under the circumstances, its meager daily motion would have been nearly unnoticeable. But if Le Monnier was not so careless after all, why did Bouvard say he was—in effect slandering him?

Leaving that question aside for the moment, we return to Bouvard's calculations. Possessing an unprecedented trove of observations covering both the forty years since the discovery of Uranus as well as seventeen prediscovery observations spread over the previous ninety years, Bouvard appeared to be in a favorable position. The procedures were clear enough, and for an expert computer like Bouvard they would, on the face of them, have seemed relatively straightforward, although the work involved drudgery on a grand scale. It is sobering to think that the series of computations that would have taken a human computer like Bouvard months, if not years, to finish would require less than a millisecond for a modern electronic computer.

From the outset he carefully distinguished two series of observations. The first comprised the prediscovery observations, which, though generally imperfect due to the instruments used and the circumstances in which they were made, were "very precious all the same because of the extension they give to the arc traversed by Uranus since their time." The second was made up of the numerous observations dating from 1781, which formed an uninterrupted series made with much more accurate instruments.[13]

He continues:

> The natural idea was to use the two series for the elliptic elements. I then made the calculation of the ancient observations. Then I took from the modern observations all the oppositions and quadratures which are in the precious collection of Maskelyne and those of Pond continued until 1815, and lastly those of the Observatory of Paris registers 1800–1819. I used the formulae of perturbations from the *Mécanique Céleste* and the new masses of Jupiter and Saturn. Provisional tables were constructed and 77 equations of condition formed.

(There was one equation of condition for each observation.)

The result of all this careful and backbreaking labor? One can almost hear a sigh as he tells us: "The values I obtained for the elements have failed to answer to the hopes with which I began the work."[14]

He tabulated the residuals (the theoretical values minus the observed ones) resulting from combining the two series of observations: they are very large, ranging in of arc from +98."7 in 1690 to −124" in 1750. The plus sign means the planet was lagging behind, the minus running ahead of the predicted positions. Finding it hard to believe that the modern observations could result in such large errors, he put the blame on the old observations. He had reasons to be suspicious. Thus, he claimed, "Bradley's observation was a single one and recorded only in degrees and minutes; the same for Mayer's. It is known that Flamsteed's instruments were not well made or exactly placed on the meridian. As to [Le Monnier's, we already know] what to think." (Now we see something of the motivation behind his "sloppy Pierre" slander.)

Bouvard could not reconcile both the old and the new observations, so Bouvard decided to toss the old ones. Admittedly, this meant imputing an error of 65.9" to Flamsteed's observation of 1690, even though his observations could generally be relied upon to 10"; he also assumes the observations of Bradley, Mayer, and Le Monnier to be in error by 40" or 60", again badly off their usual standard of 5" or 6". Perhaps this legerdemain weighed on his conscience; perhaps he even lost some sleep over it. But at least now he could finish.

He repeated the whole laborious calculation (though it was not quite so laborious—tossing the ancient observations meant that he now had only sixty-seven equations of condition to evaluate). Though the residuals for the modern (post-1781) observations were better than he had found for the ancient observations, they were still much greater than the several seconds of arc he had hoped for: since 1813 the residuals had been 40", 50", and—for one observation in 1819—as much as 76". Not only were they large, they appeared to be increasing. An unseemly gap remained. With the following remarks, Bouvard introduced his *Tables* to the world:

> If we combine the ancient observations with the modern ones, the first will
> be adequately represented, but the second will not be described within their
> known precise tolerances; while if we reject the ancient positions and retain
> only modern observations, the resulting tables will accurately represent the
> latter, but will not satisfy the old figures. We must choose between the courses.
> I have adopted the second as combining the most probabilities in favor of
> truth, and I leave to the future the task of discovering whether the difficulty of
> reconciling the two systems results from the inaccuracy of the ancient obser-
> vations, or whether it depends on some extraneous and unknown influence
> which may have acted on the planet.[15]

Reading between the lines, we see two questions implied. First, Bouvard was asking whether Uranus would in fact follow the motion the tables prescribed. Second, he was asking what the real problem was with the pre-1781 observations. By suggesting that the old observers had been unreliable, Bouvard was ignoring another possibility: the known architecture of the Solar System might not be complete. With his "extraneous and unknown influence," Bouvard was throwing out a thread that if followed might lead to nothing less than the existence an unknown trans-Uranian planet.

4

"With the Tip of a Pen"

ALEXIS BOUVARD'S TABLES were published in 1821. Throughout the following decade, Uranus's track was closely monitored from several European observatories; it continued to diverge from its predicted course. Bouvard continued to tinker with the orbit, without success. Meanwhile, a new leader emerged, Cambridge University Observatory astronomer (and soon to be Astronomer Royal at Greenwich) George Biddell Airy.

Airy was born in 1801 at Alnwick, Northumberland, the son of a collector of excise taxes. From age twelve he was raised by an uncle and achieved such distinction at the Colchester Grammar School that he was sent to Cambridge. At Trinity College—the college of Isaac Newton—he swept all the mathematics honors by the time he graduated in 1823. In 1824 he was appointed Fellow of Trinity, and two years later Lucasian Professor of Mathematics (the chair that Newton had once held, though Airy was a year younger when he ascended to the position). In two more years, Airy added yet another prestigious position, Plumian Professor of Astronomy and Experimental Philosophy. Among his duties was taking charge of the recently completed Cambridge University Observatory, and he did so with a longer view. As Cambridge University Observatory historian Roger Hutchins has written,

Figure 4.1 George Biddell Airy at about the time he moved from Cambridge University Observatory to the Royal Greenwich Observatory. Credit: William Sheehan Collection.

Airy was determined to establish himself as the only candidate to succeed the elderly John Pond as Astronomer Royal [at Greenwich]. To this end he organized his observatory as an exemplar. Instead of the old English method of trying to detect instrument errors and then adjust them to zero, he followed [the German astronomer] Friedrich Bessel in leaving the instrument stable, rigorously measuring instrument errors, then applying numerical corrections of devised standard formulae to correct for the physical variables of clock, temperature, humidity, and personal error. Differentiating between expert and routine work, Airy also devised simplified standard procedures on printed forms for each step of the reductions. These could then be performed even by unskilled school-leavers, and readily checked for errors.[1]

At a time when reductions of observations often lagged years behind their making, Airy completed them within days. This greatly increased their usefulness. At a meeting of the British Association for the Advancement of Science (BA) at Cambridge in 1833, he called for all the Royal Observatory's observations of planets made between 1750 and 1830 to be reduced to a uniform system—a huge task that he would later accomplished himself. He was also personally responsible for the erection of the Northumberland 11.75-inch (30-cm) refracting telescope that, though matched by a well-to-do amateur's instrument in London, was far more useful than its rival because of its innovative double-yoke equatorial mounting. Designed by Airy "down to the last nut and bolt," it was a forerunner of the mountings used for the great twentieth-century telescopes at Mount Wilson and Palomar in California.

Airy pushed himself hard at the Cambridge University Observatory during his seven years there, and even he—despite his comparative youth, his unusual ability, his high level of motivation, and his sheer genius for organization—found the experience "grueling." Hutchins concludes, "His example there could not have been sustained for long with the resources available. Only after promotion did he admit that 'the overwhelming mass of reductions was incompatible with active lecturing.'"[2]

Among this "overwhelming mass of reductions," those of Uranus were of particular importance and put him in a position to comment on the Uranus problem in a report on progress in astronomy to the BA in 1832. For the first few years after 1821, he announced, Bouvard's *Tables* represented the observations "pretty well," to within an accuracy of 9" or so. Until 1829–30, the planet's actual positions were running slightly ahead of the tabulated positions. In 1829–30, the tabular and observed longitudes briefly coincided, followed by a reversal of sign, and the planet began falling behind. As measured at Airy's own observatory in 1832, Uranus was lagging by an intolerable 30", leading him to conclude, "With respect to this planet a singular difficulty occurs. . . . It appears impossible to unite all the observations in one elliptical orbit, and Bouvard, to avoid attributing errors of importance to the modern observations, has rejected the ancient ones entirely. But even thus the planet's path cannot be represented truly; for the Tables, made only eleven years ago, are now in error nearly half a minute of space."[3]

Then and later, Airy would refuse to speculate as to the cause of the planet's divagations. There were several explanations. One came from Bessel, who for a time was associated with Schröter at his Lilienthal Observatory and since 1813 had been director of King Friedrich Wilhelm III of Prussia's new observatory in Königsberg (now Kaliningrad). Bessel had taken a keen interest in the Uranus problem since the 1820s and thought that the fault might lie with the law of gravitation itself.[4] Another idea was that Uranus's motion was being impeded by a resisting medium. Most astronomers, however, still had full faith in the Newtonian law, which had so far emerged triumphant in every test, and they hoped to see a solution emerge from within the familiar paradigm. In other words, they anticipated the eventual identification of Bouvard's "extraneous influence" with an unknown planet.

The first person to look for such a planet was an English amateur astronomer who is almost forgotten today, the Reverend Doctor Thomas J. Hussey. Hussey was the Anglican rector of Hayes, Kent (the rectory still exists; it is now a public library) and had a very wide acquaintance, including Charles Darwin, who lived nearby at Down House. (In his only reference to Hussey, in a letter to his sister, Darwin noted only that his neighbor "talked grand nonsense albeit about church and local matters"). Hussey lived at a time when Anglican rectors had comfortable livings and the leisure to pursue outside interests, which he did, emphatically and without apology.[5] With a private observatory that was among the best equipped in England, boasting a 6.5-inch refracting telescope by legendary optician Joseph von Fraunhofer, a 7-ft Herschelian reflector, and a 9.3-inch Gregorian-Newtonian reflector, he achieved contemporary fame as one of the first observers in Britain to catch Halley's Comet on its return in 1835. He also contributed a star chart, *Hora XIV*, to the Berlin Academy's great star-mapping project, which was then getting underway. A successor of von Zach's plan to map the stars of the zodiac to assist in the discovery of asteroids, this project, international in scope, owed its original impetus to Bessel but was currently being supervised by the hard-working director of the Berlin Observatory, Johann Franz Encke. It would amply live up to the original hopes for it of helping identify asteroids. It would also lead to a far more substantial quarry.

Hussey's own observations showed that Uranus was lagging increasingly behind the positions prescribed in Bouvard's *Tables*, and voiced his suspicions that an unknown planet might be the cause while visiting Paris. Putting the matter to Bouvard himself, he found the master calculator not only encouraging but willing, if he found the time to carry out the necessary calculations to determine a position for the planet, to assist Hussey in a telescopic search. Unfortunately, Bouvard, by then the director of the Paris

Observatory, was always greatly overworked with endless calculations and administrative duties, and nothing more came of it.

Hussey did not give up—at least not yet. On November 17, 1834, he wrote to Airy, the leading expert on Uranus in England but also possibly the one man who was busier than Bouvard. Airy was still at Cambridge University Observatory but about to be appointed seventh Astronomer Royal at Greenwich (succeeding the Reverend John Pond) when he received Hussey's letter. "The apparently inexplicable discrepancies between the ancient and modern observations," Hussey wrote,

> suggested to me the possibility of some disturbing body beyond *Uranus*, not taken into account because unknown. My first idea was to ascertain some approximate place of this supposed body empirically, and then with my large reflector [the 9.3-inch Gregorian-Newtonian] set to work to examine all the minute stars thereabout: but I found myself totally inadequate to the former part of the task. . . . I therefore relinquished the matter altogether; but subsequently, in a conversation with Bouvard, I inquired if the above might not be the case: his answer was, that, as might be expected, it had occurred to him. . . . Upon my speaking of obtaining the places empirically, and then sweeping closely for the bodies, he fully acquiesced in the propriety of it, intimating that the previous calculations would be more laborious than difficult. . . . If the whole matter do [*sic*] not appear to you a chimera, which, until my conversation with Bouvard I was afraid it might, I shall be very glad of any sort of hint respecting it.[6]

No matter how busy, Airy could almost always be counted on to be conscientious in his correspondence. He posted his reply from Cambridge on November 23. What he had to say was not encouraging. "It is a puzzling subject," he wrote, "but I give it as my opinion, without hesitation, that it is not yet in such a state as to give the smallest hope of making out the nature of any external action on the planet. . . . I am sure it could not be done till the nature of the irregularity was well determined from several successive revolutions."[7]

Since Uranus's period of revolution was eighty-four years, this came out—roughly!—to several centuries.

Hussey seems not to have been entirely satisfied with Airy's reply, for he took the matter up again with Bouvard. The French astronomer confided that he disagreed root and branch with Airy's opinion. He not only thought that the irregularities in Uranus's motion could be explained by the existence of a perturbing planet, he now suspected two.[8] Perhaps with this encouragement Hussey would have pursued the matter further, but in 1838 he suffered a severe accident (of unspecified nature), after which he sold off his instruments to Durham University.[9] At this point he seems to have given up his serious interest in astronomy, and he here leaves our stage.

Despite Hussey's departure, further developments soon followed. The return of Halley's Comet to perihelion in 1835 had shown the magnitude of the refinements that had occurred in celestial mechanics since its last return in 1759. Whereas Lalande, Alexis Clairaut, and Nicole-Reine Lepaute had won universal praise for succeeding in calculating the comet's previous 1759 return to within thirty-three days, several astronomers pinpointed the 1835 return to within a few days. The most accurate result was achieved by Philipe Gustave Le Doulcet, count of Pontécoulant, who included the perturbative effects of Jupiter, Saturn, and Uranus between 1682 and 1835 as well as that of Earth near the comet's 1759 perihelion passage. In the last of several revisions, he was off by only three days.

Small as that difference was, it was enough to lead Jean Élix Benjamin Valz at the Marseilles Observatory and Friedrich Bernhard Gottfried Nicolai at the Mannheim Observatory independently to speculate that it might be due to a retarding force on the comet's motion. Valz added, "I would prefer to have recourse to an invisible planet beyond Uranus. . . . Would it not be admirable thus to ascertain the existence of a body which we cannot even observe?" Nicolai, invoking Bode's law, was even more specific: "One immediately suspects that a trans-uranian planet (at a radial distance of 38 astronomical units, according to the well-known rule) might be responsible for this phenomenon."[10]

Perhaps the culprit had already been sighted. The previous May, Niccolo Cacciatore, Piazzi's one-time assistant who had been present at the discovery of Ceres and was now director of the Palermo Observatory, claimed to have sighted an object of the eighth magnitude moving among the stars of Virgo.[11] Horrible weather then intervened, making it impossible for him to catch it again before it was lost in the evening twilight. Though Cacciatore's object remains an enigma—it has never been satisfactorily explained—the announcement teased out a hitherto unpublished document from Louis François Wartmann, a Belgian astronomer at the Geneva Observatory, concerning observations he had made in 1831.

In the summer of that year, he explained, the observatory was being refurbished; all the main instruments were placed in storage while the observatory's director, Jean Alfred Gautier, retreated to his chateau at Vinzel. Wartmann was left with only a single instrument, a small Fraunhofer refractor, and as a diversion he decided to see if he could use it to "discover" Uranus, having only a general knowledge of its location. As a preliminary he used the telescope's 10× finderscope to construct a chart of faint stars in Capricornus near the planet's path. While thus engaged, on September 6, 1831, he noted that one of his stars, a pale white star of seventh or eighth magnitude, seemed to be moving west, an indication of retrograde motion. He recorded a position for that date: 315°27' right ascension and −17°28' declination (Uranus at the time was actually at 313°). After a spell of bad weather, Wartmann recorded it again on September 25, confirmed its movement and examining it with the Fraunhofer refractor using a magnifying power of 60×. He could not conclusively make out a disk. He obtained a few further positions in October and November, but by December twilight began to interfere.

Subsequent attempts to find "Wartmann's planet" were unavailing. The whole matter was strange, and it would continue to nag at investigators even as the noose around an actual planet began to tighten a few years later.[12]

Of Calculation and Procrastination

That noose was being tightened by the celestial mechanicians as they continued to revise their calculations. In 1837, Bouvard planned yet another revision of his tables of Jupiter and Saturn. However, overworked and affected by the loss of close friends that inevitably happens with age, he decided to delegate the task of revising those of Uranus to a younger man, his twenty-five-year-old nephew, Eugène Bouvard. Eugène began to prepare himself systematically and methodically, and on October 6, 1837, he requested from

Airy (now at Greenwich) the latest reductions of Uranus observations. For some reason, Eugène believed that the errors in the tables of Uranus were chiefly in latitude. Airy responded six days later, pointing out that all the observations made at Cambridge University Observatory in 1833–35, as well as those made at Greenwich in 1836, had already been reduced; in addition, he (or his computers) were involved in an even more ambitious project: the reduction of all the Greenwich planet observations beginning with Bradley's in 1750. Airy advised Eugène that under the circumstances it might be best to delay any efforts on the tables of Uranus until he had finished these preliminaries. He added, moreover,

> With respect to the error of the tables of Uranus, I think you will find that it is longitude which is most defective, and that the excess in latitude is not at present increasing. . . .
>
> In the memoirs upon these subjects, I have included, in the expression for the error in longitude, a term depending on the possible error of the radius vector [the line connecting the planet to the Sun].
>
> I cannot conjecture what is the cause of these errors, but I am inclined, in the present instance, to ascribe them to some error in the perturbations. There is no error in the pure elliptic theory (as I found by examination some time ago). If it be the effect of any unseen body, it will be nearly impossible ever to find out its place. I beg you to offer my remembrance to your Uncle, whom I never saw in England, but whom I have visited in Paris in two different years.

Airy's response seems to have had the same effect on Eugène that his earlier one had had on Hussey: paralyze further initiative. Indeed, apart from a few letters that add little of substance to those just cited, the correspondence between Eugène Bouvard and Airy effectively ended. Eugène, perhaps seizing an excuse for procrastination in a thankless project, was happy to wait for the reductions that Airy promised. To the extent he continued to work on the revision of the Uranus tables, he did so in a way that must be described as desultory.

Though for the moment Eugène slips out of view, others were meanwhile coming to the fore. Indeed, as the American astronomer Benjamin Apthorp Gould Jr. recalled of that time, "Numerous mathematicians . . . conceived

Figure 4.2 The Flamsteed Building of the Royal Greenwich Observatory on Greenwich Hill as it looks today. Photo by William Sheehan, 2011.

the purpose of entering into laborious and precise calculations, in order to
determine whether the assumption of an exterior cause of disturbance were
absolutely necessary, and, if so, to determine from the known perturbations
their unknown cause."[13] One of these was Bessel, who had finally convinced
himself that the only plausible solution of Uranus's wayward motion was
an unknown disturbing planet. As usual, he embraced the research with
enthusiasm, and as a preliminary assigned his twenty-six-year-old assistant,
Friedrich Wilhelm Flemming, the laborious task of rigorously reducing all
the available observations of Uranus from 1781 to 1837 from the ledgers of
the observatories at Greenwich, Paris, and his own Königsberg. Flemming
completed the task with utmost skill. His residuals—comparisons of the
observed positions of Uranus with those calculated from Bouvard's *Tables*—
showed excellent agreement for 1781 to 1821, as was to be expected from
observations for which Bouvard's theory had been customized to satisfy.
However, during the 1820s and 1830s, the errors began rapidly to increase:
from 0."01 in 1820, the error in right ascension had increased to 8" by 1825, to

24" in 1830, to 49" in 1835, and to 62" in 1837. (To put this in perspective, the apparent diameter of Uranus's disk is about 4".) The next step would presumably have been to set up equations of condition along the same general lines that Adams and Le Verrier were later to use and to calculate a position for a disturber assumed to be an exterior planet. Unfortunately, Flemming fell victim to "fever of the brain" (typhus) just after finishing the reduction, dying at Christmas 1840. The promising start was for naught, and his substantial—and very impressive—raft of calculations remained unpublished until 1850.[14]

The loss of his assistant dealt Bessel a double blow, for in the same year his son, who was about the same age, had also died. During a visit to England in 1842, Bessel met John Herschel, the son of the discoverer of Uranus and the doyen of English astronomers at the time, and assured him that Uranus had not been forgotten, but by then it was too late—his own health was beginning to fail. After two years of intense suffering, probably from colon cancer, he died in March 1846 at the age of 62. He did not live to see the discovery of the planet whose existence long hovered on the edges of his imagination.

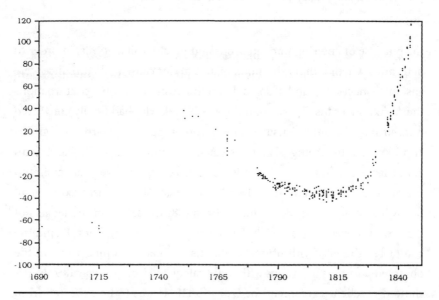

Figure 4.3 The "residuals" of Uranus—the difference O-C between Uranus's observed position and its calculated position based on Alexis Bouvard's *Tables* of 1820, plotted from Flamsteed's first "prediscovery" observation in 1690 to 1845. Based on the data in Le Verrier's 1849 paper in *Connaissance des temps*. Residuals are in arc seconds.

Meanwhile, by then Eugène Bouvard, still in charge of the Paris Obser-vatory's Uranus department, had made some progress—albeit slow. He had found that the errors in longitude, which Airy had described in 1837 as increasing "with fearful rapidity," had reached 90" by 1840 and 120"—a scandalously large 2' of arc—by 1844; this at a time when the results of posi-tional astronomy could routinely be trusted to seconds of arc. Eugène had also arrived at a result that must have surprised him greatly. In contrast to his uncle, who had thrown out the "ancient" observations, Eugène found that the ancient ones were not the problem after all: with the perturbations of Jupiter and Saturn subtracted out, a perfectly regular elliptical orbit for Uranus satisfied them. The problem was not with the ancient observations but the modern (post-1781) observations, which his uncle had presumed beyond reproach. The observations made at Greenwich under Maskelyne between 1785 and 1796 were particularly troublesome: they were both the most difficult to reconcile with an elliptical orbit and the least reliable given the significant errors in collimation that existed for the quadrant instrument Maskelyne used to make them.

Although most summaries of the Uranus problem emphasize that the planet's observed positions were running ahead or lagging behind the tabu-lar positions in heliocentric longitude, this was not the only discrepancy. As Airy himself had found from his analysis of the observations made at Cam-bridge in 1833–35 and at Greenwich in 1836, the radius vector was also con-sistently too small by an amount greater than the Moon's distance to Earth, which he announced in 1838.[15] He gave no explanation for this at the time. It might have been from an exterior planet, though Airy's intuition pointed in another direction: he suspected that the Newtonian inverse-square law of gravitation itself might not be perfectly exact at such great distances from the Sun.[16] This was a minority view, as he was perhaps the only prominent astronomer alive, as astronomical historian A. F. O'D. Alexander notes, "who still hoped that the motions of Uranus could be explained without recourse to the hypothesis of an unknown planet."[17]

Although Eugène never doubted that his uncle's supposition of an exterior planet was correct, he made no attempt to work out its position. Probably the task was beyond him. Instead, he continued in established paths, trying to reconcile the observations and improve his uncle's tables by tinkering with

Figure 4.4 The Paris Observatory around 1870. At the left is the large dome of the 38-cm equatorial, to the right that of the 31.6 cm. The low wing to the left contains the meridian instruments, while Le Verrier's luxurious apartment is on the right. Courtesy of William Sheehan Collection.

the elements of Uranus's elliptical orbit. By May 1844, he concluded that "considerable adjustments" were needed in two elements—the mean motion and the longitude of the perihelion. He was also juggling different values for the masses of Jupiter and Saturn. His goal, in fact, was rather modest: he would be perfectly satisfied, he said, if only he could produce new tables of Uranus with errors nowhere exceeding 15" of longitude.[18]

When Alexis Bouvard died in 1843, he was succeeded as director of the Paris Observatory by François Arago. Though Arago tried at first to be supportive of Eugène, his confidence in the younger man had been shattered by the poor quality of his measurements during an expedition to observe the total eclipse of the Sun on July 8, 1842, and of course by the slow progress he was making on the revision of his uncle's *Tables*. At last he ran out of patience. In June 1845, Arago decided to put the problem into another's hands—those of Urbain Le Verrier, a thirty-four-year-old mathematical astronomer at the Bureau of Longitudes, a man of boundless ambition and a seemingly inexhaustible capacity for long calculations.

Le Maître Mathématicien

A native of Normandy like Laplace, Le Verrier was born March 11, 1811, in St.-Lô. His father was a bureaucrat, employed by the State Property Administration. Le Verrier received two years of education in mathematics at Caen, where Laplace had studied, and graduated at the top of his class. However, this provincial education did not provide sufficient preparation for the difficult entrance examination to the prestigious École Polytechnique in Paris. He was not admitted, whereupon his father promptly sold his house to secure the funds needed to send him for further preparation to the Collège de Saint Louis in Paris. The stratagem proved a success, and Le Verrier duly gained admission to the École Polytechnique. According to one of his peers, his performance seemed "solid" and gave promise of an "honorable career." No one would have predicted his future brilliance.[19] He had already become interested in mathematical astronomy and provided Airy with an analysis of a problem on which the latter was engaged, the motions of Earth and Venus.

After the July Revolution of 1830, in which the Bourbon King Charles X, a political reactionary, was overthrown and replaced by the bourgeois regime of the "citizen king," Louis Phillipe I, Duke of Orléans, it became the policy of the French government to offer graduates of the École Polytechnique their choice of positions in the various departments of public service. Le Verrier, who at that time was equally torn between astronomy and chemistry, chose to work under the illustrious Joseph Louis Gay-Lussac as an experimental chemist in the Administration des Tabacs, and in short order published two creditable papers on the chemistry of phosphorus. He also began spending a good deal of time in the company of Lucile Marie Clothilde Choquet, the daughter of one of Le Verrier's former teachers. A requirement of Louis Philippe's civil servants was to carry out field work in the outlying provinces, so in 1836 Le Verrier faced a difficult choice: either leave Paris and presumably break off with Choquet or resign his position. He chose to resign, and spent a year as a tutor at the second-rate Collège Stanislaus. Mademoiselle Choquet became Madame Le Verrier.

The following year, two positions opened at the École Polytechnique: one in chemistry and one in astronomy. Le Verrier's preference would have

Figure 4.5 Urbain Le Verrier. Engraving by Maurin of the portrait by Charles Dav-
erdoing (1846). Courtesy of Wikipedia Commons.

been to stay in chemistry, but Gay-Lussac had to choose between him and
another excellent candidate, and knowing Le Verrier's interest in advanced
celestial mechanics, he proposed the other man for the chemistry position
and Le Verrier for that in astronomy. "Without dividing his attention and
without looking back Le Verrier detached himself from chemistry and rapidly became an astronomer."[20] He was hired by the Bureau of Longitudes,
located in the same building as the Paris Observatory, and devoted himself
to mastering perturbation theory. He published a series of papers on weighty
subjects, such as "Sur les variations séculaires des orbites des planètes" (On

the secular variations of the orbits of the planets) in 1839 and "Détermination nouvelle de l'orbite de Mercure et de ses perturbations" (New determination of the orbit of Mercury and its perturbations) in 1843. After attempting to straighten out the motion of the innermost planet, he turned to the calculation of comet orbits. He was just getting underway when Arago tapped him on the shoulder and diverted his attention to the Uranus problem.

There were to be two presentations on Uranus given to the Paris Academy of Sciences in the late summer and fall of 1845. The first, on September 1, 1845, was by Eugène Bouvard, who at long last presented a summary of his attempts so far at improving the tables of Uranus. This was to be his last word on the subject: the tables on which he worked so long would never appear. Moreover, no longer supported by Arago or the Academy, he would soon resign from the Paris Observatory and leave astronomy altogether.

Eugène's long-awaited but quite forgettable paper was completely overshadowed by Le Verrier's: "Premier mémoire sur la théorie d'Uranus" (First memoir on the theory of Uranus), presented to the Academy on November 10, 1845. Le Verrier's performance was masterful and thorough, and dealt with its subject decisively. After first examining carefully all the Uranus observations up to 1845, he repeated Alexis Bouvard's calculations—he seems not to have paid any attention whatsoever to Eugène's—and in doing so discovered certain perturbation terms that had been neglected, as well as several outright errors. This required him to redo parts of the calculation.

Even then, discrepancies in Uranus's motion remained. Le Verrier therefore considered the following alternatives: the exterior planet hypothesis, the possibility that Newton's inverse square law of gravitation does not hold exactly at distances from the Sun as great as that of Uranus, or the existence of a resisting medium of some kind. An adjustment in Newton's law was to be considered as a "last resort." The resisting medium, though plausible, did not lend itself to exact mathematical treatment. Hence Le Verrier decided to devote all his efforts to the unknown-planet hypothesis.

He now set himself to tackle nothing less than a problem of inverse perturbations: starting with the discrepancies in Uranus's motion, he would work the perturbation problem in reverse to find, if possible, the elements of the perturbing body responsible for producing those discrepancies. As far as he knew, no one had ever tried to make such a calculation. As 1845 drew to a close, he might have believed, with good reason, that he had the field entirely to himself.

Without simplifying assumptions, the problem was impossibly difficult, for it required the solution of equations in no less than thirteen unknown terms because the errors in Uranus's motion derived from two sources—the elements of Uranus itself and the perturbations of an unknown body of unknown mass and elements. Being inextricably intertwined, they could not be treated independently.

Le Verrier immediately began considering how he could simplify the problem. He realized, first, that given that the orbits of Jupiter, Saturn, and Uranus were negligibly inclined to the ecliptic (all by less than 2°30′), he could assume that the unknown planet was likely to be orbiting almost in the plane of the ecliptic. Second, he argued that the unknown planet could not lie between Saturn and Uranus because it would then produce disturbances in the movements of Saturn that could not have gone unnoticed. It could only lie beyond Uranus. As a necessary starting assumption, he used Bode's law, despite its empirical basis, to suggest the unknown planet's distance: about twice that of Uranus.

By these means, Le Verrier collapsed a problem with thirteen unknowns to one with only eight. To solve for these unknowns algebraically, he set up the usual equations of condition in which expressions from perturbation theory (containing as variables elements of Uranus's orbit and of that of the unknown perturber) were set equal to the residuals at various points along Uranus's orbit. In principle, there ought to have been as many equations of condition as observations, but again Le Verrier simplified. Except for the prediscovery observations where only one observation is available, he took averages for all observations. This reduced their number to only eighteen. The points around Uranus's orbit for which these equations of condition were formed corresponded to observations for the following epochs: 1690.98, 1712.25, 1715.23, 1747.7, 1754.7, 1761.7, 1768.7, 1775.7, 1782.7, 1789.7, 1803.7, 1810.7, 1817.7, 1824.7, 1831.7, 1838.7 and 1845.7.

We cannot follow the details of the mathematics, but suffice it to say, starting with a polynomial equation derived from perturbation theory he attempted systematically to eliminate seven of his eight variables until what remained was an expression in which the only unknown was the heliocentric longitude ε' of the unknown planet. (The heliocentric longitude, the position relative to the Sun, is simply related to the geocentric longitude, giving its position in the sky—and this, of course, was what he was after.) Though he could easily eliminate six variables, the seventh—the mass of the unknown planet—depended heavily on the observational errors: even a slight change would cause the mass of the unknown planet to vary widely. He found two roots of the polynomial that seemed to be acceptable and produced values of the heliocentric longitude ε' between $96°40'$ and $189°59'$ and between $263°8'$ and $358°41'$. That was still a lot of territory to cover, so these solutions were of no use in locating the planet. Even within these broad limits, however, he could not get a satisfactory solution because inserting these values of ε' into his perturbation equation gave the mass of the unknown planet a negative value—an obvious physical impossibility. On reaching this seeming impasse, his investigation ground to a halt for several months.

As Le Verrier suspended his efforts, a new planet had been found from Germany. It was not the trans-Uranian planet of Le Verrier's and others' quest but a minor planet—the fifth discovered (after a long hiatus) moving in the gap between Mars and Jupiter. Karl Ludwig Hencke, a retired postmaster at Driesen (now Drezdenko, Poland) and avid amateur astronomer, had for some time been using a small refractor in a rooftop observatory to search among the stars near Vesta for a new planet. Checking the positions of each against the Berlin Academy's recently published *Hora IV* star chart, on the night of December 8, 1845, he found an uncharted ninth-magnitude object. Its planetary nature was soon confirmed, and the year ended with astronomical headlines blaring the discovery of a "new planet," called by its discoverer Astraea. No one would have guessed then that another planetary discovery, of far greater consequence, would follow before Earth had completed one more circuit around the Sun.

Adams

With Le Verrier stymied, and a new planet discovered in Germany, interesting developments were occurring off the main stage. In England, John Couch Adams, a twenty-seven-year-old mathematician of St. John's College,

Figure 4.6 John Couch Adams in the combination room of St. John's College, Cambridge. Mezzotint by Samuel Cousins of the portrait by Thomas Mogford, 1851. Courtesy of William Sheehan Collection.

Cambridge, who was virtually unknown elsewhere and working in near iso-lation, had taken up the same formidable problem as Le Verrier and indeed managed to get somewhat ahead of the French astronomer.

Adams was born in June 1819 in an isolated farmhouse at Lidcott in the moorland parish of Laneast on the edge of the north Cornwall Downs, "an area that to this day is very quiet; the hustle and bustle of city life is a world away, and people visit it for that very quality. People deal with one another in a supportive non-confrontational way. If asked a question you give your response and that is the end of the matter."[21] His father, Thomas Adams, was a tenant farmer on property owned for many years by the Barons of Trageare but was then in the hands of a Launceton lawyer, John King Rennall Lethbridge.

At an early age, Adams was sent to the village school at Laneast. The teacher, inauspiciously named Mr. Sleep, taught himself algebra so he could teach it to his student, but the pupil soon outstripped the teacher. By eight years old, Adams was already being regarded as a prodigy by at least one acquaintance of the family, who told Adams's father, "If he was my boy, I would sell my hat off my head rather than not send him to college."[22] Adams's father seemed to agree, thinking him "too absent-minded to make a farmer. Send him to fetch in the cows for milking and you'd find him lying on his back star-gazing. He seems to know as much as the village schoolmaster, so he'd better go to his uncle's school in Devenport."[23]

At age 12, Adams was sent to Devenport as a boarder at the school run by the Reverend John Couch Grylls. Like most schoolmasters at the time, Grylls emphasized classical languages and offered very little mathematics. Fortu-nately for Adams, the school was located near the town hall, which housed the Devenport Mechanics Institute. This institution, like the more famous establishments at Liverpool, London, Ipswich, and Manchester, provided free education in technical subjects for working people and professionals,[24] and young Adams regularly attended lectures there and taught himself a good deal of mathematics from the Institute's books. His interest in astron-omy was already evident. On the return of Halley's Comet in October 1835, he wrote excitedly to his parents about seeing this comet, "which at its visit 380 years ago threw all Europe in consternation."[25] He made lines or notches on the window ledges or walls of his uncle's house to mark the position of shadows at noon, and calculated the best times to see the May 15, 1836, solar eclipse at Lidcott farm.[26] According to local legend, he also stargazed while leaning against an ancient Celtic cross near Lidcott farm.

About that time, Grylls decided to give up his school and emigrate to Australia. Adams seems to have been encouraged—and thought seriously—about doing likewise. Lethbridge, the landlord, who lived in an imposing Palladian-style house a mile from the Adams farm, seems to have been in favor of emigration because he strongly opposed the idea of Adams (or anyone else of his social class) attending college because "scarce resources should not be spent on college bills and that education was getting more prevalent and was not well paid." Adams's family, however, stood firm against the emigration plan and unanimously agreed that "he should go to the University."[27]

Once the decision had been made, even Lethbridge seemed to have had a change of heart, and he offered the promising young scholar his support. Otherwise, Adams might have been, in the words of the poet Thomas Gray, a flower destined "to blush unseen, And waste its fragrance on the desert air." The Industrial Revolution also played a role in his advancement after a manganese nodule was found on the Adams farm. Since manganese, when added to iron, hardens the iron without making it more brittle, it was coming into great demand in the burgeoning steel industry. It is pleasant to record that Lethbridge generously allowed Adams's family to control some of the profits realized from the Lidcott mine, which largely were to fund his college education, and also recommended to him a tutor to help him prepare for his admission examinations.

Adams was duly entered at St. John's College, Cambridge, in October 1839. Almost immediately after his arrival, he made a strong impression on at least one fellow student, also a first-rate mathematician, who sat for the same entrance examinations and afterward recalled, "He and I were the last two at a viva-voce examination; he broke the ice by asking me to come to his rooms to tea. I went and we naturally had a long talk on mathematics, of which I knew enough to appreciate the great talent of my new friend. I was in despair; I had gone up to Cambridge with high hopes and now the first man I meet is something infinitely beyond me and whom it was hopeless to think of my beating."[28] The examiner was of the same opinion, telling Adams at the end of the examination, "If you work, the highest honors are in your reach." His first year was spent, as for all freshmen, in rooms licensed by the authorities away from college; they cost £10 a year. In a letter home to his brother during that first year, he complained about the flatness of the countryside around Cambridge and described his routine of prayers, dinner in the hall, lectures, and preparations for examinations. He distinguished himself well enough to obtain a sizarship—a great assistance as it carried a partial allowance toward college expenses. It was the first of many scholarships and prizes he was to win, for he consistently took first place in the college examinations—not only in mathematical subjects but also in the Greek testament, in which he won first prize every year he remained at Cambridge.

In June 1841, he had just finished his second year and was looking forward to the freedom to pursue subjects of choice over the summer vacation. He was browsing in a bookstore on Trinity Street when he happened across a copy of the proceedings of the Oxford meeting of the BA in 1832—the volume containing Airy's report on the progress of astronomy that called attention to the nettlesome deviation of Uranus from its predicted path. In 1832, as noted above, this was 0.5′ in longitude and increasing; by 1841 it had reached 1.5′. Adams's interest was piqued, and before the week was out he had jotted down the following memorandum:

> July 3, 1841
> Formed a design, in the beginning of this week, of investigating, as soon as possible, after taking my degree, the irregularities of the motion of Uranus, which are not yet accounted for, in order to find whether they may be attributed to the action of an undiscovered planet beyond it.

At his graduation in 1843, he received all the honors he could have hoped for, including Senior Wrangler in the prestigious Mathematics Tripos and recipient of Smith's Prize. (When news of Adams's triumphs reached Launceton, Lethbridge rode his horse all the way back to Laneast, and on entering the yard of Lidcott farm, took off his hat and shouted, "Adams for ever!") He was named a fellow of St. John's, in which he was kept busy with his official studies and tutoring. The Uranus problem still beckoned, but could only be taken up during college vacations. During the summer of 1843 he made a preliminary investigation, sketching the case of a planet external to Uranus moving in a circular path with a mean distance of 38.4 AU, as given by Bode's law. On including such a planet in his perturbation-theory calculations, he found that the residuals had been reduced. Though Adams hardly believed that the circular orbit and Bode's law assumptions were valid, he was sufficiently encouraged to take the analysis to a higher level. First, however, he needed better data. Using the Reverend James Challis, a former Senior Wrangler himself and director of the Cambridge University Observatory (about a mile's walk uphill from St. John's), as a go-between, Adams obtained observations from Airy at the end of 1844, which included all the Greenwich observations from Bradley's prediscovery observation of 1754 to 1830.

In the autumn of 1845, Adams threw himself into a much more detailed calculation. Without awareness of one another, and each supposing himself to be in exclusive possession of the problem, he and Le Verrier were following the same track.

Adams proceeded to set up the perturbation equations with all the unknowns and to write equations of condition for the residuals at points around Uranus's orbit. He retained the Bode's-law assumption of a planet at a mean distance of 38.4 AU from the early sketch but abandoned the circular orbit. Instead, he gave it a high eccentricity, 0.161, which was ten times that of Earth. He realized that the Bode's-law assumption was to some extent arbitrary, and did not have any faith in an orbit of such high eccentricity, but at least he had a starting point for his calculations. He also did something that Le Verrier had not yet attempted: he calculated an approximate position for the planet, putting it close to the Capricornus-Aquarius border. Almost all the retellings of the discovery of Neptune—and there are many—insist on the significance of this autumn 1845 position. However, no one—Adams included—would have expected this first position to be anything more than approximate: it might put the planet in the right sign of the zodiac, but it was unlikely to be close enough to point a telescope to within a few degrees of the planet's actual position in the sky. (After the discovery, the actual planet was found to have been lurking only 2.5° from Adams's calculated position.)

In the latter part of September 1845, Adams drew up a short summary of his investigation's results so far, though without any hint of the theory he used to obtain them, and gave it to Challis, who as director of the Cambridge University Observatory took charge of the 30-cm (11.75-inch) Cauchoix refractor donated by Hugh Percy, the 3rd Duke of Northumberland, in 1833 and put in working order by Airy before he left for Greenwich. Challis was a typical Victorian: honest, duty-bound, conscientious, and chronically overworked. Under the circumstances, even though he and Adams were on friendly terms, Adams's project had, at first blush, little allure for him. He justifiably had no great faith in the position that Adams gave him. There was certainly nothing in it to impel Challis to the telescope to look for the planet. Rather, as Challis himself would later explain, it seemed "so novel a thing to undertake observations simply upon theoretical deductions that while the labor was certain, success appeared to be uncertain."[29]

Challis did exactly what any overworked person would do—he punted. Instead of doing anything with Adams's calculations himself, he recommended that Adams take them to Airy. Airy, after all, was the greatest authority on the Uranus problem in the realm. So, in early October 1845, Adams left Cambridge for a vacation in Cornwall. Though he must have known that the odds were long of finding Airy available at short notice, he stopped by the Airy residence on Greenwich Hill. Airy's wife, Richarda, answered the door, and informed him that the Astronomer Royal was away on business in Paris. He can hardly have been surprised. (Afterward, Richarda had no recollection of this visit by Adams; for all his brilliance, Adams seems to have been a rather faint personality.) On his return from Cornwall to Cambridge, on October 21, Adams again tried to call on the Astronomer Royal—not once but twice. The first time Airy was out on one of the daily walks that Richarda insisted he never missed in the interests of his health. Adams tried again in the afternoon. The butler answered the door and explained that Airy was at dinner and not to be disturbed. (It was only 3:30 in the afternoon, but astronomers' hours are often irregular, and Airy was a creature of strict routine.) Adams was sent away. He could do no more than drop a small scrap of paper containing a summary of his calculations into Airy's letterbox. He certainly left Greenwich discouraged, and no one could entirely blame him if he felt snubbed—though he later denied it.

Airy found Adams's paper and was quite interested. After all, the subject was near and dear to his heart. One question nagged at him. Another apparent discrepancy accompanied Uranus's unexplained drift in longitude: Uranus's Kepler ellipse radius (radius vector) appeared to be slightly larger than it ought to be according to gravitational theory. Since Airy himself had discovered this, his interest was rather paternal. Naturally, he considered it to be a matter of the first importance.

Even though Richarda had just given birth on October 28, and a day earlier the police had taken into custody one of the senior assistants at the Greenwich Observatory for a crime that would soon develop into a major scandal, Airy, on November 5, 1845, took time to ask Adams the question of whether his exterior-planet hypothesis would correct both the drift in longitude and the error in radius vector. Adams, famously, did not reply. We now know, from sleuthing in the Cornwall Record Office, that on November 13 Adams did indeed begin to draft a response, but he did not finish it.[30] Apparently Adams had an unfortunate tendency to seize up whenever he attempted to write a letter to anyone except members of his own family. From what he said later, it can be inferred that he considered the matter of the radius vector as something needing to be cleared up but hardly urgent. It was a mere detail. In this, he was too dismissive, for it hardly appeared that way to Airy. Indeed, the Astronomer Royal later wrote that because of Adams's nonresponse, "I did not even urge Challis to observe, because . . . I regarded the whole matter as doubtful."[31]

Though Adams continued to make a few private calculations, by the end of December he apparently gave up. (This was almost the same moment that Le Verrier's calculations were breaking down.) It would be September 1846, before Adams would write Airy again. By then it would be too late.

A Double Prediction

After several more months of inactivity, Le Verrier managed a breakthrough in his calculations. On June 1, 1846—seven months after Adams had first contacted Challis; seven months during which the shy and retiring young Fellow of St. John's all but disappeared from the scene—the French astronomer published a new paper, "Recherches sur les mouvements d'Uranus" (Researches on the movements of Uranus). Here he gave the elements of a hypothetical planet moving in a circular orbit with a Bode's-law mean distance of 38.4 AU. He also for the first time gave a position (for epoch

January 1, 1847). It was, like Adams's, near the Capricornus-Aquarius border—in fact, it as within 4° of the position Adams had dropped into Airy's mailbox the previous October.

Although Le Verrier, unlike Adams, had published his calculation, his paper did not immediately stir his French colleagues to point a telescope to the sky to search for it any more than Adams's private calculations had prompted his English colleagues to do so. In part, this was because the Paris Observatory's instruments were old and out of adjustment at the time. However, as Challis would later emphasize, Le Verrier had framed his investigation as a mere preliminary.[32]

However, when Airy received the issue of the *Comptes Rendus* of the Paris Academy of Sciences in which Le Verrier had published his paper, he immediately recalled the data that Adams had dropped on him in October. Both astronomers had put the planet near the Capricornus-Aquarius border. Airy would later claim that the positions were "the same, to one degree," though this is not the case: they were actually 4° apart. Regardless, they were clearly in reasonably close—if not quite stunning—agreement, and the effect on Airy was immediate. He later wrote, "To this time I had considered that there was still room for doubt of the accuracy of Mr. Adams's investigations. . . . But now I felt no doubt of the accuracy of both calculations, as applied to the perturbation in longitude." But he added the usual caveat: "I was, however, still desirous, as before, of learning whether the perturbation in radius vector was fully explained."[33]

Since his radius-vector query to Adams the previous November had gone unanswered, Airy now decided to put the same question to Le Verrier. The French astronomer's response was emphatic. Writing on June 28, he put the blame for the discrepancy squarely on Alexis Bouvard's miscalculations and showed that once the elements of Uranus's orbit were corrected, the error in the radius vector was corrected as well. "Excuse me, Sir," he added, "for insisting on this point."[34] In the same letter, Le Verrier invited Airy to search for the new planet: "If, as I hope, you have sufficient confidence of my work in order to look for this planet in the sky, I would, Sir, speedily send you its exact position, as soon as I have obtained it."[35]

Le Verrier seemed as strong-willed and decisive as Adams seemed procrastinating and indecisive. A famous psychologist has remarked, "Whereas in the irresolute, all decisions are provisional and liable to be reversed, in the resolute they are settled once and for all and not disturbed again. . . . These two opposing motives twine round whatever other motives may be present at the moment when decision is imminent, and tend to precipitate or retard it. . . . One says 'now,' the other says 'not yet.'"[36]

Airy, perceiving Le Verrier to be the "now" type, was persuaded as he had never been by Adams, and events began to happen with great speed. On June 29, 1846, at a meeting of the Royal Observatory's Board of Visitors at the Admiralty in London, Airy mooted "the extreme probability of now discovering a new planet in a very short time, provided the powers of one observatory could be directed to the search for it." The twelve members at the meeting were mainly Cambridge graduates or professors forming what historian Robert W. Smith has called "the Cambridge Network," explaining that all were "socially and intellectually close-knit, all boasted impeccable Cambridge credentials."[37] Presumably these men hoped to exploit their monopoly on knowledge of the double prediction of the planet's position and win one for the old school tie. Though this could never be proven, it certainly explains some unusual aspects of what went on that summer.

What is beyond doubt is that Airy now proceeded to draw up parameters for a planet search. In concept, it was similar to the search by the Celestial Police for the possible planet between Mars and Jupiter, but far more extensive. Its prosecution would fall to Challis, who as Cambridge University Observatory director had command of the 30-cm (11.75-inch) Northumberland refractor. Despite its ungainly appearance, it was a superb telescope, but the kind of search Challis meant to undertake with it would have been difficult without an assistant. Airy offered him a "senior" Greenwich man. Challis—despite the clear need—declined, presumably because he did not wish to cede control to another. Next Airy offered someone who would be clearly subordinate, the Reverend James Breen, a twenty-year-old Armagh native who had been working for Airy as a human computer since age fourteen and who Airy said was "a rough genius, but [someone] who can get through any quantity of work." Challis was more amenable that time, and agreed to take Breen "on trial" for one month. However, Breen did not actually arrive at Cambridge until well into August; even then, as we shall see, Challis did not immediately use him. In the opening stages of the search, and fatefully as it would prove, Challis was completely on his own.

Meanwhile, one omission clearly stands out: Airy's failure to respond to Le Verrier's request to search for the planet. Later Airy would claim that he was just then about to leave for his annual trip to Europe. He did not actually depart until early August. One can only assume that he wanted to keep the English search secret to maintain advantage-in for Cambridge.

Airy's instructions to Challis were to commence sweeping a broad band of the zodiac, 30° long by 10° wide, centered roughly on the planet's position as indicated by Le Verrier. Though Adams at one point had mentioned to Challis that he expected a ninth-magnitude planet, as usual he made no impression, and instead Challis—with admirable if excessive thoroughness—decided to examine all the stars in the zone down to the eleventh magnitude—some three thousand stars.

In Challis's defense, though in retrospect such a thorough search would prove unnecessary, it did reflect the consensus at the time. Le Verrier, in his June 1, 1846, paper, had recommended a broad search of the ecliptic, from 321° to 335° of heliocentric longitude. Adams apparently had similar thoughts, as we know from a hitherto little noticed document in the Challis Papers at the Cambridge University Library. In this document, Adams presented geocentric coordinates (the data needed by a telescopic searcher— i.e., Challis—including the right ascensions and declinations at which the planet would appear in the sky) for a Bode's-law orbit of zero eccentricity and no inclination. The longitude is calculated for increments of 5° on either side of the 325° figure, while on a torn-off sheet evidently added as an afterthought, Le Verrier's instruction to search between 321° and 335° of heliocentric longitude is repeated. The existence of this document proves following: despite his public silence that summer, Adams was well aware of what was going on; he agreed with everyone else the search net should be cast widely, and thus had little confidence in the exactness of the calculations; he had not begun his own final calculations; and he remained a faint background figure at that time, even to his English colleagues.

Overworked as always, Challis was hoping to finish reductions of comet observations from earlier that year—it was, in fact, a year of comets, including the famous Biela's Comet, which broke in two—before starting the planet search. He certainly did not feel any urgency about the latter, supposing that the planet, if it existed, would yield only to a protracted siege. The shutters of the dome of the Northumberland refractor remained closed until July 29. That night, Challis dutifully began recording the position of each star in his zone. On August 12, he decided to compare the observations of that night with those of July 30, just to be sure of his methods. He stopped his cross-check at star number 39. In fact, he had twice recorded the planet without knowing it, on August 4 and 12, which he later learned, with considerable heartache, after its discovery. Had he continued for just ten more stars, the planet would have been within his grasp! Ironically, though Breen had joined him by then and could have been employed in just this kind of cross-checking, for some reason Challis did not immediately make any use of him.

It has often been pointed out, most notably by Challis himself, that his search would have been facilitated immensely if only he had had an up-to-date star map of the region. Without one, his work was laborious and effectively meant producing his own star catalog of the region in which the planet was presumed to be lurking. But he *did* have such a star map: the planet, near opposition that summer, was moving between the Berlin Academy's star map *Hora XXII*, published in 1833, which could have been consulted in the Cambridge University Observatory library, and *Hora XXI*, which though complete had not yet been mailed out from Berlin. The transition from the first map to the second did not occur until late August. In other words, the planet *was* on the map in his possession. Another miss!

As Challis continued ploddingly along, the pace of events was quickening elsewhere. Le Verrier published a new paper at the end of August that contained a revised orbit. Instead of a circular orbit, as in his June 1 paper, he revised the eccentricity to 0.107 and reduced the mean distance to 36.2 AU. He also published the revised position that he had promised Airy in the letter Airy had never answered. At heliocentric longitude 318°47', this position was the closest of any of the prediscovery predictions—within a degree of the planet's actual position. As an incentive to searchers, he added the further suggestion that the planet might be most readily recognized not by painstakingly plotting its position among the stars but by the recognition of its disk.

Adams, too, had been busy. Doubtless spurred by Le Verrier's efforts, he was winding up his own intermittent and much-interrupted calculations.

His behavior during the year previous had been, frankly, baffling. One can only say that he seems to have been so focused on what he was doing in the all-consuming world of his mathematics that he had dropped from the picture for long periods. Only on September 2, 1846, did he finally break his silence by attempting to communicate his latest—and last—results to Airy at Greenwich. By now Adams's preference was for a more circular orbit ($e = 0.120$) in which the mean distance had been reduced to 37.2 AU. Even this he did not find wholly satisfactory, for his calculations were showing that by still further diminishing the distance, the agreement between theory and observation might be further improved. The way things were trending, he thought the planet might lie closer to longitude 315°, which would have been fully 10° from Le Verrier's June 1, 1846, position. In other words, Adams's revised calculations were putting the position farther, not closer, to that of the actual planet. Airy did not receive this letter because he was traveling in Germany; in any case, Adams' new calculations were entirely academic, finished far too late to influence any planet searcher.

By then the circle of searchers was beginning to widen. In addition to Challis, John Russell Hind at George Bishop's private observatory at Regent's Park, London, the first Englishman to read Le Verrier's paper suggesting the planet be searched for by its disk, had launched a search at his own initiative. John Hartnup at the Liverpool Observatory had become interested, while Hervé Faye at the Paris Observatory apparently carried out a brief search earlier in the summer. Even an American, Sears Cook Walker of the U.S. Naval Observatory (USNO) in Washington, DC, proposed looking for the planet to his supervisor but was turned down.

Le Verrier, completely unaware of these developments and not having heard anything from Airy since the end of June (the silence must have seemed deafening), on September 18, 1846, penned a letter to an old friend Johann Gottfried Galle, a thirty-four-year-old assistant astronomer at the Berlin Observatory, in which he expressed his wish "to find a persistent observer, who would be willing to devote some time to an examination of a part of the sky in which there may be a planet to discover." Galle, receiving the letter on September 23, immediately requested permission from the observatory's director, Johann Franz Encke, to search for the planet that very night. Encke, a man of stern appearance and martial discipline (he had served with the artillery in the Prussian army against Napoleon), was at first reluctant to allow a departure from the observatory's routines. However, Galle perhaps found him in an unusually good mood, for that night he planned to celebrate his fifty-fifth birthday with his family and was eager to get away from the observatory. He thought for a moment, then replied to Galle, "Let us oblige the gentleman in Paris."

That night, Galle was assisted by young Heinrich d'Arrest, whose role in the discovery would not become widely known until long afterward. After first taking up their positions at the observatory's 9.5-inch (24-cm) refractor—Fraunhofer's masterpiece—they searched for a small disk (as suggested by Le Verrier) without success. At that point, d'Arrest shrewdly suggested that they try again with the help of a star map covering the area where the new planet was presumed to be lurking. They ransacked the cabinet outside Encke's office and found the map they needed—*Hora XXI* (Aquarius) of the Berlin Academy series, completed by Carl Bremiker but not distributed to observatories elsewhere. To save postage, its distribution had been delayed so it could be sent with the next map in the series. d'Arrest spread the star map out on the table in front of him as Galle seated himself at the eyepiece.

The Planet . . . Really Exists

In less than an hour of observing, Galle came across a "star" d'Arrest could not locate on the map. At once the younger man's voice rang out—words that have reverberated in the annals of astronomy ever since: "That star is not on the map!"

Encke was awakened from his postprandial birthday-celebration slumbers and summoned to the dome. With excitement such as can only be imagined, the three astronomers tracked the curious object until it set about 2:30 in the morning.

The following night, the sky remained clear, and the three astronomers, Galle, d'Arrest, and Encke excitedly pointed the telescope to the sky. Galle later recalled what appeared in the finder scope:

Figure 4.7 The 9.5-inch (24-cm) Fraunhofer refractor of the Berlin Observatory, used by Johann Galle and Heinrich d'Arrest in discovering Neptune on September 23, 1846. The telescope is now on exhibit in the Deutsches Museum, Munich. Photo by William Sheehan, 2010.

Figure 4.8 A small section of the Bremiker map showing the position of Neptune relative where Le Verrier predicted it (marked by *X*).

Four stars of the eighth magnitude occupied the field. One of them was brought into the field of the large telescope and critically examined by my assistant [d'Arrest] and rejected. A second star was in like manner examined and rejected. A third star rather smaller and whiter than either of the others was brought to the center of the field of the great telescope, when my assistant exclaimed: "There it is! there is the planet! with a disk as round, bright, and beautiful as that of Jupiter."[38]

Galle then took d'Arrest's place at the eyepiece, and on scrutinizing the disk, exclaimed, "My God in heaven, this is a big fellow!"

Galle and Encke measured the diameter of the disk as 3.2″—almost identical with the 3.3″ Le Verrier had predicted on August 31. The fact was clearly significant: to show a disk at all from such a great distance signified it must be a giant planet, as indeed had been inferred from its ability to perturb another giant planet, Uranus. Finally, on September 25, Galle took up pen and wrote to Le Verrier, "The planet of which you indicated the existence *really exists!*"[39]

The world was now securely in possession of a new planet—one far more consequential than the small one Hencke had brought to light the previous December. The latter had been found after a brute-force canvassing of small stars, but the former's existence and even position had been deduced before it was seen in the telescope. This seemingly magical event captured the public imagination as nothing else and elevated the prestige of science to unprecedented heights. The accuracy of Le Verrier's calculation—the planet was found within 52', less than a degree, of where his August 31 paper had put it—was little short of astounding, and despite being one of the most oft-told and familiar stories in the history of astronomy, still stirs the blood. Encke would call it "one of the most beautiful triumphs that theory has ever achieved."[40]

As we shall see, the unparalleled triumph would cast a very long shadow before it, continuing to inspire later generations of astronomers. At least a few of them would dream of using Le Verrier's methods, emulating his success, and earning a place in the pantheon of astronomical immortals who discovered a new planet, as Arago expressed it, "with the tip of a pen."[41]

5

Post-Discovery Controversies

What's in a Name?

THE TRIUMPHANT DISCOVERY in Berlin soon led to controversies. One involved the naming of the planet. Galle suggested "Janus," for the Roman god of gates and doors, though Le Verrier—evidently confusing Janus with Terminus, the god of boundaries—objected that "the name would have the disadvantage of making one believe that the planet is the last of the Solar System, which there is no reason to believe."[1] He was at least temporarily in favor of "Neptune," for the brother of Jupiter and god of sea—a name that, he claimed, had been put forward by the Bureau of Longitudes. Another somewhat nautical name, "Oceanus," was briefly mooted in England by Adams and Challis. Needless to say, it was never popular on the other side of the English Channel (which the French, of course, called La Manche).

Within a few days of Neptune being suggested, Le Verrier seems to have had second thoughts. Instead he came to favor a name officially proposed to the Paris Academy of Sciences by Arago: "Le Verrier" or "Le Verrier's Planet." It has always been suspected that Arago was merely a stalking horse for the name and that it had actually been suggested by Le Verrier himself. "Le Verrier" survived in French circles into early 1847 and made its way into the tables of the *Connaissance des Temps* for 1849 before being dropped in the face of overwhelming international opposition. "Neptune," though it had never been particularly popular, was now adopted as the least bad alternative. (In retrospect, it seems a perfectly good name both on mythological grounds and because of its sea-blue color as seen in the telescope.)

Another controversy involved a dispute over priority for the discovery as information about Adams's role, hitherto unknown outside a small circle in England, gradually came to light. A great deal has been written about this seemingly important argument—indeed, it briefly threatened to provoke an international incident between England and France—but within a year tempers had died down. By the time Adams and Le Verrier met at the BA meeting at Oxford in June 1847, they could do so on terms of mutual respect. John Herschel, who was present on the occasion, remarked, "We could make no distinction between them."[2] The carefully crafted consensus that Le Verrier and Adams were co-predictors of the planet would hold up for at least a century and a half. In recent years Adams's claim to equal credit with Le Verrier has once more been challenged by some historians, though the whole issue has become rather academic.

The "Happy Accident" Theory

From our modern perspective, the most important controversy—and the only one pertinent here—involved the question of whether the predictions of Adams and Le Verrier, seemingly such a stunning success, had been valid at all. Since this controversy involved technical matters, it was far harder for the public to grasp than the naming of the planet or the apportionment of credit for its discovery. However, in the long run it was to prove far more consequential.

The controversy chiefly turned on Neptune's orbital characteristics. Within weeks of the discovery in Berlin, Adams calculated a first orbit using the two fugitive August positions, those of August 4 and 12, that had been recorded by Challis in his unsuccessful search. This orbit made clear that the planet was much nearer the Sun than expected, just 30 AU. (Remember the Bode's-law distance had been 38.4 AU.) Since Neptune's perturbations of Uranus had been the basis of the mathematical calculations, and these were critically dependent on Neptune's mass, the lesser distance also implied a lower mass. Le Verrier, from his perturbation analysis, had expected a planet three times as massive as Neptune, a great overestimate.

The first somewhat reliable determination of Neptune's mass awaited the discovery of its satellite. As early as October 10, 1846, William Lassell, a Liverpool brewer-turned-amateur astronomer who boasted a 24-inch (61-cm) reflecting telescope on an equatorial mount, announced the discovery of a possible ring and a large satellite. The ring proved to be illusory, but the satellite—later named "Triton"—was real enough, though Lassell did not definitively settle the matter until he had made further observations in July 1847.

Neptune's mass could in principle be accurately determined from observations of the satellite's motion by applying the Newtonian form of Kepler's third law.[3] Given the difficulty of accurately placing the wires of the micrometer on so faint an object as the satellite, however, estimates varied. Thus Harvard mathematician Benjamin Peirce, from observations made by W. C. Bond with the new 15-inch Merz-Mahler refractor at the Harvard College Observatory, found Neptune to be about four-fifths the mass of Uranus, but Friedrich Wilhelm Struve at the Pulkovo Observatory in Russia found it to be only about three-fifths. Regardless of the precise value, Neptune seemed to be less massive than Uranus and much less massive than Le Verrier or Adams had expected. In fact these early estimates of Neptune's mass were much too small; according to the latest results, Uranus's mass is 14.536 times that of Earth while Neptune's is 17.148 times.

Meanwhile, as Neptune continued along its track and was kept under close observation, its orbital elements were defined. By January 1847 the American mathematical astronomer Sears Cook Walker of the USNO in Washington, DC, who had been denied his request to search for the planet on the basis of Le Verrier's predictions before its discovery in Berlin, published a nearly circular orbit with a mean radius of 30.200585 AU and a period of 165.9703 years.

As in the case of Uranus, a much more precise orbit could be calculated if the arc of the orbit along which it had been observed were extended by chance observations by earlier astronomers, and Walker took up the challenge of searching for them. He briefly considered the possibility that Wartmann's star or Cacciatore's object might be prediscovery observations of

Neptune, but quickly ruled them out. (In fact, Wartmann's star was eventually shown to be nothing more than a mistakenly recorded observation of Uranus, which was then in the vicinity.) He then turned to old star catalogs, of which only one held any promise of success: Lalande's *Histoire Céleste Française*, published in 1801, which contained the positions of fifty thousand stars. Indeed, Walker found that Lalande—or, since the titular author had not actually observed any of the stars in the catalog, his nephew Michel Lalande, the actual observer—might have unknowingly come across Neptune on May 8 and 10, 1795. In fact, not only had Michel Lalande recorded the planet as a fixed star on those two nights, he had even noted a slight discrepancy in position and indicated it as uncertain. As soon as the given area of the sky could be examined with one of the Washington telescopes, it was found that this "star" had indeed vanished! It had been none other than Neptune. (It was later found that Hussey, in the 1830s, had absent-mindedly entered Lalande's star onto the Berlin Academy map he was then preparing, where it continued, for a while, to lead a ghostly existence.)

Eventually other prediscovery observations of Neptune would turn up, though Lalande's were the only two of any use in extending the arc of Neptune's observed motion. For instance, the Scottish astronomer John Lambert, who spent most of his career at the Munich Observatory, had unwittingly recorded it on October 25, 1845, and again on September 7 and 11, 1846. Even John Herschel had passed over it once, as he revealed to Struve of the Pulkovo Observatory. "The discovery of the new Planet is indeed in every way a most spirit-stirring event," Herschel wrote. "I had myself a narrow escape of it. On the night of July 14th 1830 I swept over the zone. . . . But it is better as it is. I should be sorry if it had been detected by an accident or merely by its aspect. As it is, it is a noble triumph of science."[4] But by far the most remarkable prediscovery observations of all—though they did not come to light until 1980—were found among the observing records of Galileo Galilei at the dawn of the telescopic age. He was tracking the newly discovered satellites of Jupiter in December 1612 and January 1613, when Neptune was in the field, and recorded it on two nights. He even suspected a change in position, but failed to follow up.[5]

With the two Lalande observations from 1795, Walker quickly recomputed Neptune's orbital elements. The table below compares his elements, which are reasonably close to the modern ones, and the final theoretic elements based on the perturbations of Uranus by Le Verrier (his August 31, 1846, calculation) and Adams (his unpublished Hypothesis II of September 1846):

	Walker	Le Verrier	Adams
Mean distance (AU)	30.25	36.15	37.25
Eccentricity	0.00884	0.10761	0.12062
Inclination	1°54'54"	—	—
Longitude of ascending node	131°17'38"	—	—
Longitude of Perihelion	0°12'25"	284°45'	299°11'
Period in years	166.381	217.387	227.3
Mass, Earth = 1	22.0	35.4	33.0
Helio. long., Jan. 1, 1847	328°7'57"	326°32'	329°57'

Clearly, the authority of Bode's law, always a bit shaky, had been completely undermined, while the scant resemblance of Neptune's actual orbit to those of the co-predictors led Benjamin Peirce—already gaining a reputation as America's foremost mathematician—to a conclusion he first stated in an address to the American Academy of Arts and Sciences in Boston on March 16, 1847. Neptune, said Peirce, "is not the planet to which geometrical analysis had directed the telescope; its orbit is not contained within the limits of space which have been explored by geometers searching for the course of the disturbances of Uranus, and its discovery by Galle must be regarded as a happy accident."[6] He based his argument on two propositions in Le Verrier's solution of the Uranus problem: (1) the planet's mean distance must be between 35 and 37.9 AU, corresponding to a period of from 207 to 233 years, and (2) its mean (heliocentric) longitude on January 1, 1800, must be between 243° and 252°. Peirce insisted that while one or the other of these propositions was compatible with the observations of Neptune since its discovery, both could not be. He also showed that the problem of perturbations as considered by Le Verrier and Adams had not only led to the solution they found but to two others, one 120° ahead and the other a 120° behind the actual position of Neptune. Thus, Peirce concluded, "if the above geometers had fallen upon either of these solutions, Neptune would not have been discovered."

The "happy accident" conclusion was by far the most sensational part of Peirce's re-investigation, but of greater importance in the long run was his additional work on the commensurabilities in the periods of Uranus and Neptune—their "Laplacian librations." Laplace had famously shown that because Jupiter's orbital period is to Saturn's nearly in the ratio of 5:2, their motions are in resonance,[7] which produces very large effects on the relative motions over a period of nine hundred years. Peirce noticed that something similar applied to Uranus and Neptune: Le Verrier's calculated lower limit for his hypothetical planet (35 AU) happened to be just below the distance (35.3 AU) where its period would have been exactly 2.5 times that of Uranus. But at this resonance position, the effects of the mutual perturbations would, he noted, become "peculiar and complicated" and invalidate "the continuous law of inference" that Le Verrier had assumed. Even more significant was the commensurability at the distance of 30.4 AU. Here Neptune's period would be 168 years—exactly twice that of Uranus. "The influence of a mass revolving in this time," wrote Peirce, "would give rise to very singular and marked irregularities in the motions of this planet."[8]

Peirce got much of this right, for as we now know, Neptune orbits just inside the resonance position. According to modern values, its mean distance from the Sun is 30.06 AU and its sidereal period of revolution (the period of one revolution with respect to the fixed stars—or equivalently, as observed from some fixed point outside the Solar System) is 164.82 Earth years, while the mean distance of Uranus is 19.19 AU and its sidereal period is 84.07 Earth years. (In recent years, the so-called Nice model, developed at the Côte-d'Azure Observatory in Nice, France, has revealed some of the deeper implications of these relationships for the dynamical evolution of the Solar System. We will say more about this later.)

Peirce's critique unleashed a firestorm of controversy. Many astronomers were perplexed; some were enraged. At a single blow the gloss had seemingly been taken off what had so recently been hailed as the greatest scientific discovery of the age. Several champions rose in defense of the co-prediction, including Le Verrier himself. Adams, characteristically, was silent. The most

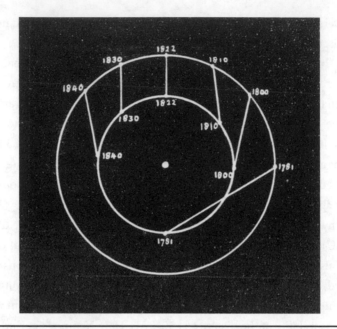

Figure 5.1 The "pull" of Neptune on Uranus. The residuals of Uranus's motion (the difference between its theoretical and observed positions) indicated that Uranus accelerated in its orbit up to 1822 and thereafter retarded. According to John Herschel's (ultimately invalid) method of graphing the residuals, the change in sign should have sufficed to give the position of the perturbing planet without the need to wade into the complicated celestial mechanics used by Le Verrier and Adams. From T. E. R. Phillips, ed., *Hutchinson's Splendour of the Heavens*, vol. 1 (London: Hutchinson & Co., 1923).

cogent argument against the "happy accident" theory was presented by John Herschel, who maintained that Neptune's orbital elements were not important to the discovery at all: it was the fact that Uranus and Neptune had been in conjunction late in 1821. Resolving the force of the exterior planet in the usual manner of vector analysis into radial and tangential components, he showed that at the time of conjunction, the radial force was nearly twice as powerful as the tangential. He further showed, by graphing the residuals of Uranus over time, that the change of longitude near the conjunction was almost entirely because of the action of the tangential force. But if this were so, then there would be a large possible set of orbits that could have approximated the gravitational force vector of Neptune on Uranus in the period between 1800 and 1850. Precise elements were not needed because the apparent position of the hypothetical planet would not have deviated far from Neptune's actual position as seen from Earth at the time, and this was all that was necessary to make a close prediction. Surprisingly, according to Herschel, the inflection in the rate of increase of longitude around the time of the conjunction as revealed by the graphical method could alone have been used to locate the planet. The complicated inverse-perturbation method of Le Verrier and Adams was not apparently necessary after all.

Though this seemed entirely convincing at the time, there was a serious flaw in Herchel's reasoning. The maximum deviation in the longitudes is not actually reached until the perturbing force has had time to annul its previous effect, and this, it turns out, is a long time after its own reversal. However, the flaw was not discovered until the end of the nineteenth century. In the meantime, the graphical method would be embraced as a welcome shortcut to the discovery of new planets: apparently all one had to do was plot the residuals of a planet's motion over time and look for the inflection points. Compared to the laborious perturbation-theory calculations of Le Verrier and Adams, this was child's play. As we shall see, the graphical method would be used by several later planet seekers, including David Peck Todd, William H. Pickering, and, for a time, Percival Lowell.

New Directions

After the discovery of Neptune, Le Verrier stood at the pinnacle of his large ambition. He was now the director, not to say virtual dictator, of the Paris Observatory, and he prioritized what had long been his grand ambition: to mathematically analyze the perturbations of all the planets in the Solar System. A particularly brilliant piece of analysis involved the innermost planet Mercury, on which he had started in 1843, even before his investigation of Uranus. He did not finish until 1859, when he announced that he had found a small but significant discrepancy—an anomalous advance of the perihelion of Mercury of some 38" of arc per century. He could only explain it by assuming the existence of a small amount of additional mass—either a planet, which he called "Vulcan" after the Roman god of the forge, or more likely, a ring of asteroidal material. Whatever it was, it needed to be present in sufficient quantity to account for the excess motion of Mercury's perihelion without affecting the planet's nodes or the motions of Venus or Earth.

Le Verrier's announcement briefly captured the imaginations of astronomers, who hoped to discover Vulcan (or "vulcans") either in transit across the Sun or as uncharted stars in the solar neighborhood during eclipses. The story is an interesting one and has been fully told elsewhere.[9] Vulcan is now known to not exist, though the anomalous advance of Mercury's perihelion is real enough: its value is 42.98" rather than the 38.0" Le Verrier calculated, and it has now been satisfactorily explained by Einstein's general theory of relativity.

Meanwhile, astronomers searched for—and found—an exponentially growing list of new minor planets in the zone between Mars and Jupiter. Hencke's discovery of Astraea in December 1845 was merely the first lightning strike of a massive storm. By the time Le Verrier died, on September 23, 1877, the anniversary of the discovery of Neptune, 174 asteroids had been identified by observers using strictly visual search methods. In 1891, Max Wolf made the first photographic asteroid discovery, and thenceforth they began turning up in such great numbers that they were being referred to as "vermin of the sky." (Apparently the expression was first used by Edmund Weiss, director of the Vienna Observatory from 1878, because asteroid trails spoiled his long-exposure photographic plates.)[10]

Beyond Neptune?

Meanwhile, what about the dark, cold region beyond Neptune? Might other planets be harboring there?

Le Verrier always thought so. Soon after Neptune's discovery, he wrote to Jean-Alfred Gautier of the Geneva Observatory, "This success allows us to hope that after thirty or forty years of observing the new planet, we will be able to use it in turn for the discovery of the one that follows it in order of distance from the Sun."[11]

For a moment, it seemed that it might already have been seen. In 1850 James Ferguson, a USNO astronomer who is remembered as the first to discover an asteroid (Eurphrosyne) on American soil, was tracking the asteroid Hygea (now known as Hygiea) through the thickets of stars in Sagittarius, using a ninth-magnitude star for comparison. A year later, the English astronomer John Russell Hind, a late entrant in the search for Le Verrier's planet and later a noted discoverer of asteroids in his own right, found that the star had apparently disappeared. After examining Ferguson's records, Hind went so far as to suggest that he had seen nothing less than a trans-Neptunian planet at about 137 AU and with a period of 1,600 years. Ferguson himself attempted to recover it, but he did not succeed—nor did anyone else. A quarter century later, Christian Heinrich Friedrich Peters, director of the Hamilton College Observatory in Clinton, New York, discovered a mundane explanation: Ferguson had apparently mislabeled one of the wires of the micrometer he was using, and thus recorded an ordinary star in the wrong position. Like Wartmann's star, it was a figment produced by observational errors.[12]

By the time of Peters's revelation, a more rigorous investigation of the possibility of trans-Neptunian planets was beginning as a byproduct of the massive calculations of the Canadian American astronomer Simon Newcomb. After a troubled childhood in Nova Scotia, which he had escaped by running away as apprentice to a quack doctor and emigrating to the United States, Newcomb rose, largely through self-education and sheer drive, through a series of positions, including computer at the U.S. Nautical Almanac Office (NAO), located at the time in Cambridge, Massachusetts. When the Civil War broke out, many positions at the USNO became vacant as Southern sympathizers left to take up commissions with the Confederacy. Newcomb was nominated for and accepted a position as professor of mathematics and astronomy. From there he embarked, with the insolence of youth—for he was not yet thirty—on the same vast project that Le Verrier had made his province: working through the mutual perturbations and reconstructing the tables of all the planets.

An early fruit of this effort was the calculation of new tables of the motions of Neptune, which Newcomb published in 1866. Therein he briefly considered whether Neptune's motions gave any indications of an unknown planet beyond it. He concluded that there were no such indications. He added that unless a trans-Neptunian planet were coincidentally in conjunction with Neptune, or much more massive than Uranus, it would not disclose itself.[13]

By 1873, Newcomb stood at the center of the enormous American effort to observe the approaching transit of Venus to refine the estimate of the Earth–Sun distance, served as a consultant to the Lick trust (the fortune of the late eccentric millionaire James Lick earmarked to build a telescope in California "superior to and more powerful than any telescope yet made"), and he tested the newly commissioned 26-inch (66-cm) Clark refractor of the USNO in Washington, DC. On top of these miscellaneous and arduous efforts, Newcomb again published revised tables of Uranus. He developed the subject with his typical thoroughness. First, he carefully considered the reductions of the old observations, and re-reduced them wherever he thought it necessary to do so. He also ruled out the various possibilities of errors in the theory of the planet's motion and considered, again, the possibility of a trans-Neptunian planet. As in 1866, he found no indications of such a planet, and instead decided that all that was called for was a small adjustment in Neptune's mass.[14]

For convenience, we summarize in the table below a few of the most important estimates of the mass of Neptune (the current value is 17.148 Earth masses).

Year	Authority	Method	Mass (Earth = 1)
1847	O. Struve	motion of Triton	7.52
1848a	Peirce	motion of Uranus	16.5
1848b	Peirce	motion of Triton	17.6
1849	Adams	motion of Triton	18.4
1851	G. P. Bond	motion of Triton	17.0
1874	Newcomb	motion of Uranus	16.75
1875	Newcomb	motion of Triton	16.97

In his never-satisfied quest for perfection in his tables, Newcomb was always keen to obtain better values of the planets' masses, so in November 1873 he began a series of observations of Neptune with the newly installed 26-inch refractor.

Though Newcomb was never particularly interested in observational work, he took it seriously, and in *The Reminiscences of an Astronomer*, he recalled what drew him to the visual scrutiny of the remote and inscrutable world of Neptune:

> In my work with the telescope I had a more definite end in view than merely the possession of a great instrument. The work of reconstructing the tables of the planets, which I had long before mapped out as the greatest in which I should engage, required as exact a knowledge as could be obtained of the masses of all the planets. In the case of Uranus and Neptune, the knowledge could best be obtained by observations on the satellites. To the latter attention was therefore directed. In the case of Neptune, it was very desirable that in addition to the one known, a more distant one should be found if it existed. I therefore during the summer and autumn of 1874 made most careful searches under the most favorable conditions. But no such satellites were found.[15]

From his further observations of Triton he published in 1875 a new mass of Neptune differing only slightly from that found from his analysis of Uranus's motions the year before. Results obtained from different lines of investigation thus reinforced each other. At the end of his study of the satellites of the outer planets, Newcomb turned the telescope over to a colleague, Asaph Hall, who would use it in 1877 to make its most momentous discovery, Phobos and Deimos, the two tiny satellites of Mars. Newcomb, meanwhile, moved up to the position he had long craved: superintendent of the NAO.[16]

Newcomb's exhaustive memoirs on the motions of Uranus and Neptune lent little encouragement to those who sought planetary quarry in trans-Neptunian space. Nevertheless, several astronomers set out in the chimerical but beguiling quest for them.

The first of these was David Peck Todd, who after graduating from Amherst College in 1875 accepted Newcomb's offer of a position at the USNO. He did so in preference to an offer from Thomas Edison, whom he had met in Newark, New Jersey, where Edison was working by day as a Western Union telegraph operator and by night on electrical experiments in the cellar of his office. Only much later in life—too late—did Todd realize that his true genius had been in mechanical engineering and regret what might have been. At the USNO, Todd became Newcomb's protégé, and as such was soon charged with heavy official responsibilities—not least the mind-numbing work of reducing all the observations of the 1874 transit of Venus for deriving a solar parallax, which occupied him for years. But in 1877 he at last found sufficient spare time to pursue some of his own investigations, including a secretive search for a trans-Neptunian planet.[17]

Rather than follow the complicated perturbation theory that Newcomb had used to recalculate the tables of the two outermost planets, Todd instead adopted a graphical method similar to John Herschel's. Todd claimed to have hit upon it independently. He began by plotting the residuals of Uranus, the difference between the observed and theoretical positions as calculated by Le Verrier in his 1873 tables. Just as Herschel had claimed that Neptune's direction—and hence its approximate location in the sky—could have been ascertained from the maximum of its residuals around the time of its 1822 conjunction, Todd reasoned that a planet beyond Neptune, if one existed, might also be determined from a plot of Uranus's residuals, assuming the peaks to represent its conjunctions with Uranus. So far, Neptune had not been observed long enough for such a purpose. He thought he found tentative indications of a planet with a period of 375 years, giving it a mean distance of 52 AU. He estimated a longitude of the planet for epoch (year) 1877.84 of 170°, plus or minus 10°, which put it in the constellation Gemini. He gave its magnitude as approximately 13+. Though he knew that his analysis was far from conclusive, and must have entertained little chance of success, he nevertheless pointed the great Washington telescope to the indicated region of the sky and spent thirty nights between November 3, 1877, and March 6, 1878, sweeping the stars of Gemini in his quest for a planet. Nothing turned up, and Todd kept the search secret for the next three years. When he finally broke his silence, it was only because another planet seeker, George Forbes, a professor of natural philosophy at Anderson's University, Glasgow, using similar methods, had announced his own pair of putative planets lurking at 100 and 300 AU and having periods of about 1,000 and 5,000 years, respectively.[18]

By then, other indications, albeit inconclusive, appeared of one or more trans-Neptunian planets besides the appearance of peaks in the graphs of Uranus's residuals. In 1880, Camille Flammarion published the first edition of *Astronomie Populaire*, which made him a famous and very wealthy man. Toward the end of the book he discussed comets that seemed to belong to families. Their family membership was indicated by the fact that their aphelia were located near those of the major planets presumed to have perturbed and captured them. Noting that Jupiter, Saturn, Uranus, and Neptune all had such families, he threw out, almost casually, the suggestion that indications of a possible comet family beyond Neptune might point to a planet beyond:

> Every comet which, coming from the outside, passed sufficiently near a planet to be subject to its attraction and be captured by it, would continue its voyage to the vicinity of the Sun, and would afterwards return to the point where it had been diverted from its primitive course, and would thus continue to revolve round the Sun. All the periodical comets have their aphelia near the orbit of a planet. Now, the third comet of 1862 and the swarm of shooting stars of August 10 follow an orbit of which the aphelion is at the distance 48 [AU]. There should exist there a large planet.[19]

When, a few years later, a comet (1889 III) discovered by E. E. Barnard at the Lick Observatory was found to have an aphelion distance near 48 AU, the cometary fingerposts appeared to point ever more insistently toward Flammarion's "large planet" at 48 AU.

These comets would soon draw the attention of Percival Lowell—a one-time student of Benjamin Peirce at Harvard who by 1894 had taken possession of a hilltop in Flagstaff, Arizona Territory, to establish an observatory dedicated to the investigation of the Solar System, especially Mars. Captured at first by the glamorous possibility of a civilization of intelligent beings on Mars, evidenced by the planet's supposed canal system, he would later be drawn to the equally glamorous possibility that a trans-Neptunian planet might exist and await discovery. Lowell's embrace of this possibility would gradually grow into a central obsession of his singularly eventful and controversial career.

6

The Search for Planet X

A Boston Brahmin

THE MOST OBSESSIVE of the mathematical explorers of trans-Neptunian space, and in the end the only indispensable one, was Percival Lowell. Though best remembered for his association with the canals of Mars and speculations regarding intelligent life on that planet, he devoted much of the last decade of his life to a massive but secretive quest for an unknown planet beyond Neptune—"Planet X."

Lowell was born on March 13, 1855, on Tremont Street, across from Boston Common. He dated his earliest memory to age three, when he was living in a vertically stacked mansion on Park Avenue: from the window of a winding staircase he caught a view of Donati's Comet, with a "magnificent scimitar-like curve" of "the most majestic celestial object of which living memories retain the impress."[1] He never forgot it.

A man with a taste for the finest things money could buy, he was prone to develop all-consuming interests that challenged his abilities and allowed him to express his individual personality. The family means were such, moreover, that he could afford to follow those interests as far as he wished—and so he did, around the globe and beyond it.

Figure 6.1 Percival Lowell poses for what would be his favorite portrait, taken in 1910 at a studio on Tremont Street, Boston. The studio was on the same street, located across from Boston Common, where he had been born in 1855. Courtesy of Lowell Observatory Archives.

The Lowells had sunk roots in Massachusetts almost from the time of the *Mayflower*. An earlier Percival Lowell, a Bristol wholesale trader in imports and exports, emigrated to the Massachusetts Bay Colony in August 1638 at the advanced age of sixty-seven, and established himself in Newbury. Subsequent members of the family played outsized roles in New England for generations, but the family's material wealth was mainly founded in the activities of Francis Cabot Lowell, an industrialist who brought the power loom he saw in operation in the cotton factories of Manchester, Birmingham, and Leeds to the banks of the Charles River. After his death in 1817, the operation was moved from its original site, Waltham, to East Chelmsford, on the Merrimack River. The textile-mill city, renamed "Lowell" in his honor, sprang up almost overnight, said the poet John Greenleaf Whittier, "like the enchanted palaces of Arabian Tales."[2]

The city's most striking feature was its complex network of canals, the achievement of James B. Francis, who engineered the greatest of them, the Northern Canal. By 1850, Lowell had grown to become the second-largest city in Massachusetts after Boston, with six miles of canals coursing through the city. The canals operated on two levels to drive water wheels powering 320,000 spindles and almost 10,000 looms, and providing employment to more than 10,000 workers. The canals depended on water drawn from the Merrimack, but to increase efficiency the mill owners dammed the river and ponded the water overnight for use the next day. Anticipating seasonal dry spells, they turned the river's watershed into a giant millpond and aggressively acquired water rights as far away as New Hampshire, storing water in lakes in the spring and releasing it into the Merrimack in the summer and fall.

Anyone familiar with Percival Lowell cannot fail to realize the importance of canals to the imagination of the future astronomer.[3]

By the time Percival was born, the Lowells had become the leading capitalist family of New England and were variously referred to as "economic royalists" or "robber barons." Together with other capitalist families, including the Lawrences, Appletons, Cabots, and Jacksons, they controlled a fifth of America's cotton production by 1850 and had built an economic, social, and political empire. The Lowells were Cotton Whigs, who favored the continuation of the South's "peculiar institution." (In contrast, the Conscience Whigs favored its abolition.) But their business interests were diverse and tentacular, and would outlast the Civil War. They helped develop the Boston and Lowell Railroad, owned controlling stock in a host of Boston financial institutions, and underwrote and insured companies of their own. They also took a leading role in developing Boston's cultural and educational institutions — forming the nucleus of the elite class that Oliver Wendell Holmes famously described as the "Boston Brahmins."

At the head of the most prosperous branch of the family was Augustus Lowell. His father, John Amory Lowell, had kept the family wealth concentrated within the family through two marriages with cousins.[4] Augustus accomplished the same end through marriage to Katherine Bigelow Lawrence, the daughter of a family business partner, Abbott Lawrence, co-founder of Lawrence, Massachusetts, and head of another wealthy textile family.

Percival Lowell, eldest son of Augustus and Katherine, was thus a privileged scion indeed. His father was a typical Brahmin, whose eldest son described him as a man who "acted for results" and, while he "never seemed to care to theorize . . . enjoyed highly the theories of others, when they did not collide with the puritanism which . . . he inherited doubly distilled. He was shrewdly right and reliably effective. But he had a sternness of countenance which made him seem 'hard.'"[5] His mother, who spoiled "Percy" — and to whom he would remain singularly devoted — was a partial invalid; she suffered from chronic renal disease, likely brought on by the strain of too many childbirths, including, between 1855 and 1862, Percival, Abbott Lawrence, the twins Roger and Katharine (of whom only Katharine survived), and Elizabeth. After an interval of twelve years followed Amy, known in the family as "the Postscript" and who became a successful poet. During the Civil War, the family went to Europe, where Percival studied French at a dame school and his mother sought to regain her health at various spas. On their return they settled in a house on ten acres on the western slope of Heath Street, a country road, in Brookline; it was not too far from Boston but at a safe enough distance not to be absorbed by the city. Augustus later called it "Sevenels" (for the seven Lowells living there). Here, in 1870, Percy set up on the flat roof of the house a telescope given to him by his mother and used it to observe Mars.

Naturally, he went to Harvard, where then-president Charles W. Eliot advocated a harmonious and balanced curriculum. Percy, who graduated in 1876, was an exemplar of Eliot's ideal, distinguishing himself in both English composition and math. His math professor was Benjamin Peirce, from whom he learned his celestial mechanics. Peirce called him "the best student mathematician of any who have come under my observation"—high praise, given that Simon Newcomb had been among his previous students. No doubt Percy was aware of Peirce's "happy accident" thesis following the discovery of Neptune. After his graduation, Percy was invited to stay on as a mathematics instructor but, reluctant to tie himself down, he decided instead on an eight-month Grand Tour through Europe and as far as Syria with cousin Harcourt Amory. His letters to friends back home suggest he might have been suffering a prolonged adolescent identity crisis. One senses a real oppression—a sense of dark heavy furniture and draperies, rather like the furnishings of Sevenels—smothering his soul. He wrote to a friend, the future literary historian Barrett Wendell, at the beginning of the fateful year 1877 (that in which the Italian astronomer Giovanni Schiaparelli would announce the discovery of the canals of Mars): "I hope that Chastity Hall is, as in old times, the scene of many a good time and that you like my old qualities. . . . [As for myself], I have . . . become decidedly misanthropic and, with the exception of a few friends, should not feel many pangs at migrating to another planet—or ceasing to exist—were either plan practicable."[6]

On his return from the Grand Tour, Percy worked in his father's office on State Street, where he would later immerse himself in his Planet X calculations. He had the kind of head for figures that could have made him wealthy beyond the dreams of avarice—but he was already wealthy beyond the dreams of avarice and wanted, as he once put it, to pursue "the good" rather than "the goods." Seized with panic both at the prospect of a life in business and an arranged marriage—he was engaged to a Brahmin woman only recently identified as Alice Lee, the daughter of Boston banker George Cabot Lee and sister-in-law of an up-and-coming New York assemblyman named Theodore Roosevelt—he resigned from business and broke off the engagement.

In taking such steps—especially the latter—Percy had committed what was, in the elite, straitlaced society in which he lived, an unforgivable sin.

Boston society was scandalized, and his family was beside itself. Augustus in particular was so livid at his eldest son's irresponsible and self-centered decision to "flutter in the breeze," to use a phrase from one of sister Amy's later poems, that he moved at once to exert an emasculatingly close surveillance over Percy's financial affairs until well past his thirtieth birthday. Nevertheless, he was never in any danger of being disowned by the family, and Augustus still bestowed upon him (as he did all his children) a sum of $100,000. This sum would be worth millions in today's terms, and Percy was shrewd enough to enlarge it with investments—by 1900, the year of Augustus's death, Percy's fortune had grown to $500,000 (about $13 million in today's dollars) and by the time of his own death in 1916 it had reached $2 million ($44 million today). He was a millionaire when the term still meant something. At no point in his life was money ever an object; he could always do as he pleased.

From the Far East to the Far Out

What pleased Lowell was to get as far away from Boston as he could. In fact, after the scandal over the engagement, he could hardly have continued to live there in any comfort. Instead he moved to the other side of the world. Electrified by an 1882 lecture series on feudal Japan given by the zoologist Edward Sylvester Morse at the Lowell Institute (endowed by another Lowell and whose sole trustee at the time was none other than Augustus Lowell), he set out the next year for the Far East. He would spend much of the next decade there and write four books about his observations and experiences: *Chosön: The Land of the Morning Calm* (about Korea), *The Soul of the Far East*, *Noto*, and *Occult Japan*.

When he set out on his last trip to Japan in December 1892, Lowell's interest in the Far East was already fading. He did not yet have a clear sense of where his next enthusiasm might take him, but he did take the trouble before shipping out from San Francisco to meet with the celebrated Lick Observatory astronomer E. E. Barnard. He also brought with him to Tokyo a 6-inch Clark refractor telescope with which he observed Saturn. When in November 1893 he sailed out of Yokohoma Bay back to San Francisco, he had not yet decided what to do next—he was tentatively planning an Easter jaunt to Seville with a wealthy friend Ralph Curtis. However, soon after that the die was cast. The decisive event came at Christmas 1893 when he received from his aunt Mary (sister of poet James Russell Lowell) a copy of Camille Flammarion's book *La Planète Mars et ses conditions d'habitabilité* (The Planet Mars and its conditions of habitability), a huge volume published in 1892 that summarized the history of Martian studies and included extensive translations from Schiaparelli's memoirs about the canals and Flammarion's own strong endorsement of the view that the planet was thriving with life, even intelligent life.

From then on, Lowell was utterly smitten with Mars (as a native Bostonian, he would of course have pronounced it "Mahs"). At the end of January 1894 he met with William H. Pickering, a junior astronomer at the Harvard College Observatory and the younger brother of the director, Edward Charles Pickering. At the time, the younger Pickering was recognized as a leading expert on Mars in his own right, and he and his assistant Andrew Ellicott Douglass were just back from observing the planet in Peru. Lowell and Pickering agreed on an expedition to Arizona Territory in the American southwest, where Pickering expected the atmospheric seeing would be as good for planetary observing as he had experienced in Peru. Pickering would take a one-year leave of absence from the observatory to join the expedition; Douglass's services were also retained. The latter was sent west early that spring, with the same 6-inch telescope Lowell had taken with him to Tokyo, to scout for Arizona sites with the most stable air and hence the most favorable seeing conditions. He reached Tombstone in early March; moved on to Tucson, where it was his bad luck to encounter stormy weather, then north to Tempe, Phoenix, Prescott, and Ash Fork. Finally, in early April, he arrived at Flagstaff, a lumber town located at an elevation of 7,000 ft on the Colorado Plateau and conveniently located on a line of the Atlantic and Pacific Railroad. On April 5 and 6, Douglass rated the seeing conditions as eight on a ten-point scale at an opening in the woods on the hill just west of town, dubbed "site eleven." It was a better result than he had found anywhere else. Lowell, who had remained in Boston and was busy correcting the page proofs of *Occult Japan*, was becoming increasingly impatient to get underway. He was keenly aware that time was of the essence because Mars's next favorable approach to Earth would occur in October, only six months away. Decisively, Lowell telegrammed Douglass to locate the observatory in Flagstaff.

With borrowed telescopes, a 12-inch Clark refracting telescope from Harvard and an 18-inch Brashear refractor diverted temporarily from its eventual destination at the Flower (later Flower and Cook) Observatory in Philadelphia, and with a prefabricated dome designed by Pickering and sent west in parts, the observatory was hastily assembled. On the eve his setting out for Arizona Territory, May 22, 1894, Lowell gave a talk to the Boston Scientific Society in which he explained his purpose: he intended to investigate the canals of Mars during the planet's next favorable approach to Earth. In fact, he already had definite views as to what he would find. "We are looking," he said, "upon the result of the work of some sort of intelligent beings. . . . The amazing blue network on Mars hints that one planet besides our own is actually inhabited now."[7]

Mars Watcher

Lowell disembarked in Flagstaff on May 28, 1894. Pickering and Douglass had been temporarily quartered in the Bank Hotel, across from the railroad station, but on that first night an eager Lowell and Pickering roughed it by camping in the still-uncanvassed dome to be ready for early-rising Mars, which was still a very long way from Earth. Unfortunately, that night the clouds moved in and it began to rain. On May 31 the sky was more cooperative, and Lowell got a first view of the planet with the 12-inch. The following night he first used the 18-inch.

So began an adventure that still seems among the most romantic ever undertaken in the history of astronomy. Lowell and his colleagues, Pickering and Douglass, kept under closest scrutiny the world that was agreed to be the most like our own from the large canvas-covered dome on the elevation that Lowell was soon calling "Mars Hill." Lowell was only a part-time observer— perhaps spurred by anxiety over his mother's failing health (she died the following April), he returned to Boston during July and much of August, then travelled back to Flagstaff in late August, returned to Boston again in early September, and finally came back to Flagstaff for the period of Mars's closest approach in October and November. During his absences, Pickering and Douglass continued to monitor the planet. For the first month or so of peering through the eyepiece, Lowell struggled to make out any of the expected canals. In time, however, he trained himself to capture fine details revealed in flashes of superior seeing. His "learning curve"—combined no doubt with the changing phenomena on the planet—can be read in his observing logbooks: in contrast with his bland early efforts, the drawings that fill the later pages are replete with spindly features.

Mostly in the library at Sevenels rather than in the observatory on Mars Hill, Lowell formulated over that summer his celebrated—and controversial— "theory" of intelligent life on Mars. He confidently placed it before the world in a six-part series of articles published in *Popular Astronomy* between September 1894 and April 1895; in a lecture series, given to a capacity crowd at Boston's thousand-seat Huntington Hall in February 1895; in four articles in the *Atlantic Monthly* that appeared between May and August 1895; and finally in his best-selling book, *Mars*, which was based partly on his earlier articles. The book appeared in December 1895, just as he was about to sail to Europe (as he would continue to do every other year through 1914).

His ideas about Mars, expressed in flowery prose, electrified the public but provoked a critical reaction from leading astronomers who emphasized the difficulty of correctly perceiving detail on another planet under the trying conditions of Earth-bound observation.[8] Indeed, even as Lowell and his associates had been busy sketching canals from Flagstaff, just a few hundred miles farther west, at the Lick Observatory on Mt. Hamilton, Barnard was studying the planet with that observatory's 36-inch Clark refractor, an instrument with twice the aperture as Lowell's and favored both by the thinness of mountain air and the laminar airflow off the Pacific Ocean. Under these conditions, Barnard saw a tremendous amount of surface detail on Mars, including mountains and elevated plateaus. But there were no canals. As he wrote to Simon Newcomb:

> To save my soul I can't believe in the canals as Schiaparelli draws them. I see details where he has drawn none. I see details where some of his canals are, but they are not straight lines at all. When best seen these details are very irregular and broken up—that is, some of the regions of his canals; I verily believe—for

all the verifications—that the canals as depicted by Schiaparelli are a fallacy and that they will so be proved before many favorable oppositions are past.[9]

Lowell, nevertheless, pressed on. In the summer of 1896, he spent twenty thousand dollars (the equivalent of more than half a million dollars today) to acquire a new permanent telescope, a 24-inch (61-cm) refractor with a jewel of a lens that, according to its maker Alvan Graham Clark, was the most optically perfect of any of the large lenses his celebrated firm had made.[10] It was housed in a wedding-cake dome built by Flagstaff man-of-all-work Godfrey Sykes that was intended to be readily taken apart, shipped, and reassembled elsewhere. Indeed, that was to be the fate of both telescope and dome: they were shipped to Tacubaya, outside Mexico City, in time for the December 1896 opposition of Mars. Though Pickering had gone back to Harvard, Douglass was still on board and took charge of the Tacubaya installation. Two other assistants, T. J. J. See and Wilbur A. Cogshall, had also joined the staff. They were already using the telescope weeks before Lowell himself arrived at the end of December, by which time it was late to do much with Mars. Opportunistically, Lowell used the telescope to make observations of Mercury and Venus, and immediately waded deep into controversy.

Venus, the most impressive of all the planets to the naked eye, had long been a bitter disappointment to telescopic observers. Most astronomers had found its dazzling disk virtually featureless and regarded the planet's surface as forever hidden beneath a highly reflective and never-parted cloud deck (as is still the case today). Lowell, however, immediately drew a spoke-like system of markings on the disk, a design in black and white, unchanging in aspect and quite unlike anything reported by anyone else. He believed himself to be looking through a thin haze directly onto the planet's surface, which he thought probably consisted of exposed ribs of bare rocks sticking out of the sand. Further, he deduced that the planet always kept the same face toward the Sun. (Though all his assistants recorded vestiges of his system of markings, it was his secretary, Wrexie Louise Leonard, whose drawings reproduced the most Lowellian effects.)

As a case of overinterpretation of uncertain observations—not to say of a runaway imagination—Lowell's work on Venus has been rarely surpassed. The reaction of the astronomical world was immediate—and fierce. Though

Lowell emphasized that his spoke system appeared completely natural (it lacked the appearance of artificiality he discerned in the Martian system), to most astronomers it was a distinction without a difference: if the streaks on Venus were not canals, then they were their lineal descendants. Several astronomers could not resist referring to them as canals. High-strung at best, and not accustomed to criticism, Lowell was unhinged and suffered a complete nervous collapse. On returning from Mexico, he made a brief visit to Boston before setting out for Flagstaff intending to resume his work, but made it no further than Chicago. Astronomically speaking, he was to be out of commission for four years.[11]

A Career Reconstituted

Lowell's recovery was gradual and proceeded by fits and starts. His father's doctor recommended the rest cure. Lowell tried it for a month, found it intolerable, and abandoned it for a regimen of travel, exercise, and sunshine. The one thing most needed was avoidance of any subjects that would upset his nerves. First and foremost, that meant Venus. Finally, by 1901 he had recovered enough to return to Flagstaff, eager for vindication. His first order of business was to dismiss Douglass, who ran the observatory during Lowell's illness and had developed an independent streak regarding the markings on Venus and Mars. He thought them mostly illusions. With Douglass's departure, Lowell resolved to carry on, alone if necessary. "I am so much at home here," he told his sister Elizabeth, "and yet no one I know knows it."[12]

Lowell's next decision was to acquire a spectrograph from Pennsylvania optician John Brashear, with which he hoped to launch a new investigation of Venus, prove its long rotation period, and add credence to the spoke-system observations that had implied such a result. Then, despite his disillusionment with assistants, to honor an earlier promise made to his former assistant Cogshall, now at Indiana University (IU) in Bloomington, to hire recent IU graduate Vesto Melvin Slipher to operate the instrument. Slipher arrived in 1901 and began working on a temporary basis. In the end, he would remain at the observatory for fifty-three years, until his retirement in 1954. He was the first of several assistant astronomers hired from IU—others included Carl Otto Lampland and Slipher's brother Earl. Their scientific skills, caution, and sheer durability would help assure the observatory's future long after Lowell himself was gone.

By 1903, Slipher had achieved sufficient competence with the spectrograph to announce, to Lowell's great satisfaction, that Venus's rotation was indeed slow and probably equal to its period of revolution around the Sun, 225 days. Unfortunately, this success did little to restore Lowell's reputation among his professional astronomical peers or quiet the din of criticism to which he was subjected.[13] Against this backdrop, Lowell's quest for the trans-Neptunian planet took shape, a quest that would eventually lead to the discovery of the world hereafter to be our main if not exclusive concern. From modest beginnings at the turn of the twentieth century, it would grow into one of the principal preoccupations of Lowell's last decade. Though he no doubt found grappling with the intricacies of celestial mechanics appealing in and of itself, he also saw in mathematics a way to achieve results that would be immune to the kind of criticisms his Mars and Venus observations had elicited. Clyde W. Tombaugh later recalled, "Carl Lampland told me, Lowell wanted desperately to improve his credibility among other astronomers. So, Lowell thought, if he could predict the location of a ninth planet, beyond Neptune, and then find it, it would surely improve his status."[14]

Planet X

Lowell traced his interest in the possibility of trans-Neptunian planets to his perusal in 1901 of a catalog of orbits of all observed comets compiled by Galle, the optical discoverer of Neptune.[15] From this he picked up on the notion of comet families, and suspected on this basis the existence of two planets at distances of 48 AU and 74 AU from the Sun.[16] He briefly alluded to these findings in a series of lectures on the Solar System he gave at the Massachusetts Institute of Technology in December 1902. After mentioning Jupiter's comet family, he added,

Jupiter is not the only planet that has a comet-family. All the large planets have the like. Saturn has a family of two, Uranus also of two, Neptune of six; and the spaces between these planets are clear of comet aphelia; the gaps prove the action.

Nor does the action, apparently, stop there. Plotting the aphelia of all the comets that have been observed, we find, as we go out from the Sun, clusters of them at first, representing, respectively, Jupiter's, Saturn's, Uranus,' and Neptune's family; but the clusters do not stop with Neptune. Beyond that planet is a gap, and then at 49 and 50 astronomical units we find two more aphelia, and then nothing again till we reach 75 units out.

This can hardly be accident; and if not chance, it means a planet out there as yet unseen by man, but certain sometime to be detected and added to the others. Thus not only are comets a part of our system now recognized, but they act as finger-posts to planets not yet known.[17]

In this line of argument he owed much, of course, to Flammarion. So far, however, Lowell gave no hint that he was personally interested in pursuing the problem further. To the contrary: no further references to trans-Neptunian planets appear in either Lowell's published works or in the Lowell Observatory Archives until early 1905.

By then he was pursuing—as he usually did—several independent investigations at once. He and Lampland were preparing to use a new planetary camera to try to photograph the canals of Mars at the planet's favorable opposition of May 1905. Achieving that success would also improve his status among astronomers. As he told Lampland, "We will get out something to make others sit up!"[18] He was also writing a new book on Mars, *Mars and Its Canals*, which he would finish in June 1906. As is apparent from the numerous drafts and revisions preserved in the Lowell Observatory Archives, it caused him far more pains than any of his other books. He intended it as his magnum opus on the red planet, and so it would prove to be.

In addition to his usual Mars-related activities, he took the first steps toward a systematic search for the possible planet to which the fingerposts of comets pointed. It was an investigation with a difference: though his astronomical activities usually had been high profile and well publicized, information about "Planet X," as he called it, was to be shared discretely.

Even members of his own staff received Planet X information on a strict need-to-know basis.

In February 1905, Lowell wrote to Edward Pickering at the Harvard College Observatory and to Camille Flammarion in France requesting astrographic charts of star fields along the "invariable plane." The invariable plane, sometimes also called Laplace's invariable plane, is the weighted average of all planetary orbital and rotational planes. As Jupiter is by far the most massive of these bodies, the invariable plane lies within a half degree of the orbital plane of Jupiter. This, rather than the plane of the minuscule Earth's orbit (the ecliptic), was the most probable zone in which to find a lurking planet.

As a further step, he hired a human computer, William F. Carrigan of the NAO, to compute residuals (the difference between the observed and theoretical positions) of Uranus and Neptune. Getting a good set of residuals would furnish, as with the nineteenth-century mathematical planet seekers, the groundwork for any mathematical investigation of perturbations due to unknown planets.

The First Invariable Plane Search

Mars remained Lowell's first priority to the end of his life, though his fundamental views about it never changed. Every other year—including in May 1905, when he began to explore the trans-Neptunian project—Mars was in a favorable position for observation and took up much of the time of Lowell and his assistants. Realizing that neither he nor Slipher or Lampland could take on the additional (and doubtless time-consuming) duties associated with any serious initiative to find Planet X, in April 1905 he founded the Lawrence Fellowship (named after his mother, whose maiden name was Lawrence), coordinated with IU and purportedly aimed at giving selected IU astronomy graduates experience with practical observing and a chance to pursue their thesis research while receiving a modest stipend (fifty dollars per month). In practice, they would have little time to pursue their thesis research, instead devoting almost all their available time in Flagstaff to the tedious work of photographing the invariable plane. Lowell hired three Lawrence Fellows: John C. Duncan (1905–6); Earl C. Slipher (1906–7), Vesto's younger brother, who would be retained as permanent staff; and Kenneth P. Williams (1907).

After some test exposures by Lampland with the 24-inch Clark telescope, whose 1° square field was too small for the purpose, a lens with a 5-inch wide field, crafted by the Viennese optical firm of Voigtländer, was tried. Duncan began testing it in September 1905. It yielded a field 15° wide, with sharp definition over some 10°. Duncan sent the first invariable-plane plates obtained to Lowell's Boston office. However, Lowell was in England announcing Lampland's apparent success at photographing some of the Martian canals, and did not actually get around to examining them until early December. By then, though Duncan had grown dissatisfied with this lens, V. M. Slipher urged him to use it to obtain another set of plates covering the same sky regions. On one of these, taken on November 29, Slipher found an image that he took to be that of a comet. He sent the plate to Lowell, who announced the discovery to the Harvard College Observatory. On December 29, Lowell further announced that he had found another comet, with "two tails," on the same plate. By then too much time had passed for the discoveries to be confirmed elsewhere, but Lick Observatory astronomer Robert G. Aitken wryly remarked, "The discovery of a comet by photography is unusual enough to be noteworthy, but to find two on a single plate is a unique achievement." He could not resist pointing out that photographic defects often look remarkably like comet tails, and added—with a touch of sarcasm—"It is of course assumed that Professor Lowell took precautions to guard against such deception before announcing the discoveries."[19] Lowell Observatory historian William Graves Hoyt concludes, "No other astronomers seemed to have sighted the purported comets."[20]

A different wide-field lens of 6.375 inches was briefly tried in place of the Voigtländer, but proved to be no more satisfactory. Only in January 1906 did the observatory procure an instrument that was deemed more suitable, a 5-inch doublet lens manufactured by the Pittsburgh optician John Brashear, which reached a limiting magnitude of 16 and boasted a sharply defined field of about 5°. Between then and September 1907, Duncan, Earl Slipher, and Williams used it to expose pairs of photographic plates along the entire invariable plane. At that point, the Lawrence Fellowship ended. The program resulted in asteroids turning up en masse, but it found no new planets.

This first survey was just a first chance. Clearly, Lowell himself had many distractions, and being competitive and fiercely independent, he rarely sought advice. That meant it took time for him and his assistants to work things out for themselves. The initial photographic survey with inadequate equipment had squandered several months of valuable time before they decided on the 5-inch Brashear lens. Also, the means of examining the plates were primitive: Lowell simply placed one plate on top of the other and examined them with a hand magnifying glass. This was highly inefficient, to say the least.

Bad luck also played a role. As Tombaugh pointed out after the discovery of Pluto—the closest thing to Planet X astronomers would ever find—Pluto's highly inclined orbit during 1905–7 placed it far from the ecliptic and outside the belt covered by plates with the 5-inch Brashear. Also, it was then a sixteenth-magnitude object, much fainter than the eleventh- or twelfth-magnitude object Lowell was expecting and at the very limit of the search plates.

Despite its negative result, this first photographic search provided, as Hoyt points out,

> some sound lessons on how to conduct a search. . . . It also showed the practical futility of searching at random through trans-neptunian space for a body whose presence was merely presumed and about which, if indeed it did exist, nothing whatever was known. What was needed, Lowell was well aware, was some indication of where to look, however general in nature, some "fingerpost" that would point out the more probable places where a hitherto unseen planet might be hiding among the millions of stars in the sky.[21]

Residuals

Lowell was already avidly pursuing such fingerposts. At the time, except for one or two highly able experts in celestial mechanics who realized the flaws in the approach, the most direct way to find the position of an unknown planet was still thought to be John Herschel's graphical method. In the graphical method, the residuals (the differences between the theoretical and observed positions of a planet) were plotted, and the peaks were identified with presumed conjunctions with the unknown, perturbing planet.

Errors could enter either from the theoretical or the observational side. Neptune had still not been observed long enough to be of any use; therefore the search for an unknown planet still depended on Uranus. On the observational side, astronomers believed the handful of observations made accidentally before William Herschel's discovery in 1781 to be subject to large errors, so much so that Newcomb—in computing his own tables of Uranus in 1873—had been reluctant to rely on them at all. The post-1781 observations were much more accurate, but even they were not beyond reproach. Newcomb had written,

> To make all the data of reduction rigorously homogeneous and uniform, it would be necessary to completely re-reduce the greater part of the observations made before 1850, using the modern values of the constants of reduction, and to compare each observation separately with the geocentric place determined from the provisional theory. Such a reduction and comparison would be extremely desirable. The execution would, however, involve an amount of labor far greater than it is now possible for the author to bestow upon the problem. We must, therefore, adopt the reductions which have been already made, applying such systematic corrections of reductions to a uniform system of star places as we have the means to readily determine.[22]

Of course, the theories of the motions of the planets also contained uncertainties. As Lowell himself would later point out,

> The residuals left by the theory are not at all the outstanding perturbations, but only such small part of them as cannot be got rid of by suitable shuffling of the cards. We have then no guarantee that our supposed elements are the real ones, but only the best attainable under the assumption *that no unknown exists*. Every theory of a planet is thus open to doubt, seeming more perfect than it is. It has been legitimately juggled to come out correct, its seeming correctness concealing its questionable character.[23]

As noted earlier, when he first began to lay the groundwork for his planet search, Lowell had hired William F. Carrigan to do the very thing that Newcomb had said would involve so much labor—recompute the residuals for both Uranus and Neptune. Carrigan was then forty years old, and as an employee of the NAO devoted his regular business hours to grinding out long computations for the *American Ephemeris and Nautical Almanac.* Despite the rigors of his day job, he was eager enough (or desperate enough) for the extra income to agree to devote as much time as he was able to make additional computations on Lowell's behalf. Hoyt calls him the "forgotten man in Lowell's Planet X" search. He was not a drudge but a skillful mathematical astronomer who published papers in leading astronomical journals. Lowell, however, did not want to consult him on perturbation theory. He only wanted his services as a drudge.

Lowell's correspondence with Carrigan (in the Lowell Observatory Archives) shows the human computer's fitful progress in carrying out the project. It also shows Lowell's conflicting criteria of wanting everything done carefully and rigorously versus his increasing impatience for results and desire to keep costs down. Only in March 1908 did Carrigan finally finish the first stage of his protracted work and present Lowell the recomputed residuals of Uranus from 1780 to 1820. At the moment, Lowell was eager to have him press on with those from 1820 to 1900 but was also advising a more streamlined approach, telling Carrigan, "I should not take every observation but a few of the best ones only for each [year]."

As always, Lowell was stretching himself thin. He was busy arranging for publication of the results of his 1907 Mars observations, publicizing V. M. Slipher's January 1908 spectrographic attempt to detect water vapor in Mars's atmosphere (which at the time seemed to have succeeded), and working up his most recent course of Lowell Institute lectures for publication as *Mars as the Abode of Life.* He also was negotiating with the firm of Alvan Clark and Sons for a new 40-inch (102-cm) reflector he hoped would vindicate the Martian canals to skeptics who claimed that they could not be discerned in large apertures. Moreover, in June 1908 he finally acceded to endless entreaties and went from bachelor to benedict. His betrothed was his Boston neighbor, interior decorator and Back Bay real-estate speculator

Constance Savage Keith. After their New York City wedding, the couple set sail for a summer honeymoon in Europe—a literal highlight was Lowell's balloon ascension with his cousin A. L. Rotch, the founder of the Blue Hill Meteorological Observatory, into the skies over London to photograph the crisscrossing paths in Hyde Park for comparison with the Martian canal network. When Lowell returned from Europe in early October, he made no move to reopen his theoretical trans-Neptunian exploration. It didn't seem urgent. However, that would change when his onetime associate in the founding of the Lowell Observatory, William H. Pickering, came forward with a startling announcement.

A Competitor Appears

Pickering's surviving notebooks, preserved in the archives of Tulane University in New Orleans, show that his documented interest in a trans-Neptunian planet (or planets), like Lowell's, also dates back to 1905. As Lowell had done, he had scanned lists of comets to identify possible families belonging to distant planets, and had identified three prospects. From this early—and never published—inquiry, Pickering soon extended his researches, but instead of attempting to do as Newcomb had suggested and as Carrigan was doing for Lowell, he made no attempt to recompute the residuals. Instead he did as Todd had done and simply plotted the residuals as given by Le Verrier in his 1873 memoir. This was a risky approach—Le Verrier's theory of Uranus was now well out of date, having been supplanted by the revised theories of Uranus of Newcomb (in 1873 and again in 1900) and Le Verrier's protégé Jean-Baptiste Aimable Gaillot (1903). These theories used different elements—in particular, different masses for the planets, the adoption of which would necessarily produce conflicting residuals. As the entire problem depended on the residuals, differing masses would give different solutions for the unknown planet.

Pickering ignored these difficulties and deduced from the Le Verrier residuals the existence of a planet, "O" (so-called because O was the next letter in the alphabet after N, for Neptune). Given that Pickering had used the same residuals as Todd had used in 1877, his result was similar in some

Figure 6.2 The 5-inch Brashear refractor used by the Lawrence Fellows John C. Duncan, Earl C. Slipher, and Kenneth P. Williams during Lowell's first search for Planet X, 1905–7. The plates obtained with this telescope were simply superimposed on top of each other and examined by Lowell with a handheld magnifying glass. Courtesy of Lowell Observatory Archives.

respects. On the assumption of a circular orbit for O, Pickering concluded that its mean distance was 51.9 AU, its period 373.5 years, and its mass twice that of Earth. He also gave for its approximate position (for epoch 1909.0) R.A. 7h 47m, Dec. +21°. This would have put it, had it existed, in eastern Gemini at the time. Pluto was then one constellation over, in Taurus. Neptune was in Gemini, not far away from the position of the putative O.

Before publishing his results, Pickering organized a secret search for the planet, but nothing turned up. On November 11, 1908, at a meeting of the American Academy of Arts and Sciences in Cambridge, with Lowell in the audience, Pickering presented his paper. After speaking, he went so far as to drop Lowell a note, suggesting that perhaps the latter might search for his planet from Flagstaff. Lowell responded tersely, "I am looking up the whole subject myself, and have not yet got far enough along to undertake any visual search. When I get a position I will let you know."[24]

Despite the nonchalance of Lowell's response, Pickering's paper actually provoked something of a crisis because it revealed that Lowell had at least one serious rival. The immediate effect on Lowell seems to have been to suggest that perhaps the signature of the perturbing planet in the residuals of Uranus was greater than he had supposed. After his longstanding attitude of (usually) patient understanding toward Carrigan, he suddenly began berating his computer almost daily with letters demanding new data and residuals. He was now determined to try the rule-of-thumb graphical method for himself. As a byproduct of this galvanized effort he drew up several graphs of residuals—not only for Uranus, but also for Jupiter and Saturn. One graph—showing the residuals of Jupiter from Bouvard's tables—showed three peaks corresponding to conjunctions with the then-unknown Neptune. On making this discovery, Lowell exulted to Carrigan, "Neptune could have been found apparently from Jupiter's residuals."[25]

This phase of Lowell's investigation continued until the spring of 1909, when Pickering published a somewhat expanded version of the paper he had given the previous November.[26] Lowell read through it carefully and made annotations in the margins of his copy. He saw at once that Pickering had already covered much of the same ground as he—for instance, Pickering had also noticed the peaks corresponding to conjunctions of Jupiter in the residuals of Neptune. Lowell also saw that Pickering, using the same data and following the same methods as Todd, had essentially reached Todd's conclusions. The chief difference between Todd and Pickering, as Pickering himself had realized, was the assigned positions of the unknown planet, some 84° in longitude apart. According to Hoyt, "Pickering's full statement of his work, along with Lowell's own experience with Pickering's graphical method, seems to have convinced Lowell that the problem of a trans-neptunian planet admitted no easy solution."[27] Lowell's last note on Pickering's paper read, "This planet is very properly designated O [and] is nothing at all."

Soon after he read Pickering's paper, Lowell terminated Carrigan's services. Increasingly irritated by the latter's thoroughness (which he had at first asked for) and by how much money it was costing him, Lowell informed Carrigan,

> I have now completed my investigation (I regret without being able to use any of your computations) and I find evidence of an exterior planet at 47.5 astr. units from the sun, magnitude 13+, hel. longitude Jan. 1, 1909, 287°±. My paper will soon be published. In view of my completion I shall not need any more computations at present, though I shall hope to call on you in the future.[28]

In fact, he would never call on Carrigan again, nor did his promised paper appear, though several drafts of it exist in the Lowell Observatory Archives, from which it can be inferred that he had been planning to publish the results he obtained using the graphical methods.[29]

A yearlong hiatus in the Lowell Observatory Archives follows, from July 1909 to July 1910, during which Lowell finally, if reluctantly, concluded that the shortcut methods such as those used by Todd and Pickering were hopeless. He now faced a crucial fork in the road: he could either give up the problem as too difficult or commit even more energy and resources to its solution. Lowell, without apparent hesitation, opted for the second path. By 1910, he was immersing himself in deep researches into perturbation theory and was mobilizing the same mathematical methods Le Verrier had used to capture Neptune to prize another planet from the tiny residuals in Uranus's motion that remained after Neptune's contribution had been factored out.

Lowell's Second Search and
Memoir on a Trans-Neptunian Planet

Lowell's rather obsessional personality and preoccupation with order were attested by his wife, Constance, who recorded an incident that occurred shortly after she and her husband returned from their European honeymoon in 1908:

> In October, soon after our return from Europe, I discovered that the scientist's motto is—"Time is sacred." I was to meet him on the train for Flagstaff leaving the South Station [Boston] at 2 p.m.; anxious to impress him with my reputation for being punctual, I boarded the train about ten minutes before two. Percival came into the car, holding his watch in his hand, just about two minutes before two. He turned to me: "What time were you here?" I answered triumphantly: "Oh, I got here about ten minutes ago." His reply was: "I consider that just as unpunctual as to be late. Think how much could have been accomplished in ten minutes!" I have never forgotten that remark. Percival never wasted minutes.[30]

But if a mere train was to be held to a punctuality of minutes, planets were to be held to an even higher order of punctuality. The observed motion of Uranus, after the corrections of Neptune's pull had been deducted, was still off by a few seconds of arc from calculated motion. Lowell intended to squeeze those seconds for all they were worth, and to make them yield—if possible—the position of an unknown planet.

The trail in the Lowell Observatory Archives of Lowell's preoccupation with Planet X, which had petered out in 1909, resumes in the summer of 1910. We have a glimpse of the man at the time, secure in his Mars Hill redoubt, in the notes kept by Edward Pickering, who in the late summer of 1910 was wending his way west to the International Solar Conference to be held at Mount Wilson. Pickering arrived at the Flagstaff train station late in the evening of August 24, and after resting—presumably in the adjacent Bank Hotel—was up the next morning at seven, feeling refreshed, and "walked up hill to Observatory, about two miles off, and five hundred [ft] up, to the surprise of other members of the party, most of whom rode." He continued,

Much of interest at the Lowell Observatory. Fine collection of illuminated photographs, mainly of planets and comets. Large equatorial and reflector. Mrs. Lowell very cordial. I urge her to bring Mr. Lowell to Mt. Wilson, and assure them a hearty welcome. Rotch [Lowell's cousin] wants me to repeat my invitation to Lowell. The latter replies, "No, I am an astronomer!" I feel somewhat snubbed at this remark and so does Rotch, who said, "He is always rather abrupt."[31]

In insisting, "No, I am an astronomer," Lowell was opposing the term to the new and fast-developing domain of astrophysics. He elaborated later to John Trowbridge, president of the American Academy of Arts and Sciences, when Lowell proposed to endow a monetary award to "an astronomer" working in what Lowell considered the more traditional lines of the discipline: "I beg to call your attention to the fact that astronomer is used in its technical sense, and not including astrophysics which, of late years, owing to the effect that pictures have on people, has usurped to itself the lime light to the exclusion of the deeper and more profound parts of astronomy proper."[32] Though Lowell encouraged research into astrophysics among his younger assistants—and eventually gave a rather free hand to V. M. Slipher, whose spectrographic studies of nebulae issued in fundamental astrophysical discoveries—his own researches remained thoroughly devoted to what he referred to as astrophysics' "elder sister," astronomy.

This was nowhere more evident than in the mathematical search for Planet X and in his determination to follow the classic nineteenth-century paradigm of inverse perturbation calculations that led Le Verrier and Adams to the triumphant discovery of Neptune. He had, in one sense, an easier task than theirs; they had already shown the way. But in all other respects his task was much more daunting. Le Verrier and Adams were like the first climbers of the sheer face of Yosemite's El Capitan, who had scaled it using ropes. Lowell was like one intent on climbing the wall, smooth as alabaster and nearly vertical, with nothing but handholds and footholds. The analogy is not entirely fanciful. Whereas in 1845, the residuals of Uranus had reached the precipitately steep figure of 133" of arc—an error more than twice the diameter of Jupiter—when Lowell set out in 1910 to employ the analytical methods of inverse perturbation theory to the small residuum in the remaining residuals, they had dwindled to next to nothing. As Lowell himself admitted, "To-day its residuals do not exceed 4.5" at any point of its path."[33] Even this figure was an exaggeration of what he had to work with because 4.5" was the maximum; most of the residuals were 2" or less—within the reasonable margins of observational error. This was especially the case for those between 1781 and 1840. Moreover, Lowell had little clear evidence of any systematic changes over time. He was facing the now-familiar problem of signal to noise—and he was at least somewhat aware of the fact that, as Hoyt puts it, "even the best of methods is no better than the data to which it is applied, and that a constant, coherent body of data for the trans-neptunian body was lacking, and that much time and labor would be necessary to acquire such data."[34] But by then Planet X—like the Martian canals, which presented a similar signal-to-noise conundrum—had become an obsession. In both cases he was destined to flounder because he based his conclusions on data that did not rise convincingly above the margins of error.

After the discovery of Pluto, Ernest W. Brown of Yale revisited with cool and dispassionate analysis Lowell's calculations, and concluded, "Practically the only significance which can be attached to the residuals between 1750 and 1840 is their approach to a straight line. . . . If a planet does not produce sensible oscillations it is of course impossible to do anything in the way of prediction."[35]

As we have emphasized repeatedly in the preceding discussion, the residuals themselves are an amalgam compounded of uncertainties in the theories of the planets' motions and those in the observations themselves, which are impossible to eliminate. The errors, especially in the pre-1781 observations, were bound to be far from negligible, and had already been thoroughly discussed by previous investigators, beginning with Bouvard. He had given no details of his reductions, but the mean times Bouvard obtained for these observations were shown to be inaccurate by Le Verrier, who after correcting them published new reductions of all Flamsteed's and Le Monnier's observations. Lowell seems to have not appreciated, at first, the thoroughness of what had already been done with the reductions, or else he would never have imposed the burden of recalculating them on Carrigan. However, by 1910, he had learned enough to appreciate that the positions were unlikely to improve much by further examination of original sources or attempts at new reductions. The raw data, marginal as it might be, was as good as it was going to get.

Thus in 1910 he simply adopted Le Verrier's reductions of the observations from 1873. He proceeded to estimate the observational errors by grouping those taken around the same times and applying to them a not-quite-legitimate least-mean-squares analysis. Among the pre-1781 observations, the groups consisted of Flamsteed's observations of March 4 and 10 and April 29, 1715; Le Monnier's of October 14 and December 3, 1750; Bradley's single observation of December 3, 1753; Mayer's single observation of September 3, 1753; Le Monnier's eight from December 27 and 28, 1768, and January 15 to 23, 1769, together with his one on December 18, 1771. Regarding the likely errors in these, it should be noted that Le Monnier's observations, all made within the same month, differed from each other by 16.9″. This ought to have given a clear idea of the limits of their accuracy!

As Le Verrier and Adams had done, Lowell set the residuals (or the least mean squares of groups of them) for each time interval on one side of a series of condition equations. These were complicated formulas expressing the perturbations caused by the unknown planet on the orbit of Uranus in terms of the orbital elements of Uranus and the unknown planet. Using a long series of iterative algebraic operations, the orbital elements—and the position of the unknown that would allow it to be found in the heavens—were to be solved for.

The orbital elements were the length of the semi-major axis, mean motion, longitude at epoch (the planet's place in its orbit), its eccentricity, the perihelion of the orbit, and the inclination. As Adams and Le Verrier had done, Lowell had to assume a distance. He decided on a distance sunward from what would have been expected using the no-longer-creditable Bode's law: 47.5 AU.

Lowell was assisted in this vast undertaking by several computers, including, at the head, Elizabeth Langdon Williams, distinguished as the first woman to graduate from the Massachusetts Institute of Technology and a good enough mathematician to occasionally find—and rather gingerly point out—math errors in her boss's calculations.[36] Lowell, Williams, and other computers worked not in Flagstaff but mostly in Lowell's State Street office in Boston—the same office from which Lowell, as a young man, had managed textile mills, electric companies, and trusts.

Among the state-of-the-art instruments that helped their effort was a "Thacher's Cylindrical Slide Rule," billed as the "ultimate cylindrical slide rule" and a "Millionaire Calculator," the first commercially successful motorized mechanical calculator and advertised as the "only calculating machine on the market . . . that requires but one turn of the crank . . . for each multiplier or quotient."[37] It was fast, but big and clunky, and occupied an entire desk. Despite this inconvenience, it was a step up from doing multiplication and long division by hand.

Though Lowell knew from the comet-family indications more than one trans-Neptunian planet might be awaiting detection, for practical reasons he decided to try calculating for only a single body that would account as nearly as possible for the whole of the residuals. That was problem enough: the calculations and re-recalculations of Lowell and his collaborators would eventually fill some twenty large file boxes in the Lowell Observatory Archives.

Already by the end of 1910, Lowell and Williams had constructed a curve of the perturbations from which they were finding "some strange things, looking as if Le Verrier's later theory . . . were not exact."[38] They also were on the verge of obtaining preliminary results. Thus, on March 13, 1911—Lowell's fifty-sixth birthday—he telegraphed Lampland, "Please begin to photograph ecliptic where south with forty-inch [reflector]. Hope to wire position in a few days. Calculations tremendously long."[39]

Lowell did not, in fact, have a position to send Lampland in a few days. He was, as his secretary Leonard informed V. M. Slipher, "working like a slave on Planet X and just now he has a very severe cold—but he won't stop working."[40] Despite the effort expended, he and Williams did not arrive until the end of April at a position for Planet X: heliocentric longitude 235°. It did not stand for long. A week later, Lowell found that "our residuals, from necessities in the case very uncertain, seem to indicate for Planet X a heliocentric longitude of 239°. . . . Please take plates [with the 40-inch reflector] in neighborhood for 2° or 3° on either side of the fundamental plane and devise best method of comparing them for the stranger."[41] These positions put Planet X in Libra but moving into Scorpius and the inextricable star-thickets of the Milky Way.

As Lampland examined the plates exposed to the area of sky corresponding to the latest calculated position, he realized how utterly ill-suited the small field of the 40-inch reflector was for obtaining the plates and how inefficient the method was of superposition and examination with a hand magnifying glass. Three years earlier Lampland had proposed that Lowell ought, while in Europe, to try to obtain a recently invented viewing apparatus for comparing plates called a blink comparator from the Zeiss firm in Germany. The blink comparator—sometimes called a blink microscope—had been invented by physicist Carl Pulfrich at Zeiss in 1904, and uses an electromagnet to flip a small mirror back and forth to redirect the light path successively from a small area of one plate to the corresponding area on a second matched plate. This way, the image of any planet registered in exposures taken several days apart will betray itself immediately by appearing to jump back and forth. Lowell did look into it while in Europe but had then been cool to the idea. By 1911, his thinking had changed, and now that he was planning to go to Flagstaff to assume direct oversight of the observational phase of the search, he decided to act on the suggestion.

Lowell arrived in Flagstaff on May 8 and remained through June. The blink comparator arrived in Flagstaff during that time. During Lowell's lifetime, it regularly traveled between Flagstaff and Boston to "blink" the Planet X search plates. Later, with modifications, it would reveal Pluto.[42]

On returning to Boston, Lowell asked that the blink comparator and plates be sent there so he could continue to do the work himself. However, he telegraphed V. M. Slipher in early July, "Hold comparator and star plates. Hope to be back soon."[43] Unfortunately, the telegram arrived too late; Slipher had already sent the blink comparator back to Boston. Lowell at once had it returned to Flagstaff.

Meanwhile, Williams had been recalculating the residuals using Gaillot's theory of 1903 instead of Le Verrier's of 1873. (As noted above, she and Lowell had begun to doubt Le Verrier's theory for the post-1873 observations.) This meant that the condition equations had to be reset and everything computed over again. Williams clearly worked quickly, for that July she and Lowell had a preliminary position from Gaillot's residuals, which was well to the east of the positions Lampland had received just months earlier. Within a few days, however, the position had been revised again: further tweaking put the planet at around longitude 210°. Instead of being in Libra or at the border of Libra and Scorpius, the planet was now sliding into Virgo.

Though in late July a "suspicious object retrograding" was caught on Lampland's plates, the Lowell Observatory Archives give no indication of just what it might have been. Suffice it to say, it was not Planet X. At this point Lowell's attention was diverted to Mars, then coming to the latest of its biennial oppositions, for which he remained at the observatory from mid-August till February 1912. After this he and Mrs. Lowell set out for Europe, and only on their return, in May 1912, did the search resume. Lowell was now determined not only to increase the pace of the photographic search but also to refine the theoretical work by grappling not only with the longitude but the latitude of Planet X. He explained to Lampland, "I feel convinced that the inclination is considerable, both because of the size of the residuals in latitude and because of the great eccentricity, the two going together and diminishing the mass also."[44] By then the strain of the calculations was beginning to take a toll. Leonard, returning to Boston in September after a vacation in Maine, found Williams "a mere shadow from her perplexing computations."[45] Nevertheless, Williams soldiered on. A few days later Lowell wrote to Lampland that he and Williams—now joined by additional computing assistants—were continuing to revise and extend their calculation. Even so, they did not yet have a new position, and Lowell wrote in frustration, "Every new move takes weeks in the doing."[46]

At the end of October, Lowell's own health collapsed. Not only did he have to postpone a planned visit to Flagstaff, but he was not even able to drop into the Boston office. Leonard reported he could manage "only a word now and then on the phone." She worried that he was "weak and run down and must be careful and quiet."[47] She could only hope that he would not be out of commission as long as he had been during his earlier breakdown, and confided to her Flagstaff colleagues, "He worries about the work—he wants to be *in it*!" and, "It is nervous exhaustion, and he is *up* and *down*! Some days he cannot even telephone. He gets . . . impatient for things to come from Flagstaff."[48]

Fortunately, Leonard's worst fears were not realized. Lowell was by late January again wiring Lampland the latest positions for Planet X. As usual, none of them stood for long, and in early February he sent a new set of elements with a new position at heliocentric longitude 58°4'. This latest position put it in Taurus, not far from the Pleiades. Lampland undoubtedly found these frequent revisions frustrating, and soon took his turn to collapse. Low-ell urged him to do as he had done, telegraphing, "Go off and get well. I cannot afford to lose you."[49]

By spring 1913, Lowell, says Hoyt, "apparently decided to make one final all-out attack on the trans-neptunian problem, now extending the time span of his solutions by including observations up through the year 1910 and investigating the higher powers of the eccentricities of Uranus and X."[50] His first step was to write to Gaillot for residuals in latitude and longitude for Uranus since 1903. Gaillot sent the requested data by June, and Lowell at once asked for the same for Neptune. Meanwhile, Williams and the other computers were hard at work on the rigorous and time-consuming least-mean-square solutions.

Lowell, continuing his recuperation, left Boston for the summer home of his sister, Katharine Bowlker, at Mount Desert Island, Maine. From there he kept in constant touch with the Planet X work through telegrams and letters from Williams in Boston as well as forwarding brief progress reports and probable positions to Lampland, now returned to health, in Flagstaff. On July 10, 1913, Lowell telegraphed Lampland, "Generally speaking, what fields [of sky] have you taken? Is there nothing suspicious?" and on August 21 he telegraphed, "So far best determination for first power e' [eccentricity of Planet X] for present position is two hundred thirty nine degrees and for second power ditto two hundred forty one degrees. Use these. Suspect inclination large probably south. Am personally still on the retired list. Await another excitement proving true. Any news grateful."[51]

As 1914 approached, Mars again brightened in the sky as it moved toward a January opposition. Lowell advised V. M. Slipher via telegram on August 27, 1913, to "rig up the 40-inch for the best possible visual observations at coming Mars opposition." The 40-inch had originally been acquired for Mars obser-vations. However, whenever it was not needed for Mars, Lampland used it to take plates in the search for Planet X It proved, however, a decided disap-pointment for either purpose: in part because of Lowell's decision to locate it partially underground, the image stability, or seeing, was never very good for planetary work, while, as Lampland was well aware and had already pointed out to Lowell a year earlier, it was not really a good choice for the planet search either because of its small field and the distortion of star images at

Figure 6.3 Lowell's chief rival in the search for trans-Neptunian planets, his Beacon Hill neighbor William H. Pickering. This photograph, taken in 1909, shows Pickering soon after he presented his trans-Neptunian planet investigations, based on John Herschel's graphical analysis, before the American Academy of Arts and Sciences in Boston. His presentation led Lowell to redouble his resolve in his search for Planet X. Courtesy of Wikipedia Commons.

the edges of the plates. Lampland himself recommended to Lowell to acquire a refractor with a doublet lens providing a large field. Lowell accepted the recommendation and wrote to John A. Miller, director of the Sproul Observatory of Swarthmore College in Pennsylvania, to see if an instrument such as Lampland was suggesting might be available on a temporary basis. Luckily, it was. A 9-inch Brashear photographic doublet lens mounted in a camera, giving significantly more light-gathering power than the 5-inch Brashear instrument used in the earlier phases of the search, was delivered to Flagstaff by April 1914 and set up in a temporary dome on Mars Hill.

As the first plates with the Brashear were exposed to the sky over Flagstaff, the activity in Lowell's Boston office had already reached its climax. Lowell himself, Williams, and the other assistants had been throwing themselves into yet another gigantic calculation—the last attempt to reach the goal so long and so avidly sought. The bulk of the computations were finished that April, and Lowell accordingly relieved two of the assistants, Thomas B. Gill and Earl A. Edwards, of their computing duties and sent them to Flagstaff to help Lampland with the photographic search with the Brashear doublet.

Lowell was evidently still under severe strain. His biennial trip to Europe planned for that spring was delayed when Mrs. Lowell needed surgery for an ulcer. In Boston, he monitored both Mrs. Lowell's recovery and the progress of the search in Flagstaff, with which he was growing increasingly impatient. In May he telegraphed Lampland, "Don't hesitate to startle me with a telegram 'FOUND.'"[52] The Lowells finally set sail for Europe in May, but they remained for only three months, arriving in London just a week or so before Great Britain declared war on Germany. With the beginning of hostilities, they booked an immediate return to the United States. This was to be Lowell's last visit to Europe. Lowell, back in Boston in August, immediately wanted a progress update from Flagstaff. The search had been suspended— not only by the usual July monsoons, but by a further medical issue. Now it was Mrs. Lampland's turn to need ulcer surgery. Lowell wrote rather plaintively to V. M. Slipher, "I feel sadly of course that nothing has been reported about X, but I suppose the bad weather and Mrs. Lampland's condition may somewhat explain it."[53]

There was still no news of Planet X when he returned to Flagstaff for an abbreviated visit in October. Among his other activities, he observed Venus by daylight with the 24-inch Clark refractor on October 17. That day, a visiting Philip Fox, then at Northwestern University and later to serve as the first director of Chicago's Adler Planetarium, captured the iconic image of Percival, in sartorial splendor, with a backward-turned soft cap covering his shiny bald head, seated alert and erect in his chair peering intensely at Earth's sunward neighbor. Though Venus was on his eyeball, the Planet X calculations were still running through his head, for at that moment he was putting the finishing touches to his *Memoir on a Trans-Neptunian Planet*, which he planned to present to the Academy of Arts and Sciences in Boston. In December he told Lampland, "I am giving my work before the Academy on January 13. It would be thoughtful of you to announce the actual discovery at the same time."[54]

Figure 6.4 Lowell assistant Thomas Gill, observing with the 9-inch Brashear dou-blet, borrowed from Pennsylvania's Swarthmore College and used for the second Planet X search, 1914–16. Courtesy of the Lowell Observatory Archives.

Memoir on a Trans-Neptunian Planet

No such announcement preceded the brief summary he gave to the Academy on January 13, 1915. He spoke in the same venue in which Pickering had presented his trans-Neptunian graphical predictions seven years before. Nor was there any reaction—terse or otherwise—from academicians present or astronomers not present. Adding insult to injury, the Academy declined to publish it. When the memoir did appear, nine months later, it was as a private publication of the Lowell Observatory. In contrast to Lowell's other accessi-ble and popular books, much of it is impenetrable, with even his brother and first biographer acknowledging that it

> consists of many pages of transformations which, as the guide books say of mountain climbing, no one should undertake unless he is sure of his feet

and has a perfectly steady head. But anyone can see that, even in the same plane, the aggregate attractions of one planet on another, pulling eventually from all possible relative positions in their respective elliptical orbits with a force inversely as the square of the ever-changing distance, must form a highly complex problem. Nor, when for one of them the distance, velocity, mass, position and shape of orbit are wholly unknown, so that all these things must be represented by symbols, will anyone be surprised if the relations of the two bodies are expressed by lines of these, following one another by regiments over the pages. In fact the *Memoir* is printed for those who are thoroughly familiar with this kind of solitaire.[55]

Here Lowell summarized the results of his and his assistants' five-year mathematical quest. As with Le Verrier and Adams, he had needed to make some assumption about Planet X's distance from the Sun, and had started with an assumed distance of 47.5 AU. With this as a basis, and with the errors of the observations reduced as far as possible by the method of least squares, he proceeded to estimate the eccentricity, the place of the perihelion, and the mass of the unknown planet, and recomputed residuals for every 10° around the orbit. By means of this laborious reshuffling of the cards he found two possible positions, near 0° and 180°, which reduced the residuals to a minimum. Needless to say, each of these positions had required a huge amount of calculation, much of it done by Williams and her fellow assistants but by no means a negligible share by Lowell himself. To be safe, he repeated the same calculation assuming Planet X to be at other distances—40.5, 42.5, 45, and 51.25 AU. This recalculation showed that the residuals seemed to be most nearly accounted for by a distance not far from 45 AU. He was still not satisfied. For greater assurance, he took up terms of the second and third order of the eccentricity, but found that the additional calculation did not significantly change the result. He next attempted to determine the effect on his results if the planet moved in an orbit highly inclined to the ecliptic but abandoned the effort as soon as he realized the results would not be reliable.

His two final "best solutions" gave heliocentric longitudes, for July 1, 1914, at 84°.0 and 262°.8. The first position put it in eastern Taurus moving toward Gemini, the other in Scorpius moving toward Sagittarius. The math provided no reason to favor one over the other. However, since the latter would put the planet in a part of the sky "nearly inaccessible to most observatories," in the densest thickets of the Milky Way's stars, Lowell chose to concentrate on the former. He estimated that his planet had a mass seven times that of Earth and an orbit eccentricity of 0.202 for the first solution and 0.195 for the second. Because eccentricity and inclination were usually correlated in other members of the Solar System, he guessed that the inclination was likely to be about 10° (a high inclination that, he noted, would make it more difficult to find). He speculated that the magnitude might be 12–13, and that the apparent diameter of the disk was about 1" of arc. (This gives some idea of what kind of object he and his colleagues were looking for on their plates.)

By this point in the long and laborious affair, the complications, mounting uncertainties, and continuing failure of the observational search had profoundly discouraged him. "It must be remembered," he admitted, "that the actual as against the probable errors of observation might decidedly alter the result."[56] This turned out to be correct: the errors in the observations, especially for those made before 1781, were significantly greater than the probable errors he had assigned on the basis of the least-square analysis. He saw that terms above the squares in the eccentricities of Uranus and Planet X, though too ponderous to deal with and so "necessarily left out of account," might also alter the result.

Uncharacteristically for Lowell, whose pronouncements on Mars, for instance, were usually dogmatically stated—and once stated, repeated continuously in much the same terms—he concluded his Planet X memoir with caution and humility. He harkened back to the paradigmatic discovery of Neptune by Le Verrier and Adams, whose analytical methods had inspired his adoption of the same:

> That Le Verrier's solution gave him limits which were erroneous shows how necessary to a full comprehension of the problem is the rigorous and more complete method of solution. This does not detract from the great analytical skill displayed by both Adams and Le Verrier in their masterly attack on the problem. That alone deserved success. Why it attained it is nevertheless a cause for surprise, for Le Verrier left out terms bigger than two he retained. The explanation would seem to lie in the nearness of Neptune and the near

circularity of its orbit. Neptune turns out to have been most complaisant and to have assisted materially to its own detection.[57]

If this, however, had been true for complaisant Neptune, it was much more so in the case of Planet X:

But that the investigation opens our eyes to the pitfalls of the past does not on that account render us blind to those of the present. To begin with, the curves of the solutions show that a proper change in the errors of observation would quite alter the minimum point for either the different mean distances or the mean longitudes. A slight increase of the actual errors over the most probable ones, such as it by no means strains human capacity for error to suppose, would suffice entirely to change the most probable distance of the disturber and its longitude at the epoch. Indeed the imposing "probable error" of a set of observations imposes on no one familiar with observation, the actual errors committed, due to systematic causes, always far exceeding it.[58]

He added, with almost Olympian detachment,

Owing to the inexactitude of our data, then, we cannot regard our results with the complacency of completeness we should like. Just as [Joseph-Louis] Lagrange and Laplace believed that they had proved the eternal stability of our system, and just as further study has shown this confidence to have been misplaced; so the fine definiteness of positioning of an unknown by the bold analysis of Le Verrier or Adams appears in the light of subsequent research to be only possible under certain circumstances. Analytics thought to promise the precision of a rifle and finds it must rely upon the promiscuity of a shot gun after all, though the fault lies not more in the weapon than in the uncertain bases on which it rests. But to learn of the general solution and the limitations of a problem is really as instructive and important as if it permitted specifically of exact solution.

For that, too, means advance.[59]

By the time the *Memoir on a Trans-Neptunian Planet* was published in September 1915, Lowell himself seems to have given up the quest. Lampland

Figure 6.5 One for the ages: an iconic photograph of Percival Lowell, observing Venus by daylight with the 24-inch Clark refractor, taken on October 17, 1914, by visiting professor Philip Fox of Northwestern University. Lowell, beginning to despair of ever finding Planet X, was only months away from presenting his *Memoir on a Trans-Neptunian Planet* at the Academy of Arts and Science in Boston, and had only two more years to live. Courtesy of Lowell Observatory Archives.

did his best to shore up Lowell's flagging spirits. In August he wrote, "X is not yet in sight, though you may well believe that I am in hopes that he is not far away. . . . This is no time to be discouraged."[60] A month later he wrote to Millar in a similar vein: "The distant planet has not yet been located but for all of that we are not discouraged. . . . I suppose you are getting tired of extending the time of stay of the 9-inch. But you see we are a hopeful lot—in some things at least. Each day brings the hope that a little more work may turn the trick. . . . After so much work on a problem how one hates to give in!"[61]

At least as far as Lowell was concerned, the effort was to no avail. The subject of Planet X, long the energizing core of so much activity at the Lowell Observatory, disappears at that point from Lowell's correspondence. Gill's observing log notes that Lowell paid a brief visit to the 9-inch on October 8, 1915, while the photographic search with that instrument continued until July 2, 1916. On that date an entry appears in its usual place, "Lunch," suggesting a project only temporarily ended. It would not be taken up again during Lowell's lifetime.

Lowell was not, as Hoyt points out, "a man to admit defeat in so many words; his usual reaction was to ignore it and turn his attention to other matters."[62] Abbott Lawrence Lowell, in considering his brother's exchanges with Lampland when the blood of the search was still running high, would have perhaps the final word: "Through the banter one can see the craving to find the long-sought planet, and the grief at the baffling of his hopes. That X was not found was the sharpest disappointment of his life."[63]

The Final Year

Lowell now had less than a year to live. At the beginning of 1916 he was in Flagstaff to observe Mars during its less-than-favorable February opposition. In addition to the usual staff members, he was joined by George Hall Hamilton, who had left a position at Bellevue College in Nebraska to spend the period around the opposition at Flagstaff as a volunteer observer. He saw the canals much as Lowell saw them, and would later marry Elizabeth Williams, Lowell's chief computer on the Planet X project. Meanwhile, Lowell was continuing to add to the buildings on Mars Hill. With Mrs. Lowell's advice on architectural designs, he planned the construction of the administration building (now called the Slipher Building) into which the astronomers would move by the end of the year and in which Pluto would be discovered long afterward. After briefly returning to Boston after the Mars opposition, Lowell was in Toronto in April delivering a speech title "The Genesis of the Planets," apparently announcing the next topic he was hoping to investigate. In May he told a correspondent, "Eventually I hope to publish a work on each planet—the whole connected together—but the end not yet."[64] That autumn he barnstormed the Pacific Northwest, lecturing on Mars and the other planets at Washington State and Reed Colleges; at the universities of Idaho, Washington, and Oregon; and at Stanford and

Berkeley in California. "More exhausted than he was himself aware," wrote his brother, "he returned to Flagstaff eager about a new investigation he had been planning on Jupiter's satellites."[65] He was hoping to use the fifth satellite of Jupiter, Amalthea, discovered by Barnard at the Lick Observatory in 1892, as a probe to work out the interior structure of the giant planet and to determine whether the presumably molten inner core was more oblate than the outer gaseous envelope. With this end in view he and Earl Slipher were busy observing positions of the fifth satellite night after night during late October and early November. Their last recorded observations were on the night of November 11. The next morning, Sunday, November 12, 1916, Lowell was in his residence (the "baronial mansion") when he suffered a massive cerebral hemorrhage from which he never regained consciousness. He died about 10 p.m. that evening, Mrs. Lowell dutifully inscribing in chalk on the wall of the room, "Percival Lowell's earthly existence terminated in this chamber upon the green couch."[66]

He left the Lowell Observatory—and the unfulfilled search for Planet X—among his legacies.

7

Clyde's Planet

FROM JULY 1916, when Thomas Gill exposed the last plate with the camera bearing the 9-inch Brashear doublet lens, thirteen years would pass before another plate would be exposed at Lowell Observatory in the search for Planet X. But while the Lowell Observatory search for a trans-Neptunian became moribund in the years after Lowell's death, Lowell's chief rival, William H. Pickering, now retired from Harvard and based at a plantation in Mandeville, Jamaica, remained active, publishing with almost metronomic regularity new predictions for planets, which he called "P," "Q," and "R." In each case, the elements were based rather casually (and illusively) on graphical assessments of residuals or statistical analyses of comet aphelia. In contrast to Lowell, who by the time he finished *Memoir on a Trans-Neptunian Planet* had become aware of the inherent uncertainties in the data and the vagaries of the methods relied on to pluck another planet from it, Pickering never advanced beyond a completely nonrigorous "shotgun" approach. His predictions spawned like sturgeon eggs, and were accordingly greeted by most astronomers with deserved indifference. Only his first planet prediction (Planet O) inspired an actual search. In 1919, Milton L. Humason, using a 10-inch Cooke refracting telescope at the Mount Wilson Observatory, exposed several plates on the region suspected of holding Pickering's planet without success. (Ironically, Pluto registered as a faint image on a couple of these plates, but it escaped detection at the time.)

It was not until the mid-1920s that interest in the trans-Neptunian planet revived in Flagstaff. The remote and icy orb of Lowell's dreams had never been far from the minds of his loyal staff, but funds had not been available to do anything tangible in the direction of finding it.

The lack of funds had to do with Lowell's widow, Constance, onetime interior decorator and real-estate entrepreneur from whom Percival had bought his three-story brownstone row house at 11 West Cedar Street in Boston after his mother's death. She was apparently a devoted, agreeable, and compatible companion from 1908 until his death eight years later, but questions had always hung over her, even within the extended Lowell family. As the late Lowell Observatory Sole Trustee William Lowell Putnam III wrote, "One can only speculate as to why Percival, one of the more eligible bachelors in the nation, chose to marry a woman who was by no stretch of the term a beauty and, at age forty-four [to her husband's fifty-three], could hardly be counted on to give him an heir. Matrimonially speaking, she had been 'on the shelf' for years, and there exists no record of Percival's thinking on this matter. Was there some overriding element of family pressure?"[1]

Putnam himself blamed Lowell's two older sisters for nagging him into the marriage, and could not help wondering why he did not marry his helpful—attractive—and unquestioningly loyal secretary, Wrexie Louise Leonard, though he added that this was probably of the question because proper Bostonians of that day simply did not marry their secretaries. Whatever the case, after Lowell's death, Constance took on the role of villain, and in the years since she has had few if any defenders. Her husband's corpse was hardly yet cold when she sent Leonard packing. Shortly thereafter an unpleasantly argumentative, contentious, grasping, and pathologically litigious aspect to her personality—not apparent during her husband's lifetime—emerged. The consequence of this would be that, as Putnam put it, "seen in hindsight . . . [the] marriage was a disaster of the first magnitude for the pursuit of astronomy."[2]

The sad fact is that Mrs. Lowell proceeded to exert a stranglehold on her husband's estate that was resolved only after a costly and protracted legal struggle, during which the observatory was forced to keep its operating budget to a bare minimum. At the time of Lowell's death, a court-appointed audit of his estate revealed a very conservative portfolio, with major holdings in a variety of utility bonds and stalwarts such as AT&T, and a value of $2,336,813.77 (more than $40 million in today's terms) on February 5, 1917. By the time Mrs. Lowell was through with it, on November 30, 1925, the value had been reduced to less than half—$1,105,017.08.[3] Most of the difference, of course, had evaporated into the hands of lawyers.

This disastrous litigation left the Lowell Observatory long to be regarded as famously broke. The roaring twenties completely bypassed it. No funds were available to acquire a proper telescope for a new search for Planet X. Only after the litigation was settled did Guy Lowell, Percival's cousin and the first sole trustee of the observatory after Percival's death, acquire two partially finished 13-inch glass disks from the estate of the late amateur astronomer and optician the Reverend Joel Metcalf as a first step toward reviving the search. However, he died suddenly of an intracerebral hemorrhage early in 1927, causing another interruption. His successor as sole trustee, Roger Lowell Putnam, Percival's nephew, next took up the cudgel for the project, helped by a significant contribution of $4,000 from Percival's brother, Abbott Lawrence Lowell, the president of Harvard, that largely paid for the telescope. Using the Metcalf disks, a 13-inch doublet lens was crafted by Carl A. R. Lundin, senior optician of Alvan Clark and Sons. The telescope and the dome housing it, the latter designed by Stanley Sykes and based on his brother Godfrey's design of the Clark refractor dome, was to be located on U.S. Forest Service land ceded to the observatory for the purpose several hundred feet northwest of the observatory's main building.

At that time, only three astronomers were on the staff—Director V. M. Slipher, his brother Earl Slipher, and Carl Lampland. They were busy with their own researches, including trustee Putnam's chief priority at the time, the publication of a new book on Mars, which the astronomers seemed to have regarded more as a duty than a pleasure and on which they made slow progress. As the new telescope neared completion and the resumption of the Planet X search loomed, they realized that an assistant was needed, but there was still little money to spare. The person chosen had to be young, in robust health, and willing to work extremely hard for little pecuniary reward. As

V. M. Slipher advised another young astronomer at the time, "It is probably true that the pay in the field of Astronomy is lower than in other lines of intellectual endeavor of a more commercial relation. In other words if the matter of remuneration weighs much now or may do so in the future in your choice of work, I doubt whether one could recommend to you the pursuit of Astronomy. However, Astronomy, as well as many other sciences, holds out something in addition to pay alone."[4]

A Young Man from Kansas

Timing, they say, is everything, and just as the observatory was starting to cast around for an assistant, a young farmer from Kansas, Clyde W. Tombaugh, knowing of the Lowell Observatory's reputation for expertise on the planets, obligingly wrote asking for an opinion concerning sketches of Mars and Jupiter he had made with a homebuilt 9-inch reflecting telescope in the fall of 1928. Though his interest in astronomy had been fostered long before in views of the Moon and planets obtained with a Sears and Roebuck 2.5-inch refractor, Tombaugh soon yearned for a more powerful telescope, and did as many others have done: he decided to build a telescope himself. He made several mirrors, of which the third attempted, of 9 inches aperture, was of excellent quality and provided sharp views of the planets with magnifying powers up to 400×. As for the mechanical parts of the telescope, the base had been part of a cream separator; another component used the crankshaft of his father's 1910 Buick.

Tombaugh had chosen a good time to begin observing the planets. Jupiter was then full of complex and rapidly changing details as it was undergoing one of its periodic, and dramatic, South Equatorial Belt disturbances, and it presented much to see and draw. He made exact copies of his drawings and sent them to Lowell Observatory. V. M. Slipher was favorably impressed with Tombaugh's sketches and immediately saw that the young farmer from Kansas with an interest in astronomy might be just the person he should hire as an assistant for the renewed Planet X search. "It seemed to me," Slipher confided afterward, "that we would probably get more real assistance from this young man than we would from the highly trained variety for the reason

Figure 7.1 The young man from Kansas: Clyde Tombaugh, with the 9-inch home-built reflecting telescope he used to make sketches of the planets in 1928–29. Courtesy of Lowell Observatory Archives.

that the latter care only to take up new pieces of work for themselves rather than help us with lines the Observatory has been doing."[5]

Offered a position, at a monthly salary of $125 plus living quarters on the second story of the observatory's administration building (built in 1916), Tombaugh did not hesitate. In January 1929, he boarded a train in Larned, Kansas, for Flagstaff. Before the train pulled out, his father Muron imparted the advice, "Clyde, make yourself useful and beware of easy women." He seems to have taken both pieces of advice to heart; he certainly made himself useful. When he arrived at the platform in Flagstaff, the completed 13-inch telescope was not far behind. Several more months were needed to set up the telescope and to make the necessary adjustments, but Tombaugh had plenty to do in the meantime, acting as an all-purpose gofer. His duties included shoveling coal and wood into the administration building's furnace, cleaning snow off one of the telescope domes, and giving afternoon tours to visitors.

In those days astronomical images were registered on glass plates onto which light-sensitive chemicals (the photographic emulsion) had been applied. Those that would be used for the Planet X search were large, 14 by 17 inches, and because of their size they had to be given a slight curvature for stars to be perfectly focused across the entire plate. Because the smallest star images were only about 0.3 mm in diameter, unless the images were exactly the same size and shape on both plates of a pair, the plates would be "unblinkable" (i.e., they could not be examined successfully with the blink comparator.) To get all the star images in focus, all parts of the plate had to be within 0.13 mm (1/200 inch) of the focal surface of the instrument. The exact curvature of each plate was produced by pressure screws on the back, while a testing table, designed by V. M. Slipher, which used a long amplifying arm, was used to accurately check the curvature of each plate in the plate holder before it was loaded into the telescope.

Under Slipher's supervision, Tombaugh learned to adjust the curvature of his plates and to load them into the telescope. To take exposures of an hour or more, he needed to keep a brighter "guide star," selected from *Norton's Star Atlas*, centered in the eyepiece of the guide telescope. After several nights of working under supervision, Slipher decided that Tombaugh had "passed" and could be trusted to work reliably on his own. At this point Slipher stopped coming out to the dome.

On April 6, 1929, Tombaugh made a first test exposure centered on the star Delta Cancri in what would become the third Planet X search. Two nights later, he exposed a first plate on the constellation Gemini centered on Delta Geminorum. Despite being already far gone in the western sky, Gemini was the area Slipher was actually interested in, since it was here that, per Lowell's last calculated positions, the planet was thought to be lurking. (Ironically, this plate and a companion plate exposed on April 30 recorded faint images of what eventually proved to be Pluto. The images were not recognized at the time not only because Gemini's lowness in the sky and atmospheric refraction caused the images to be poorly matched, but also because the Pluto images were far fainter than expected for Lowell's Planet X. According to Lowell's estimates, his planet was supposed to be magnitude 12 or 13.)

Especially in cold weather, the pressure of the screws tended to shatter the plates, and the first Delta Geminorum plate suffered this fate. Several other plates eventually became casualties of this field, and in later years Tombaugh would often quip, "I ought to have had the presence of mind to know that this was the place to find the elusive little rascal."[6]

In an unheated dome at 7,000 ft of elevation, guiding the telescope for hours on end was brutal. As Mount Wilson astronomer Robert S. Richardson, who knew Tombaugh well, later put it,

Without some experience in direct astronomical photography, it is hard to form an idea of the amount of work involved in such a search. Part of the work is plain physical labor, and although it is not quite so strenuous as swinging a sledge hammer in a rock quarry, it can nevertheless be extremely tiring. One has to open the dome, set the telescope, load plateholders, and attend to a dozen other minor tasks. Tombaugh once told me that dodging wild animals on the path to the little dome was another hazard. The most boring part of the job is sitting for hours beside the telescope, guiding it to make sure it is always pointed in precisely the right direction. Speaking from personal experience, I would say that sitting up with a telescope is much worse than babysitting. The highest type of babysitting requires an alert, aggressive mental attitude on the part of the sitter, while guiding a telescope requires practically no mental attitude whatever. After an hour's work taking a plate, it may turn out to be

Figure 7.2 The 13-inch Abbott Lawrence Lowell astrograph used by Clyde Tombaugh in the third search for Planet X. Photo by William Sheehan, 2015.

Figure 7.3 Clyde Tombaugh at the guiding eyepiece of the 13-inch telescope. This photograph was taken in 1931, a year after the discovery of Pluto. Courtesy of Lowell Observatory Archives.

defective, forcing one to take it over again if time permits. Or clouds may come up suddenly and spoil the whole night's work.[7]

Tombaugh was grateful for at least some entertainment to relieve the boredom: his radio picked up a Mexican station from across the border.

After they had been exposed, the plates were examined on the blink comparator that Lowell, at Lampland's urging, had acquired in 1911. The use of this device brought an enormous improvement in the speed and accuracy over the old method of superimposing one plate on another and examining the star images with a hand magnifying glass. Using the comparator was still a tedious and mind-numbing chore, but at least sifting through the millions of star images, any one of which could be the sought-for planet, was humanly possible.

The first Gemini plates exposed by Tombaugh were blinked by the Slipher brothers. They blinked in vain. As Tombaugh continued to expose plates, he naturally assumed that the senior astronomers would continue to want to blink them, since whoever did so would bear both the responsibility for the success—or the failure—of the search. He could not believe that this awesome responsibility would ever be entrusted to a newly hired assistant. However, the Sliphers' initial enthusiasm for the project rapidly waned, and before long the plates began stacking up as Tombaugh exposed them faster than the Sliphers could examine them. Tombaugh continued to expose plates on Gemini, which in April 1929 was already 90° past opposition (the point where this region of the sky lay opposite the Sun). Working backward through the zodiac, by mid-June Tombaugh had caught up with the opposition point— the point where Planet X, if it existed, would be directly opposite the Sun. He was now exposing his plates to the star-rich fields of the Milky Way in Scorpius and Sagittarius (where the second of Lowell's theoretical solutions had placed the planet—though Lowell had decided to ignore this solution and concentrate on the one in Gemini because the thickets of stars here would have rendered the search almost impossible). As Tombaugh noted, the star images in these plates were three times as numerous and closely packed as those of Gemini. Whereas each of the 14-by-17-inch plates exposed on Gemini contained perhaps 300,000 stars, those exposed on Scorpio and Sagittarius contained upward of a million. As Tombaugh somberly reflected, "To pick out one image that shifted position . . . is an awesome task."

After a month of blinking Tombaugh's plates, V. M. Slipher was already losing interest—or at any rate coming to realize that the task greatly exceeded the time he and the other senior astronomers were able to spare for it. At the end of June, as Flagstaff's monsoon season approached, making the photographic work more intermittent, Slipher came to Tombaugh's office with an announcement: thenceforth he would not only take the plates, he would blink them as well.

Perhaps, Tombaugh speculated afterward, the older men, who had endured the disappointments of the earlier searches during Lowell's lifetime, had already unconsciously given up. Possibly they also had a far more transparent motive. "This was," as Tombaugh himself admitted, "the most tedious work ever done." Richardson writes,

> Taking the plates is only part of the job. For, after the plates are obtained, they must be minutely examined for moving objects. . . . Unfortunately, since only an area of less than a square inch [actually, only 1 cm by 2 cm] can be viewed at a time, it takes hours to scan a pair of plates thoroughly. Every object that jumps is not necessarily a new planet. It is surprising how many defects there can be in a photographic emulsion that would make first-rate planets. Tombaugh found the best way to eliminate these spurious objects was to take three plates of the same star field each week, although this procedure had the drawback of increasing the observing program by fifty percent. After two plates were compared, one would be removed from the machine and the third inserted in its place. The plate suspects would then be reexamined to see if they still jumped. On this occasion, very few of them did.
>
> Such a project would not be especially arduous if there were two people to carry it on, one to take the plates at night and the other to blink them during the day. But when the same person has to take the plates and blink them, besides finding time to sleep and shave occasionally, it develops into a full-time job.[8]

The whole project now became all-consuming for Tombaugh, and not only because of the herculean nature of the task. He was also a self-described perfectionist. On the farm in Kansas, he afterward recalled, "When I planted the kafir corn and milo maize, the rows across the field had to be straight as an arrow or I was unhappy. Later, every planet-suspect, no matter how faint, had to be checked out with the third plate—either yes or no, not maybe. I had always taken seriously one of the adages in one of my school books: 'Do a thing well, or don't do it at all.'"[9]

The combination of the undertaking's sheer massiveness with Tombaugh's determined insistence that it be done well soon led to a personal crisis. As Tombaugh later explained,

> I encountered several dozen asteroids which shifted in position in the interval between the dates the plates were taken. How would I know which one was

Planet X? Several of these plate regions had no third plate and some were taken near the asteroid stationary points. I just could not reconcile myself to conducting a planet search on such a hit-and-miss manner. I was in a state of despair. With this uncertainty, I could see no point in going on. I fell to doing some soul-searching. Whatever other scientific endeavors that I might aspire to, I was blocked by the lack of a university education and a degree.

Asteroid stationary points are those where the asteroids, in shifting direction with respect to the background stars in their retrograde loops, appear to slow nearly to a standstill. At such times they are moving so slowly against the stars that they could easily be mistaken for a slowly moving planet beyond Neptune. Recall that it was when Uranus was moving near its stationary point in 1769 that Le Monnier observed it six times in nine days and failed to notice its motion at all.

Tombaugh was still feeling depressed about the situation in July when he boarded the train for a three-week visit home. Among other things, he would have a chance to see his baby sister, whom he had not yet seen, and help with the wheat harvest. When he got back to Flagstaff, with the monsoon season continuing, he made a concerted attack on the asteroid problem. He carefully studied the *American Ephemeris and Nautical Almanac* and noticed how the daily coordinate positions of Mars, Jupiter, Saturn, Uranus, and Neptune changed over time, especially around their oppositions and various stationary points. Suddenly the solution dawned: by exposing his plates only near the opposition positions, he realized, any body orbiting beyond Earth would be caught during the time of its most rapid retrograde movement through the field. This was critical. Since the retrograde loops are actually reflections of Earth's motion produced by parallax, the nearer the body to Earth, the faster the motion through the loop. By exposing the plates only near the opposition positions, not only would the asteroid stationary points be avoided, but—as a bonus—the relative distance of any body recorded on the plates would declare itself at a glance. Ironically, Lowell had realized the same thing long before, but his attempts to impress the point on his assistants had failed to register. Thus, when Pluto's image was found long after the fact on two plates exposed by Gill in March and April 1915, it turned out that this region was photographed 90° from opposition—just past Pluto's apparent stationary point. Tombaugh now realized that this kind of procedure was simply wrong in planet searching.

The Flagstaff monsoon season finally ended in mid-September, allowing Tombaugh to resume exposing plates along the zodiac. He did so with renewed enthusiasm. He now cheerfully chased the opposition point through Aquarius and Pisces, finding them "lovely regions with hundreds of spiral galaxies to view and not so populous in stars, only about 50,000 per plate. . . . I could blink a pair of plates in three days of work."[10] Though he was now searching regions 120° from Lowell's position for Planet X, Tombaugh was not concerned—he had never had faith in the calculated positions, which shifted so drastically from one revision to the next. He now knew that if a planet, or planets, were to be found, it would be through sheer hard work using the new, very effective equipment he now had in hand.

He labored on through Aries and, by November, was reaching into western Taurus. Once more he was moving back toward the Milky Way, exposing plates that contained ten times as many stars as those exposed on the fields in Aquarius and Pisces. Spiral galaxies disappeared because of the Milky Way's obscuring interstellar dust clouds. At this point, he later recalled, "there seemed to be an increasing sense of anticipation" among the senior astronomers. He speculated about the reasons for this:

> Perhaps Slipher was getting inquiries from Putnam. Likewise, Lampland seemed to be getting anxious. I worked furiously, totally abandoning the eight-hour day. I was feeling more and more . . . acceptance, as I was included on all the doings and discussions with occasional visits from other astronomers. Many stopped off at Flagstaff for a day or two on their way to or from the astronomical mecca, the Mount Wilson Observatory. Their 100-inch telescope was the largest in the world. I got to meet and talk with the famous astronomers I had read about in books. I forgot all about homesickness. I was living it up. It was fascinating to hear what they were learning about the Universe in their researches.[11]

By January 1930, Tombaugh was once more pushing on into eastern Gemini—the region where one of Lowell's positions, calculated forward to January 1930, put the planet—back to where he had started nine months before. By now, in light of his accumulated experience, he realized that the Sliphers had blinked the 1929 Gemini plates much too fast. He therefore decided to re-photograph Gemini. He obtained three plates, on January 21, 23, and 29. The first was of only marginal quality, with the star images blurred by a fierce northeast wind. The others, however, were satisfactory. As moonlight began interfering with photographic work on February 7, Tombaugh returned to blinking, putting the plates up on the blink comparator in the administration building one floor below his apartment. He finished blinking a pair of plates of the eastern Taurus region. Then, about February 15, he put the Delta Geminorum plates up on the blink comparator and blinked them over the next few days. On Tuesday, February 18, he continued to grind through hundreds of thousands of star images. It began as a day like any other, but by the time it was over, it would be one he would never forget.

The Discovery

Tombaugh woke up at about 7 a.m., drove the mile down Mars Hill to downtown Flagstaff, and ate breakfast. He consumed most of his meals downtown, his favorite restaurant being the Black Cat Café, directly across from the train station. He was back at the blink comparator by 9 a.m., settling in for another tedious day of blinking plates. He spent three hours at the comparator's microscope eyepiece, occasionally taking breaks from the mentally fatiguing task—without these breaks, his eyes would start blurring the images together and his concentration would falter. At noon, he drove to the Black Cat for lunch, and returned an hour later for another lengthy session of blinking. What happened next is best told in Tombaugh's own words:

> Upon starting a new strip to the east of star Delta . . . I suddenly spied a 15th magnitude object. . . . "That's it," I exclaimed to myself . . . I saw the images almost instantly in a 1 by 2 cm field containing about 300 stars. . . . I had never encountered a planet suspect so promising during all the months of blinking. . . . I removed one of the plates and placed the 21 January plate [taken during miserable weather] on the comparator. Sure enough, there was the swollen image of Planet X exactly where it should be. . . . For three quarters

Figure 7.4 The January 23, 1930 plate, centered on the star Delta Geminorum, one of two that Tombaugh was blinking on February 18, 1930, when he discovered Pluto. The actual size of these plates is 14 × 17 in. Courtesy of Lowell Observatory Archives.

of an hour, I was the only person in the world who knew exactly the position of Planet X.

Carl Lampland was sitting at his desk across the hall. He heard the clicking noise of the blink comparator suddenly stop and then a long silence. He thus suspected that I had run onto something and was nearly dying of suspense (he told me later). I called him in to view the new planet images and I explained that the shift was right for a trans-neptunian planet. Then I went down the hall to V. M. Slipher's office. . . . Slipher was sitting at his desk, working with some papers. I boldly strode into his office. "Dr. Slipher, I have found your Planet X," I said. He rose up from his chair, as if propelled by a spring, with a facial expression of excitement, but reservation in his voice. . . . He was on his way to the blink comparator room so quickly, I had to step lively to keep up with him.

Lampland surrendered the comparator to Slipher, commenting, "It looks pretty good." The air was tense with excitement as I interchanged the plates. . . . Finally, Dr. Slipher said to me, "Re-photograph the region as soon as possible." I looked out of the window . . . it was pretty much overcast. "Doesn't look very promising for tonight," I said, "but I will sit up for it throughout the night." Lastly, Dr. Slipher charged, "Don't tell anyone about the discovery. It could be very hot news. We need to keep it secret for a few weeks to study the object."

I drove downtown to eat my dinner in a café and perform my usual duty of picking up the observatory mail at the Post Office. After dinner, I noted the sky was heavily overcast. I was extremely excited and had to calm myself down. So I went to the Orpheum Theater to see Gary Cooper in "The Virginian." I can never forget that night. After the gun fight, my knees were shaking more than ever. . . . After the show, I waited . . . frequently going outdoors to view the sky. I gave up a little after midnight because the third quarter Moon was due to rise . . .

The following night, the sky was fairly clear, and I took a new plate. . . . I expected the planet's image to be 10 to 11 millimeters farther west. Sure enough, there it was, just where it should be . . . The following night, [after making a finder chart], Lampland, Slipher and I walked to the 24-inch refractor for a visual look. . . . Soon [Slipher] picked it up. "I've got it," he said, "but I don't see any disk," with sadness in his voice. What a thrill it was to me, and I realized we were looking farther out in the solar system than anyone had ever looked before. It was also disappointing—the planet was a very faint, unimportant looking . . . star-like point.[12]

The lack of any disk in the 24-inch refractor gave everyone pause, so Slipher, though excited, did not announce the discovery immediately. After

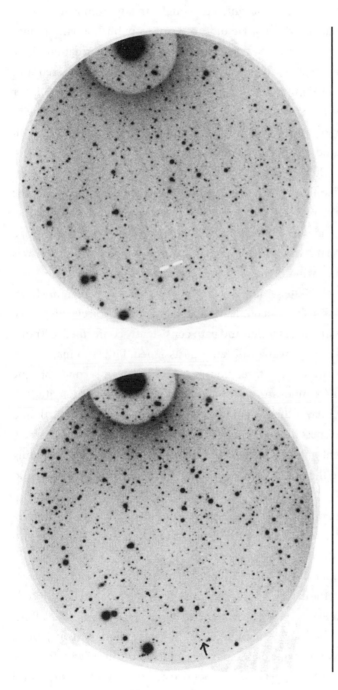

Figure 7.5 Small portions of the plates on which Pluto was discovered: (a) January 23, 1930; (b) January 29, 1930. Each of these portions is about the size of a nickel and contains about nine hundred stars. Pluto is identified with a black arrow on the January 23 plate, and is between two white bands on the January 29 plate. The shift of Pluto is about 35 mm (0.125 in) on the original plates — enough to indicate to Tombaugh that the object was beyond the orbit of Neptune. The images are enlarged about six times from one of the original plates. Courtesy of Lowell Observatory Archives.

all, the Lowell Observatory had suffered enough already from ridicule over the Martian canal controversy. He did not want to publish anything until he could be absolutely sure. Nevertheless, he passed the information to the observatory's sole trustee, Roger Lowell Putnam, who in turn passed it to Percival's brother, Abbott Lawrence Lowell. The latter exclaimed to Slipher about "the great news if it turns out to be true. To put Percy's name with that of [Le Verrier] and Adams will be a great and deserved honor, and it distinctly would be a glory for the observatory."[13]

In the weeks after the discovery, the object—though invested in great hopes, it was still only an object at this point—was kept under photographic surveillance by Lampland with the 40-inch reflector. It was indubitably trans-Neptunian, and it gave every reason, except for the troubling lack of a disk, to believe that it was Lowell's Planet X at last. On March 12, Slipher was assured enough to wire the news to the Harvard College Observatory. His telegram was brief: "Systematic search begun years ago supplementing Lowell's investigation for Trans-neptunian planet has revealed object which since seven weeks has in rate of motion and path consistently conformed to Transneptunian body at approximate distance he assigned. Fifteenth magnitude. Position March twelve days three hours GMT was seven seconds of time West from Delta Geminorum, agreeing with Lowell's predicted longitude."[14]

The following day—the 75th anniversary of Percival Lowell's birth and the 149th of Herschel's discovery of Uranus—a *Lowell Observatory Observation Circular* went out, in which Slipher gave a few more details, including for the first time the name of the assistant astronomer who actually found the object.

The trans-Neptunian object had been found only 5°.9 from Lowell's predicted longitude, or just a little more than the distance between the two pointer stars of the Big Dipper.

Slipher did not call the object a planet, but he didn't have to. The press eagerly seized on the discovery as sensational news and reached its own conclusions. "Ninth Planet, Greater than Earth Found" read a typical headline, adding a subheading suggesting the planet might be as massive as Jupiter. Obviously there were strong biases in favor of an identification with Planet X given all the money and heartache expended for the search. A successful ending of the long-standing drama solved a lot of the Lowell Observatory's problems. It would generally be agreed that it pulled the observatory's chestnuts out of the fire. At the very least, it persuaded Roger Lowell Putnam to grant the staff astronomers an indefinite reprieve from doing their long-procrastinated and perfunctorily embraced Mars book.

The discovery set off a competition among astronomers to compute an orbit. The Lowell Observatory wanted to be first, but none of the astronomers there had any particular expertise in orbit calculation, so John A. Miller of Swarthmore College's Sproul Observatory, an old friend from whom Lowell had borrowed the 9-inch Brashear doublet lens used in the second search, was invited to Flagstaff for the purpose. In the weeks after the discovery, Miller, with the Sliphers and Lampland, sat at a large table every day with tables of logarithms, plodding through the orbit calculation. Armin O. Leuschner, a world-renowned computer of asteroid and comet orbits with the Students' Observatory of the University of California, intending to do the same calculation, asked Slipher for positions. So did Frank E. Seagrave, an amateur astronomer from Providence, Rhode Island. Seagrave had his own private observatory at what is now Peeptoad Road in North Scituate, Massachusetts—it still exists and is in working order—and was a gifted amateur orbit computer who in 1910 produced the most accurate prediction of the return of Halley's Comet. He had been a friend of Lowell, and on March 14, 1930, wrote to Slipher: "Many years ago . . . the late Dr. Lowell promised me that if the Lowell Observatory ever found X that he would let me be one of the first to compute its orbit. Thank you in advance for anything you can send me in relation to this most interesting planet."[15] Apparently Slipher failed to reply because Seagrave wrote again on April 15:

About the middle of March I sent you a letter asking for some photographic positions of the outer Neptunian planet or the new Asteroid or the new Comet [or] whatever it proves to be. Up to [this] date I have never heard from you. I am wondering why as the newspapers all mention positions secured during January and February. The last time I was with the late Dr. Percival Lowell was in September 1916. At that time he was getting ready to leave Boston (I was at the Lowell Observatory Office 53 State St. Boston with him) for Flagstaff and was to stay at three or four places on the way to lecture. He showed me his computations in relation to the outer Neptunian planet, and said to me

"Seagrave, if the Lowell Observatory is the observatory that will first find this planet, you will be the first one to compute its orbit." No writing to this effect, only a verbal statement. I am wondering why you did not send me some positions as you seem to have sent some to Dr. Miller of Sproul. Dr. Van Biesbroeck of Yerkes Observatory sent me four positions.[16]

This time, Slipher responded:

It was my aim to get off to you yesterday some of the early positions of Lowell's Transneptunian object, apparently Planet X, but the day was not long enough. In fact, as you can imagine the days have for weeks all been too short for us to get done what should have been done, and we, in consequence have had to appear unfriendly or discourteous to many good people like yourself . . .

Even before making the announcement of the finding of what has been appearing to be Dr. Lowell's Planet X, it seemed to me that we here should determine for it a preliminary orbit. This is because it seemed best for Lowell Observatory to find it out and make it known if the object were thus shown to be less important than it had appeared. Dr. Lowell and the Observatory had put so much into the problem as to appear to justify this policy. We have felt that in such a case special carefulness was necessary and we have tried to be careful. (This, of course, concerns the general matter, not the orbit.)

Dr. Miller was here at the Observatory and worked with us on the orbit. Accurate positions for orbit purposes were not sent out although besides yours there were many urgent requests for early positions. The enclosed positions are being mailed to you in advance of any other person (this is, of course, in confidence). And I hope you will feel that we have tried to be fair.[17]

Though the theory of orbit calculation had advanced since astronomers grappled with the problem of computing the orbit of Uranus in the eighteenth century, the principle was essentially unchanged. Success was an uncertain proposition for an object observed over such a short arc of its orbit. The difficulty arises from the fact that any observation made from the platform of the moving Earth gives only two quantities, the angular coordinates of the body. This gives the direction along a line to the object, but not the distance. Additional observations obtained over time give other lines.

Since an orbit is defined by six elements (listed in chapter 2), at least three observations are needed. However, the orbit that is calculated, and the ability to use it to calculate where the object will be at any given time, depends on how accurately (in an absolute sense) each of the six quantities can be measured. In the months after the new planet was discovered, these quantities were known only roughly.

Thus, the first orbits were crude. Working under Leuschner, two of his graduate students at Berkeley, Ernest C. Bower and Fred L. Whipple, got there first. As published in a *Harvard College Observatory Announcement Card* dated April 7 and a *Lick Observatory Bulletin* a week later, they had arrived at "a group of solutions giving a fairly well determined distance from the earth of approximately forty-one astronomical units," and an orbital inclination of 17°. They noted that "no definite conclusion concerning the eccentricity and period can be drawn without observations covering a considerably longer interval" and that it was not even possible at that moment to determine whether the object's orbital motion was directed toward or away from Earth.[18] Meanwhile, the Lowell computers were finishing their work. On April 12, Slipher telegraphed the result to Putnam, who in turn passed it to Harvard College Observatory director Harlow Shapley. Though in overall agreement with Bower and Whipple's orbit, they assigned the object an eccentricity of 0.909, making the orbit nearly parabolic, and a period of three thousand years. At once the newspapers took up the story with headlines that can hardly have been welcomed at the Lowell Observatory, such as "Planet X Orbit Raises More Doubt" and "Lowell Observatory Estimates Put Trans-neptunian Object in Asteroid or Comet Class—Course is Long Ellipse."[19]

Slipher continued to hope that the object would prove to be the new planet of Lowell's predictions. On April 19, he wrote to Miller, "Leuschner seemed to think it an asteroid or comet—just the sort of thing he is working with every day! . . . There seems to be no doubt that the term planet to the best of our knowledge best fits the object."[20] That same day, he wrote to Putnam,

If it is a comet it is different from all other known comets and if it is an asteroid the same is to be said of it, if it is a planet it does appear to have the most exceptional orbit of all, and much the most exceptional. On the other hand, we know that in the outer satellites of Jupiter we have a similar change in the satellites as regards size and orbit . . . so we might find that there is a somewhat similar change in the planetary bodies of the sun's system between Neptune and Planet X.[21]

He continued, rather prophetically,

Your question as to size and mass is not easy to answer. . . . When the orbit is more accurately determined a year or two hence it will be possible to see how much it has perturbed the other planets, and in this way get a measure of its mass. . . . So Planet X could still be some thousands times more massive than all the asteroids rolled together. Hence only in apparent size does it compare with asteroids, owing to its enormous distance. . . . The term planet is clearly to be preferred to asteroid or comet. It took a long time to find Planet X, and so it may take a lot more time and work in determining clearly its real nature. . . . We would have wished for a brighter object and an orbit that would seem to fit better what was expected, but this, whatever may be said of it, is causing more thinking than anything that has been found in a very long time.[22]

Calling It Pluto

The orbit computations—and questions about the exact nature of what was still being called the "Lowell object"—would continue for some time. The details, of course, could only be followed by a relative handful of experts. Meanwhile, the matter of finding a suitable name for the planet was looming. Even before the official March 13 announcement, Putnam had solicited Mrs. Lowell for suggestions, and she had offered, rather lamely, "Zeus." In anticipation of the obvious objection, she told Slipher: "Mr. Putnam asked me if I had any thoughts. . . . He said he had thought of Diana. . . . Zeus is my choice. Some people believe the name to be one and the same as Jupiter but of what I have read it is not so—Jupiter is a Roman god—Zeus is a Greek god."[23] Putnam, meanwhile, leaned toward "Cronus," which might have been proposed by his mother, Elizabeth (Mrs. William Lowell Putnam II),

Percival's sister. It might well have been accepted but for the fact that another astronomer, Thomas Jefferson Jackson See—who had served a brief and controversial stint at the Lowell Observatory in the 1890s and was so disliked by the other staff that they described him as "a reptile"[24]—had used it for a planet he had once predicted. The mere association of See with the name effectively rendered it unviable. "Minerva" was another attractive choice, and was favored not only by Putnam and the New York Times, which saw it as singularly apt for "the planet that sprang to human view full panoplied from the mind of man," but by nearly half of all the letter writers to Putnam and the Lowell Observatory. (Remember, "Minerva" had also been John Herschel's choice for Neptune.) In the end, it too had to be ruled out—it had already been wasted on an asteroid.[25]

The day after Slipher's announcement, Putnam announced that suggestions from the public were welcome. After this came the deluge. Suggestions began to pour in, ranging from "Atlas," another of Putnam's early favorites, to "Zymal." Mrs. Lowell now changed her mind about "Zeus" and decided to weigh in again; she decided—even though Putnam was sure that Percival Lowell would have wanted a classical name—that "the gods of the past are worn out." On March 14, she wrote to Slipher, "In eastern newspapers and at a luncheon today unanimous demand that the planet be named Percival, and we hope that you at the Observatory who made possible its finding will be in sympathy with the appropriateness of the name."[26] "Percival" did not prevail; neither did her next choice, "Constance."

A particularly interesting suggestion—though it seems never to have been sent—was offered by H. P. Lovecraft, the horror writer, who as a sixteen-year-old had met Percival Lowell at a lecture given by the latter in Seyles Hall at Brown University in Providence, Rhode Island. He was keenly interested in developments and wrote to his friend Elizabeth Toldridge,

Asteroidal discovery does not mean much—but a major planet—a vast unknown world—is another matter. I have always wished I could live to see such a thing come to light—& here it is! The first real planet to be discovered since 1846, & only the third in the history of the human race! One wonders what it is like, & what dim-litten fungi may sprout coldly on its frozen surface! I think I shall suggest its being named Yuggoth! . . . Another thing that pleases me is that the newcomer came to light at the Lowell Observatory, & from Lowell's own calculations. Poor chap! His better known observations & speculations never have fared well in the scientific world; but now, thirteen years after his death, it is possible that his calculations may win him a major place among astronomers.[27]

Meanwhile, several writers had suggested "Pluto," including someone calling himself (or herself) "Nostradamus" and another person named "H. L. Blood." The name was also independently proposed by an 11-year-old schoolgirl, Venetia Burney, who proposed it to her grandfather, Falconer Madan, a retired librarian of the Bodleian Library in Oxford. Madan passed the suggestion along to his friend Herbert Hall Turner, the Savilian Professor of Astronomy and director of the Radcliffe Observatory at Oxford. Turner—who died suddenly only a few months later—agreed that it was an excellent choice and telegraphed it to Slipher. (Evidently the Madan family did this sort of thing rather well. Madan's brother, Henry, the science master at Eton College, had suggested the names for the two satellites of Mars, Phobos and Deimos, to Asaph Hall in 1877.)

The name was, of course, singularly apt, as Pluto was the god of the underworld and the brother of Jupiter and Neptune, and Slipher formally proposed it in a *Lowell Observatory Observation Circular* on May 1. By then, the word had already leaked out, and in fact "Pluto" was already being used informally by astronomers, especially in Europe. Slipher noted that the first two letters were the same as Percival Lowell's initials and so proposed for its planetary symbol an anagram formed of the initials P and L. In general, the name was received enthusiastically, though there were a few dissenters—including Mrs. Lowell, who still dressed in mourning black and was rumored to hold séances in which she attempted to commune with her husband's spirit. In the summer of 1930, she visited the Lowell Observatory, eager to meet the young man who had found "my husband's planet." As late as New Year's Day 1931, she wrote to Slipher, "I wish the Planet could be called Planet X and not Pluto. I do not like Pluto for the name."[28]

Despite Mrs. Lowell's preferences, the name was getting a lot of traction, and not only among astronomers. For example, public interest in the new planet seems to have been the motivation for animator Walt Disney to name his canine character, which first appeared in his cartoon features in the same year as the discovery, Pluto the Pup.[29]

Another was William. H. Pickering. Almost like a voice from the grave, Pickering—now seventy-two and in retirement on his old-fashioned plantation in Jamaica—pointed out that whereas Pluto was found 5.9° from the position Lowell had predicted, it lay only 5.6° of the place of his own Planet O. Moreover, it turned out that Pluto had even been registered on the plates Humason had taken at Mount Wilson in 1919, but had been overlooked at the time. While claiming this putative triumph, Pickering neglected to note that he had published a revised orbit for Planet O in 1928, which was in much poorer agreement with that of the actual planet, or that he had come to favor no fewer than three planets, instead of one, as the source of the perturbations that affected Uranus. Lowell Observatory seems to have fretted about Pickering for a while, though probably the final verdict on him was accurately pronounced by trustee Roger Lowell Putnam, who wrote to V. M. Slipher, "I don't think I should worry much about Pickering's predictions. I think he has predicted just about everything—from one planet to three, in varying positions. At any rate, Dr. Lowell predicted it, and you have found it, which is more than Pickering has done."[30]

Nevertheless, Pickering continued his intermittent fulminations. As early as April 10, 1930, he penned an article for *Popular Astronomy*, to which he had long been a regular contributor, complaining that the name Pluto had been adopted for the object discovered at the Lowell Observatory. By early 1931, he was arguing that, because the planet penetrated Neptune's orbit, it ought to be called "Salacia," or "Amphitrite," after the wife of Neptune, while reserving Pluto for Planet P, which as deduced from comet aphelia revolved around the Sun at 70 AU. He even claimed, as he wrote a few months later, to have mentally reserved the name Pluto for Planet P. Pluto had usurped the name, and so, he quipped, it "ought to be called Loki, for the god of thieves!"[31]

Eventually he mellowed as a series of refined computations of Pluto's orbit showed it to be truly trans-Neptunian. A definitive orbit was finally published 1931 by Ernest Bower, Leuschner's graduate student who had been involved in the early efforts. Pluto was indeed trans-Neptunian and moving in an orbit that bore only the most general resemblance to that of Lowell's Planet X. Where Lowell had surmised a planet with a highly elliptical orbit at a mean distance from the Sun of 43 AU and an inclination to the ecliptic of 10°, the object—planet, comet, asteroid, or whatever—discovered by Tombaugh was situated at a mean distance of 39.4 AU in an orbit that proved to be the most eccentric and sharply inclined of any planet. When farthest from the Sun, as it will next be in 2113, it reaches a distance of 7.4 billion km (4.6 billion miles). At perihelion, which it passed in 1989, its distance is only 4.4 billion km (2.7 billion miles) and it skims well inside the orbit of Neptune—though there is not the slightest chance of a collision since, owing to the extreme tilt of its orbit (17°), it is then more than 1 billion km above the plane of Neptune's orbit.

The sharp orbital inclination means that Lowell's 1905–7 search, which was concentrated on the ecliptic, afforded no chance of success—the planet was then so far away from the ecliptic that it was not even in the search zone covered by the plates. Though it was imaged by Gill in 1915, it was far too faint to have been recognized in what was, in any case, the waning days of hope for the search. Note that the planet takes 248 years to make each circuit of the Sun, so it will not return to where Tombaugh discovered it until 2178. For its next perihelion passage, we must wait until 2237.

Meanwhile, another giant of celestial mechanics, Ernest W. Brown of Yale, had published a devastating critique of the calculations that had led to the discovery. Brown carried out a fresh appraisal of Lowell's work and showed that the values obtained for the distance, mass, and eccentricity of Planet X all depended on three groups of observations made before 1783, observations that had appreciable errors.[32] Lowell himself, by the end of his quest, had suspected as much. As noted, he realized that if the actual errors of the observations exceeded the probable ones assumed, his whole calculation would fail. Brown noted that Lowell—following the method Le Verrier had used for the discovery of Neptune, which involved the reduction of the squares of the residuals to a minimum, and which he had carried out in great detail and with apparently high accuracy—had nevertheless been flawed in the first link of his extended analytical chain. He had given no probable errors for his results, nor had he weighted the residuals he had used. But, Brown concluded, "the weighting of the material in a problem of this nature is of fundamental importance, because the question of the validity of the hypothesis depends almost entirely on the probable error of the unknowns."

Lowell, then, had been chasing a phantom. Despite the heroic efforts expended by himself and his computers, he had, as with the canals of Mars, pushed his uncertain data far beyond what it could bear.

But if he made errors, they were at least fruitful ones, as Brown himself admitted: "In so far as [Lowell's work] has stimulated a search for an outer planet which has proved successful, one cannot regret its completion and publication." Nevertheless, once the supposed residuals were rejected, there were no perturbations to be derived from them. Moreover, as Brown noted that, Pluto was much fainter—and presumably much less massive—than expected, so it could not be Lowell's Planet X. (Recall that Lowell had expected it to have a mass 6.6 times that of Earth.) Echoing Peirce in the aftermath of the discovery of Neptune, Brown concluded, "The fact that it was found near the predicted place was purely accidental."

After reading Brown, Pickering changed course and rose to defend Lowell's calculations—and by implication, his own. He protested that both he and Lowell—with different methods—had succeeded in predicting the place of the planet, and concluded,

> One or the other of us should have the primary credit for it alone. We both computed the position and the brightness of the little planet for definite dates, by different methods. As to which method gave the more reliable result, based on the deviations recorded of Uranus and Neptune, I leave to others to decide. . . . Neither Lowell nor I ever saw any of these plates, and we therefore were not responsible in any way for finding the planet upon them, except in so far as Lowell left money to have the search made. I made no such gift. If it was a mere accident . . . that his position was so near to the actual one, then it was certainly a very fortunate accident. . . . Had it not been for his bequest, it is quite probable that the planet would not have been found for many years, and he should receive therefore full credit for this very important part in the discovery.[33]

Pickering's accolade to Lowell for instigating and providing the means to carry out the search was touching—but not quite accurate. It turns out that in November 1930 and again in May 1931, Pluto was in the same field with Jupiter, within a fraction of a degree. At Mount Wilson, Seth B. Nicholson, who had discovered the ninth satellite of Jupiter in 1914, was using the 100-inch telescope to track "his" moon. In November 1930, Nicholson wrote to W. H. Wright of the Lick Observatory, "Had you noticed that Pluto has been in the same field with Jupiter's satellites? The last plate of it showed Jupiter VI [Himalia] not far off. Now I think Pluto will be on the same field with IX [Sinope]. If Tombaugh had just waited a year we might have saved him a lot of work."[34]

As unwelcome as it might be—especially by astronomers at the Lowell Observatory—Brown's conclusion that Pluto was not massive enough to be Lowell's Planet X has stood the test of time. Pluto's mass is not remotely large enough to perturb the giant planets. In 1930–31, this was suspected— even Slipher and Putnam worried about it—but not yet firmly established. Brown's position was supported by Bower, who pointed out that even Lowell himself had placed "no great confidence" in his results.[35] The opposing camp could claim for their side the authority of the British astronomer Andrew Claude de la Cherois Crommelin, whose interest in the trans-Neptunian calculation was long-standing and who as far back as 1909 had responded to Lowell's request for residuals of Uranus from the Greenwich Observatory. While admitting that Lowell had overestimated Pluto's mass, Crommelin basically defended the validity of his work, simply refusing to believe that the accuracy of either Lowell's or Pickering's predictions could be acciden- tal.[36] In 1934, Crommelin, saying he got the idea from Sir James Jeans, sug- gested a possible way to "save" Pluto's mass while accounting for its star-like appearance in even large telescopes. Perhaps, he said, what was seen was a specular reflection of sunlight from a surface as smooth and shiny as that of a ball bearing, in which case what was seen against the black-sky back- ground was only this central bright spot, not the true breadth of the body. Though Griffith Observatory astronomer Dinsmore Alter was still toying with the idea as late as the early 1950s, it was never very convincing; even if Pluto had once had such a smooth, shining surface, it could not have maintained its luster over millions of years of comet and meteoroid colli- sions.[37] Another suggestion—which popped up within a year or so of the discovery—was to believe that it was extremely dense. It might be as small as it appeared to be and yet massive enough to be Planet X if, say, it was twelve times as dense as Earth or sixty times as dense as water; in fact, very dense objects consisting of highly crushed matter, like white dwarfs, were known. Ultimately, these were only speculations and were apt to remain so until an actual measurement of Pluto's diameter was available. For that, astronomers would have to await completion of the 200-inch Palomar telescope, still almost two decades off.

Meanwhile, important observations of Pluto were being made by the German astronomer Walter Baade. While working as an astronomer at Hamburg Observatory in 1930–31, Baade determined several highly accurate positions for Pluto and attempted to measure its magnitude accurately. In 1931 he left Germany for the United States and joined the staff of Mount Wilson Observatory. In March and November 1933, he made the first attempts to determine Pluto's true brightness using the techniques of photographic photometry with the 60-inch and 100-inch telescopes. He found the photographic magnitude for Pluto to be 15.58 ± 0.02, which is slightly fainter than Pluto when at its average distance from the Sun. Baade also measured the brightness of the planet through a set of colored filters and thereby made the first determination of Pluto's color (a color index). The result showed that Pluto is not gray but reddish in color.[38] As a very faint reddish object, Pluto's actual measured characteristics did not seem to agree with the Jeans-Crommelin theory of a highly reflective icy body. The most straightforward conclusion from Baade's work was that Pluto was a small body of probably insignificant mass.

We now know, of course, that this was right, and that Tombaugh's discovery of Pluto was serendipitous—like Wilhelm Röntgen's discovery of X-rays or Henri Becquerel's discovery of radioactivity. While searching for a planet, he discovered something else, possibly even more momentous—the brightest object of the Kuiper Belt and a harbinger of things to come, sixty years ahead of its time. It was a discovery that owed something to Lowell's inspiration but as much or more to Tombaugh's perspiration.

As for Lowell, a man of obsessive personality, whose deep-seated need was for things to be sharply defined, black and white, without ambiguity or shades of grey, he had ironically placed his utmost hopes on a "great maybe."[39] In his essay, "Is Life Worth Living?," James—while not referring to Lowell—managed to sum up perhaps better than anyone else what Lowell had done by speculating on, attempting to calculate, and finally leaving as a legacy his dream of the "great maybe" that was Planet X and which was realized as Pluto:

> "May be! May be!" one now hears the positivist contemptuously exclaim. "What use can a scientific life have with maybes?" Well, I reply, the "scientific" life itself has much to do with maybes, and human life at large has everything to do with them. So far as man stands for anything, and is productive or originative at all, his entire vital function may be said to have to deal with maybes. Not a victory is gained, not a deed of faithfulness or courage is done, except upon a maybe; not a service, not a sally of generativity, not a scientific explanation or experiment or text-book, that may not be a mistake. It is only by risking our persons from one hour to another that we live at all. And often enough our faith beforehand in an uncertified result is the only thing that makes the result come true.[40]

The Search Continues

In later years, Tombaugh often confided that Pluto was so much dimmer than what was expected for a planet at the calculated distance that he had always harbored doubts about whether Pluto truly was a full-fledged planet. When the orbital parameters were finally calculated, and the fact emerged that it had an orbit far more eccentric than any other planet's and also a steep inclination to the orbital plane of the other eight known planets, these doubts only grew. Of course, though the Lowell Observatory was hardly a disinterested party and was very motivated to get its planetary status affirmed, Roger Lowell Putnam himself had said, shortly after it turned up on Clyde's plates, "The new object, whatever it is, is a most interesting discovery. If it is not strictly planetary, it comes into a wholly new class, which is, in some ways, more exciting."[41] Putnam's remark would prove to be remarkably prescient.

The lurking doubts about Pluto led V. M. Slipher to authorize an extension of the planet search around the ecliptic and over wider areas of the sky.[42] At first, Tombaugh did as he had done in the Planet X search: he obtained the plates and blinked them. The first year's plates consisted of a single strip, centered along the ecliptic. The search was then streamlined so that two strips around the sky, one north and one south, parallel to the strips along the ecliptic, were taken at one time. When observing conditions were good, the procurement of the plates proved to be easier than their examination. By the end of the Flagstaff monsoon season in September 1932, however, Clyde had caught up: a solid belt 30–35° wide, consisting of the three adjacent strips extending all the way around the sky, had been blink-examined.

Unusually bad observing weather in the winter of 1932 interfered with the further extension of a second strip to the south, but in later years it was again possible to photograph two strips at a time, located at ever-greater distances from the ecliptic. Tombaugh noted that "such a vigorous observing program put a strain on the available observing hours with the telescope when too many nights of a lunation were not fully satisfactory. Often the [third] check plate was taken by 'forcing' the conditions. If there were passing patches of haze, the intensity and duration of the absorption was estimated, and the exposure was prolonged, sometimes up to 50–100 per cent. Yet the star images were usually well matched for blinking."[43] The mind-numbing process of guiding the plates had to be carried out in a state of at least semi-alertness. Of course, the blinking required even more mental alertness than exposing the plates. He noted afterward,

> It was found that the time required for thorough blink examination of a given area on well-matched plates was roughly proportional to the number of star images when they exceed about 400 per square inch, but when the number of star images was less, the rate of sky coverage was proportional to the plate area. The reason for this appears to be that a roughly constant number of spurious aggregations of grains [in the photographic emulsion] attract attention, whether there are any stars on the plate or not.[44]

After the discovery of Pluto, Tombaugh received a scholarship to Kansas University, in Lawrence, where he spent 1933–35 (except for the summers, when he returned to Lowell Observatory) completing his bachelor's degree. (As the only living discoverer of a planet, he was specifically forbidden from taking the introductory astronomy course, which was deemed beneath his dignity.) He met his future wife, Patricia Edson, whose brother was also an astronomy major and a close friend of Tombaugh who boarded as a student roomer in their house. Tombaugh married Patricia in 1934, and the two of them spent summers at Lowell Observatory, where he continued to work on the planet search, living on Mars Hill in a small, brown-shingled cottage in which they cooked and heated with wood fires and had no refrigerator, telephone, or washing machine.

During Tombaugh's periods of absence in Lawrence, Frank K. Edmondson, an Indiana University graduate, carried on the photographic work for a time. From 1936 to 1938, helped by two grants from the Penrose Fund of the American Philosophical Society, two persons could carry forward the work; Henry Giclas, a native of Flagstaff whose father had helped set up the 42-inch reflector on Mars Hill and who had been hired by V. M. Slipher as a "general assistant," obtained most of the plates, giving Tombaugh the luxury of devoting all his time to blinking them. He found that three to six hours a day were all he could blink with efficiency. A good day's work meant the examination of thirty thousand to sixty thousand stars for planetary motion.

In 1939, Tombaugh commenced a series of deep plates in regions not very crowded with stars, and eliminated fifteen hundred planet suspects of seventeenth and eighteenth magnitudes by checking them against third plates. He did this to rule out the existence of any zone of asteroids brighter than the eighteenth magnitude beyond the orbit of Saturn. Finally, in May 1943, he completed the main program, having photographed and blink-examined the entire sky visible from Flagstaff, from 50° to the North Celestial Pole. In all, he estimated that he had logged seven thousand hours sitting at the blink comparator and examined about 90 million star images. Tombaugh had carried out the work with such "care and thoroughness" that he was confident that no unknown planets brighter than sixteenth magnitude could exist. Further, he was sure that even planets between sixteenth and seventeenth magnitude would have had a good chance of being discovered during his search.[45]

Between 1943 and 1945, Tombaugh taught courses in navigation to Navy personnel at Arizona State College (now Northern Arizona University). He returned to Lowell Observatory briefly in August 1945, spending the next few months finishing the blink examination of some plates in northern areas of the sky that he had not yet completed. Since Lowell Observatory did not have the funds to hire him back, he left—not without bitterness—the observatory where he had made the discovery for which he and it would be remembered forever. He finished out the year as visiting assistant professor of astronomy at UCLA, and in 1946 transferred to the ballistics research laboratories at White Sands Proving Ground (now White Sands Missile Range), where he developed telescopes to track rockets and missiles during test flights. From 1953 to 1955, he led the U.S. Army's Near Earth Satellite Search, conducted at Lowell Observatory, which involved inventorying the space in Earth's vicinity in preparation for the launch of the first artificial satellites. In 1955 he transferred to the Physical Science Laboratory at New Mexico State University in Las Cruces, where he developed, in collaboration with Bradford A. Smith and others, the university's Planetary Patrol and Study Project, which lasted from 1958 to 1973, the year of Clyde's retirement.

In 1980, when Tombaugh was being feted around the globe for the fiftieth anniversary of the discovery of Pluto, he distilled the hard-earned experience of his years of planet seeking in the following document—part humorous, part serious—which seems a fitting note on which to end this chapter.

The Ten Special Commandments for a Would-Be Planet Hunter according to Clyde Tombaugh

- Behold the heavens and the great vastness thereof, for a planet could be anywhere therein.
- Thou shalt dedicate thy whole being to the search project with infinite patience and perseverance.
- Thou shalt set no other work before thee for the search shall keep thee busy enough.
- Thou shalt take the plates at opposition time lest thou be deceived by asteroids near their stationary positions.
- Thou shalt duplicate the plates of a pair at the same hour angle lest refraction distortions overtake thee.
- Thou shalt give adequate overlap of adjacent plate regions lest the planet play hide and seek with thee.
- Thou must not become ill in the dark of the Moon lest thou fall behind the opposition point.

- Thou shalt have no dates except at full Moon when long exposure plates cannot be taken at the telescope.
- Many false planets shall appear before thee and thou shalt check every one with a third plate.
- Thou shalt not engage in any dissipation, that thy years may be many, for thou shalt need them to finish the job!

8

Planetary Astronomy

Near Death and Revival

THOUGH TOMBAUGH'S DISCOVERY of Pluto arguably saved the Lowell Observatory from extinction, it did little to rescue planetary astronomy from the moribund state in which it found itself after Lowell's death. The recoil of many professional astronomers from the subject was in part due to exhaustion over the endless controversies regarding the Martian "canals" and the sense that, as Neville J. Woolf has put it, "the observations that were made did not convert into facts. . . . As a result, speculation flourished unchecked by observation. The widespread public interest has also caused difficulties, since it has tended to encourage cranks to enter this field, which is difficult enough without them."[1]

In general, Lowell's enthusiasm was acknowledged but with reservations, even during his lifetime. In 1895, James Keeler wrote to George Ellery Hale, "I dislike his style. . . . It is dogmatic and amateurish. One would think he was the first man to use a telescope on Mars . . . and he draws no line between what he sees and what he infers."[2] Eventually, Hale and Keeler, America's leading astrophysicists, simply refused Lowell's submissions to the *Astrophysical Journal*, whereupon Lowell founded his own outlet, the *Lowell Observatory Bulletins*. Later, after Lowell published *Mars as the Abode of Life* and *Evolution of Worlds*, which drew eclectically not only on astronomical but on meteorological, geological, and botanical sources in attempting to forge a new science he called "planetology," he faced savage reviews. Forest Ray Moulton, an astronomer at the University of Chicago who had coauthored the Chamberlin-Moulton theory of the formation of the Solar System, described Lowell as "that mysterious 'watcher of the stars' whose scientific theories, like Poe's vision of the raven, 'have taken shape at midnight.'"[3] After Lowell's death in November 1916, the doyen of American astronomers, Henry Norris Russell of Princeton, in a generally appreciative obituary fell short of full-throated endorsement:

> Under ordinary circumstances, Mars is a heartbreaking object for the observer. The larger and less interesting details upon his ruddy disc are indeed visible telescopically on any good night, but the finer markings are very delicate, and flash into view only by glimpses in the too rare moments when the ceaseless turmoil of the atmosphere through which we must look dies down and permits us to see the planet with relatively little blurring of its finer lineaments. . . .
> If the observer knows in advance what to expect . . . his judgment of the facts before his eyes will be warped by this knowledge, no matter how faithfully he may try to clear his mind of all prejudice.[4]

Because of the seemingly insuperable difficulties facing planetary astronomy in the visual era, professional astronomers largely decided to leave the planets (and the Moon) to amateurs. Gerard Peter Kuiper later looked back at the situation during the 1920s and 30s:

> The phenomenal growth of astrophysics and the exciting explorations of the Galaxy and the observable universe led to an almost complete abandonment of planetary studies. . . . Physical observations of planetary surfaces, particularly Mars, led to controversies and speculations that may have been appreciated by the public but hardly by the professionals. More and more this branch of planetary work, including the study of the Moon, became the topic *par excellence* of amateurs—who did remarkably well with it. The *Memoirs* of the British Astronomical Association became the chief record for the development of planetary surface markings. Astronomers with large telescopes were

so occupied with the engaging problems of stars, nebulae, clusters, the Galaxy, and the universe that astronomy became almost entirely the science of the stars.[5]

World War II would change all that. The technology forged in that cauldron of events created a new opportunity for planetary astronomy to flourish.

Modern physics—which included quantum theory, the theory of relativity, and nuclear physics, all of which would eventually shape and mold planetary astronomy—developed in the first four decades of the twentieth century. Ernest Rutherford said at the meeting of the BA in Liverpool in 1923, "We are living in the heroic age of physics." No one could gainsay that. C. P. Snow, who knew him, says that Rutherford "went on saying the same thing, loudly and exuberantly, until he died, fourteen years later."[6] "The curious thing," he added,

was all he said was absolutely true. There had never been such a time. The year 1932 [two years after the discovery of Pluto] was perhaps the most spectacular year in the history of science. Living in Cambridge, one could not help picking up the human, as well as the intellectual, excitement in the air. James Chadwick, grey-faced after a fortnight of work with three hours' sleep each night, telling the Kapitsa Club (to which any young man was proud to belong) how he had discovered the neutron; P.M.S. Blackett, the most handsome of men, not quite so authoritative as usual, because it seemed too good to be true, showing plates which demonstrated the existence of the positive electron; John Cockroft, normally about as much given to emotional display as the Duke of Wellington, skimming down King's Parade and saying to anyone whose face he recognized: "We've split the atom! We've split the atom!"

That heroic age of physics was largely a European affair. On the experimental side, it was mainly British, on the theoretical largely German and, even more narrowly, German Jewish. (The Jewish physicists included Albert Einstein, Niels Bohr [mother], Wolfgang Pauli, and Max Born, while others, like Enrico Fermi, had married Jews.)

The United States, because it was rich and a safe distance from Germany, received many Jewish scientists, creating what Snow calls "the Jewish explosion, a burst of creativity in all fields, not only science."[7] These refugees created a new dynamic in science. Another development was the way that physicists were co-opted into the war effort, on both sides. Thousands of physicists worked on radar, thousands more on rockets or the atomic bomb. Mark Oliphant, one of Rutherford's students at Cavendish Laboratory at Cambridge who went on to the University of Birmingham to develop the cavity magnetron, which made microwave radar possible, would remark when the first atomic bomb was dropped, "This has killed a beautiful subject."

Inevitably, the war ran its course. Following the Allied invasion of Italy in September 1943, the United States—as part of the Manhattan Project—established the Alsos Mission, led by two excellent physicists, Samuel Goudsmit and George Eugene Uhlenbeck, who were originally Dutch and were now American. Military, scientific, and intelligence personnel followed close behind the front lines to secure valuable resources and investigate what the German nuclear physicists had been up to. What they found was surprising. They had, in fact, done very little, apparently because German authorities, with whom decisions often went all the way to Hitler himself, weren't prepared to devote resources to projects that wouldn't guarantee results within a couple of years. Hitler's focus wasn't the nuclear bomb, it was jet-propelled flying bombs and rockets. The great achievement of Nazi Germany's scientists and engineers would be the V-2.

Among the scientists sent to Europe as part of the Alsos Mission was Kuiper, a native of the Netherlands, professor at the University of Chicago, and prominent astronomer at Yerkes Observatory. He left Yerkes for war service in 1943, going first to the Harvard Radio Research Laboratory. During a brief respite during the winter of 1943–44, he arranged for two months' leave to return to the McDonald Observatory (then jointly administered by the University of Chicago and the University of Texas). A discovery he made during that brief period of leave would, as we shall see, change the direction of his career. Afterward, he was sent to Europe as a civilian scientist attached to the Eighth Air Force based in England. Fluent in Dutch, German, French, and English, and having contacts with scientists and research facilities throughout western Europe, he was charged with reading and analyzing captured or intercepted reports and with debriefing French and Belgian scientists liberated by the advancing Allied armies. In 1945, he joined the Alsos Mission.

Figure 8.1 Yerkes Observatory, Williams Bay, Wisconsin. Photo by Dale Cruikshank.

In May of that year, as both the Russians and the Americans were converging on Berlin, American soldiers learned that the famous German physicist Max Planck was holed up in the German countryside outside Berlin, in the eastern zone occupied by the Russians. The Americans saw that there was a chance of getting to him before the Russians did, so Kuiper, who had sufficient rank to commandeer a jeep, raced across the countryside to Planck's house, where he found Planck and his wife before the Russians arrived. They were hurried away to the West and surrendered to Allied forces. Eventually, Planck settled in Göttingen. Long afterward—even after Max's death in 1947—the Kuipers received a Christmas card every year from Göttingen.

Kuiper's rescue of Planck was dramatic, but in practical terms the most important results of his Alsos work was what he learned from his interrogations of captured German scientists. From the Austrian-born physicist Erich Regener, he learned that the Germans were planning to put scientific instruments aboard a V-2 rocket to study the upper atmosphere. They had also been making great strides in developing infrared detectors, which they hoped to use to guide the jet-propelled V-1 flying bombs and V-2 rockets

The Spectrum, Wavelengths, and Spectrometers

Visible light is only one small component of the electromagnetic spectrum, which includes the x-ray, ultraviolet, visible, infrared, and radio regions. Because electromagnetic (EM) radiation behaves as waves, each term refers to a range of wavelengths. All EM radiation travels through space at the speed of light, but the wavelength, or physical distance between the crests of the waves, ranges widely. To denote wavelengths we use metric units of nanometers ($1 \, nm = 10^{-9}$ m, or one-billionth of a meter) and micrometers ($1 \, \mu m = 10^{-6}$ m, one-millionth of a meter). The *visible* spectral region covers wavelengths 300 to 1,000 nm (0.3–1 μm), although the human eye is sensitive to only a portion of that region, about 400 to 700 nm.

The region 1–5 μm is generally called the *near infrared*, while in astronomical terms the *mid infrared* covers approximately 5–35 μm and the *far infrared* covers approximately 35–700 μm. Wavelengths longer than that are in the radio region, and radio waves span an enormous range of about 700 μm to several kilometers. Some astronomers and physicists define the wavelength ranges differently depending on their special interests and the instruments they use in their studies. Most importantly, each wavelength region carries information about the composition, temperature, and other characteristics of the source of that EM radiation.

Spectrometers used in the laboratory and at the telescope are optical devices that use prisms or diffraction gratings to disperse the light from an astronomical object or a laboratory source into its component wavelengths that are recorded with detectors (sensors) that are sensitive to the wavelength region of interest. The first lead-sulfide detectors were sensitive in the range approximately 1–4 μm.

to sources of heat to maximize destruction of Allied military and civilian targets. These detectors utilized cells of lead sulfide (PbS), a mix of lead and sulfur atoms that yields an electrical signal when exposed to heat radiation. Though designed for malign purposes, Kuiper, as an astronomer, envisaged nonmilitary uses for them. The lead-sulfide detectors were sensitive to a region of the electromagnetic spectrum with wavelengths longer than the human eye, or even the best photographic emulsions, could detect. Kuiper knew that using these detectors on a large telescope and applying them to stars and planets would open a previously unseen window on the compositions and temperatures of these bodies. He knew that the infrared spectral region, as it is called, carries a tremendous amount of information on the properties of atoms and molecules that make up the atmospheres of the stars and planets.

As soon as he returned to Yerkes, Kuiper discovered from still-classified records that the Americans had been trying to build similar lead-sulfide detectors. These records were quickly declassified, and Kuiper learned that Robert J. Cashman, a physicist at Northwestern University, had been in charge of these efforts. Conveniently, Cashman's laboratory at Northwestern was a short drive from Yerkes where Kuiper lived, so Kuiper contacted Cashman at once and began to collaborate with him. At the time, as Kuiper knew, no other American astronomer had his insider knowledge of the lead-sulfide cells or his grasp of their potential for scientific research.

Who was this visionary who possessed, though still in primitive form, the technology that would eventually lead to the development of infrared astronomy, and so eventually to the ability of astronomers to detect and analyze the composition of the icy bodies of the outer Solar System, including Pluto?

Kuiper

Gerard Peter Kuiper (originally Gerrit Pieter Kuiper) was the son of a tailor. He was born on December 7, 1905, in the Netherlands, in the small town of Tuitjenhorn, in the former municipality of Harenkarspel. (In 2013 Harenkarspel merged with Schagen and Zijpe into a new municipality, Schagen.) He was an excellent grade-school student, but to continue his education, he had to leave his small town. He went to Haarlem, where he enrolled in an institution devoted to training primary-school teachers. Normally, this would have been a dead end because admission to a university, such as Kuiper already contemplated, was open only to students who went to proper high schools. Already possessed of the enormous drive, persistence, and self-assurance that he would demonstrate throughout his career, Kuiper passed an especially difficult examination that allowed him to enter Leiden University, and also passed the examination for certification to teach high-school mathematics.

Kuiper later recalled that his interest in astronomy and cosmogony—the study of the origins of the Solar System, which was to become his principal life's work—was first sparked by the writings of the French philosopher René Descartes and encouraged by his father and grandfather, who put into his hands a small telescope. He would devote a whole winter as a teenager to mapping the faintest stars in the Pleiades star cluster with the naked eye. On sending the results to Leiden Observatory astronomers, he learned that the limiting magnitude he achieved was 7.5, nearly four times fainter than what the normal human eye can achieve. Even in later years, Kuiper's visual acuity remained exceptional.

Kuiper entered Leiden University in September 1924. Already he had a clearly defined sense of direction unusual for an entering student. Bart J. Bok, who later became a leading astronomer in his own right, recalled the circumstances of their first meeting in Leiden's Institute of Theoretical Physics:

> Earlier that morning, Gerard and I had both registered as entering students at the University. Since we did not belong to any of the social clubs at the University, we had little to do on that first morning except to join the special student and reference library. . . . So it was not surprising that we should have met looking over the card catalog of astronomy books in the Reading Room. After introductory formalities were over, Gerard asked me what was my special field of interest in astronomy. I promptly replied that I was most interested in the Milky Way and the Cepheid variable stars. He responded: "That is a not uninteresting field. But I expect to study a more fundamental area, the problem of three bodies and related questions about the nature and origin of the Solar System." It was evident right at the start that Gerard was determined

to make major research contributions to astronomy and astrophysics. Gerard knew that the Solar System was to be his to explore, the physics of the planets as well as their motions and origins.[8]

Both Bok and Kuiper were fortunate in their first years in Leiden. Paul Ehrenfest, who succeeded Nobel Laureate Hendrik Antoon Lorentz as chair in theoretical physics, was among their teachers. Ehrenfest would later ask Kuiper to become tutor of his son Paul, while Kuiper would later name his own son Paul, after the senior Ehrenfest. They learned their astronomy from the likes of Willem de Sitter, Ejnar Hertzsprung, Jan Woltjer, and Jan Oort, whom Bok referred to as "the youngest of the great." Their older friends included Goudsmit and Uhlenbeck, who later played important roles in Alsos and who were then discovering the spin of the electron.

Kuiper elected to do his doctoral thesis on binary stars, under Hertzsprung. He was just getting started on his dissertation—on the statistical analysis of binary orbit periods and mass ratios—when Pluto was discovered by an American of the same age. Kuiper took note, and in an article for *Hemel en Dampkring*, the monthly magazine of the Dutch Association for Meteorology and Astronomy, pointed out that since Pluto did not fit with Bode's law, the so-called law could not be valid.

On receiving his PhD in 1933, Kuiper received a two-year postdoctoral fellowship to Lick Observatory, whose director, Robert Grant Aitken, was a specialist in double stars who had learned from the legendary double-star observer E. E. Barnard, who in turn had learned from the equally legendary Sherburne W. Burnham.[9] The Lick 36-inch refractor, the most powerful instrument in the world when commissioned in 1888, was now dated for many purposes, but it was still a superb instrument for the detection of close double stars. Kuiper, with his acute vision and relentless work ethic, discovered many new binaries and white dwarfs—small, dense objects whose nature was not yet understood.[10] Though Kuiper would no doubt have liked to stay on at Lick, Aitken retired in 1935 and the new director, William H. Wright, was not interested in double stars; he was interested in the spectroscopy of stars and nebulae. Wright had two positions to fill, and chose Nicholas U. Mayall and Arthur B. Wyse, both spectroscopists like himself and Americans, married, and products of the Lick graduate program. Kuiper was none of the above. In addition, in the small mountaintop community, the painful struggles of early astronomers like Barnard and Burnham, as well as the first director, Edward S. Holden, were still remembered. Kuiper was perceived not only as talented but as outspoken, abrasive, and self-centered, the kind of man who might not wear well over time. Kuiper, therefore, was hired for another postdoctoral year by Harvard College Observatory, after which he intended to take up the only job on offer at the time, at the Dutch Bosscha Observatory at Lembang, Java.

At Harvard, however, he had a literal change of heart. He met and fell in love with Sarah Parker Fuller, whose family had donated the land on which the George R. Agassiz Observing Station (now known as Oak Ridge Observatory) was built. The future Mrs. Kuiper—they were married in June 1936—made it known at once that she was not at all keen to move to Java. Just then, Otto Struve, director of the University of Chicago's Yerkes Observatory, invited Kuiper to join the staff there, and so instead of going to the "beautiful and happy island of Java," as he called it, Kuiper went to Yerkes. In retrospect, he was probably lucky, not only because his career at Yerkes would prove to be so productive, but also because if he had gone to Java he might not have been able to escape the Japanese prison camps that were set up after the Japanese invasion of the Dutch East Indies in 1942.

The Yerkes Observatory, built around its famous 40-inch refracting telescope—still the largest instrument of its kind on Earth, and likely to remain so—had been operational since 1897. The founder, George Ellery Hale, had demonic energy and an enviable knack for persuading millionaires to finance his grandiose schemes. Hale was spoiled by his doting father, who had made a fortune building elevators for skyscrapers in Chicago after the Great Fire and who funded George's first observatory, equipped with a 12-inch refracting telescope, and a laboratory in the family mansion in Chicago's affluent Kenwood neighborhood. In his early twenties, Hale managed to wangle funds for the 40-inch telescope from Chicago elevated streetcar magnate Charles Tyson Yerkes. (Of Yerkes, known as the "Boodler," it was said that he didn't invent corruption in Chicago, he only perfected it.) Since Yerkes agreed to pay only for the telescope, Hale had to get millionaire and University of Chicago benefactor John D. Rockefeller to pay for everything else. The architecture of the building is superb—complete with gargoyles and featuring exquisite terracotta figures on the columns meant to represent Yerkes, Rockefeller, and University of Chicago president William Rainey Harper. Rockefeller is in profile; there was originally a bee on his nose, presumably showing him being stung for the money. Someone decided this might not possibly be in the best taste, so before the grand opening the bees were chiseled off all the noses on all the columns but one, though their outlines are still plainly visible to this day. The observatory backs onto woods along Lake Geneva and a golf course, and is a picture of rural tranquility— too much so for the restless Hale, who soon left to found bigger observatories equipped with even larger telescopes in California and left Yerkes in the hands of Edwin B. Frost. Frost was a Dartmouth-trained spectroscopist who, though a humane and widely read individual, turned out to be a traditional, rather plodding researcher who was completely blind during the last decade and a half before he reached the mandatory retirement age of sixty-five in 1932. Though Frost was never a strong researcher, he did make some key hires, none more consequential than Struve.

Otto Struve had been born in 1897 to a family of astronomers. He was a grandson of the Russian astronomer Otto Struve and a great-grandson of Friedrich Wilhelm Struve, both of whom had directed the Pulkovo Observatory. He was also a nephew of Hermann Struve, the director of the Berlin-Babelsburg Observatory. Young Otto had planned to follow an astronomical career himself in Europe but the First World War intervened, and Otto, after a crash course in the artillery, was co-opted for service on the Turkish Front, where he found himself fighting with the White Russian Army against the Bolsheviks. With the defeat of the White Army, he escaped to Turkey and endured eighteen months in miserable exile at Gallipoli and Constantinople. At that time he managed to get in touch with the widow of Hermann Struve. Though Hermann himself had died in 1920, his successor as director of the Berlin-Babelsburg Observatory obtained for him a letter of introduction to Frost, and without hesitation, Frost—considering Struve's lineage alone— offered him a position as a spectroscopist at Yerkes. He arrived at Williams Bay in 1921, within two years finished his PhD thesis on short-period spectroscopic double stars, and proceeded to rise meteorically through the ranks, becoming instructor, assistant professor, and full professor at the University of Chicago and—when Frost retired in 1932—director of Yerkes Observatory.

If, during the Frost years, the Yerkes Observatory experienced what historian Donald E. Osterbrock describes as a "near death,"[11] Struve accomplished its resurrection, partly by keeping the best young PhD students (such as William W. Morgan and Philip C. Keenan) on staff, but mainly by hiring brilliant young astronomer-astrophysicists from abroad. Among these was the Indian Subrahmanyan Chandrasekhar, who had studied in England under the great Cambridge theorist of the atmospheres of stars Sir Arthur Eddington. (While on board the ship that was bringing him to America, Chandrasekhar carried out the calculation for which he is famous, that of the limiting mass of a star—Chandrasekhar's limit—which determines whether it will supernova or form a white dwarf.) Another was Bengt Strömgren, the son of Elis Strömgren, the director of the Copenhagen Observatory. Third was Kuiper. This group, with Struve, was to be the brain trust at Yerkes in the years before World War II.

Struve was physically large and this, together with his outsize personality, could make him an intimidating figure. He worked extremely hard and expected others to do the same. However, he was also a tremendous inspiration. During the Struve years, the scientific output at Yerkes was at the highest intellectual level, and the students educated there went on to become the senior astrophysicists of the next generation.

At first, Kuiper immersed himself in problems in stellar astronomy, sometimes in collaboration with Struve and Strömgren. For instance, in 1937 he and Struve published a pioneering study of the eclipsing binary Epsilon Aurigae, in which they proposed that a large star surrounded by a partly transparent gas halo was eclipsing an F supergiant whose ultraviolet radiation had ionized part of the larger star's tenuous atmosphere.

Though temporarily busy in other areas, Kuiper never gave up his early dream of understanding the origin of the Solar System. Asked to review a book on the subject, he wrote that the state of astronomy did not yet permit its solution; he instead devoted himself to the closely related—and more soluble—problem of the origin of double stars. He continued the statistical studies of binary stars he had begun as a PhD student, eventually arriving at the important result that at least half the stars in the sky were binaries. Meanwhile, he and W. J. Luyten of the University of Minnesota, a fellow Dutchman and often a bitter rival who had first coined the term "white dwarf" in 1922 (and took a rather proprietary view of the field), were between them responsible for cataloging most of the white dwarfs known until the early 1960s.

Another highly productive initiative to which Kuiper contributed was the establishment of the new McDonald Observatory. Its 82-inch telescope, when it became operational in 1939, was the second largest in the world at the time, eclipsed only by the Mount Wilson 100-inch. The site testing for the telescope had begun early in Struve's tenure, with Mount Locke near Fort Davis in west Texas being eventually selected.

Possessed of almost unbelievable energy and drive and able to thrive on only a few hours of sleep a night, Kuiper was indefatigable as an observer at the telescope. When, in November 1942, he independently discovered Nova Puppis, he obtained four spectra and a position measurement of the nova before the Sun rose, despite having already completed a full night of observing. Coauthor Cruikshank recalls his experience as Kuiper's student assistant on many long observing runs at McDonald and Kitt Peak observatories in the 1960s:

> During these periods, sometimes amounting to fourteen consecutive clear days and nights at McDonald, Kuiper would function on three to four hours of sleep on some days, while another assistant and I would work ten-hour shifts. When heavily fatigued at some time during the day [when he was observing bright stars and planets], he would occasionally lie down on the observing platform only to awaken twenty minutes later appearing fully refreshed and ready to press on for another four to six hours.[12]

Kuiper's determination to use every scrap of available time is well illustrated by the fact that, during a two months' respite from war service in the winter of 1943–44, he arranged for leave to return to McDonald Observatory. Though still chasing stars of large proper motion and parallax at the time, which included many white dwarfs, he noticed one morning that several planets and the ten brightest satellites in the Solar System were lined up in a part of the sky where he had run out of "program stars," so he decided to study each one with the spectrograph. The subjects included the giant planets; Jupiter's four largest satellites, Io, Europa, Ganymede, and Callisto; Saturn's four largest satellites, Dione, Tethys, Rhea, and Titan; and Neptune's large satellite, Triton. It was a rather ambitious agenda, and it produced a sensational result: in the spectrum of Titan, he found the signature absorption band of methane gas, thereby achieving the first detection of an atmosphere of a satellite.

In making this discovery, Kuiper had recorded the spectrum of Titan on ordinary photographic emulsions. As we have seen, when the lead-sulfide detectors fell into his hands soon afterward, he seized his great opportunity and never turned back. The new detectors opened a previously inaccessible part of the spectrum, which would enable him—while not entirely abandoning his stellar researches—to devote most of his time to studies of planetary atmospheres and the origin of the Solar System.

Though the existence of the near-infrared had first been noted by William Herschel in 1800—it was a serendipitous discovery, occurring when Herschel decided to investigate why a glass filter shattered while he was observing a transit of Mercury—not until the early 1900s were sensors sensitive to wavelengths greater than about 5 micrometers (μm) available for laboratory use. These sensors showed chemists that the spectra of nearly all molecules and compounds have distinct emission and absorption bands. These emission and absorption bands contain an enormous amount of information about molecular structure, and eventually led to a huge literature on molecular chemistry and physics based largely on infrared spectroscopic studies. However, the detectors used by chemists were designed for use in the laboratory; they proved to be virtually useless at the telescope and for astronomical purposes, mainly because the region of the near-infrared that is optimal for planetary studies is at about 1 to 5 μm and the early detectors only worked well at longer wavelengths. Prior to the advent of lead-sulfide detectors during World War II, simply nothing existed that was suitable for astronomical near-infrared studies.

For infrared astronomy to become viable, two important obstacles had to be overcome: (1) Earth's atmosphere, which is largely opaque at infrared wavelengths, thus blocking incoming light from stars and planets; and (2) the faintness of the infrared radiation of planets and stars.

When he returned to Yerkes in 1945 and began to collaborate with Cashman, Kuiper could see that the just-declassified lead-sulfide cells would be sensitive to the highly informative but hitherto unavailable region of the

Figure 8.2 Kuiper at Yerkes Observatory with the first-generation near-infrared spectrometer using lead-sulfide detectors. Circa 1947. Courtesy of Special Collections Research Center, University of Chicago Library.

near-infrared (1–2.5 μm). Here stars of all temperatures have distinct and diagnostic spectral lines and bands, and the gases of the giant planets display their component molecules in exquisite detail. Researchers in this part of the spectrum are also spared the complications of the longer-wavelength infrared, since not only is Earth's atmosphere effectively transparent in the near-infrared, the objects themselves are significantly brighter there. Thus, at a stroke, both obstacles to infrared astronomy were pushed back, if not removed.

Obtaining from Struve a grant of $500 for the purpose, Kuiper built a spectrometer around one of Cashman's cells for use on the 82-inch telescope. With this spectrometer he could record radiation in in the 1 to 3 μm region of the infrared, well beyond the cutoff even of specially sensitized photographic emulsions used at the time. The first paper describing the detector and spectrometer was jointly authored by Kuiper, who designed the spectrometer; Cashman, who designed the cell; and W. Wilson, who built the electronics for an amplifying system. In an early test on the McDonald 82-inch telescope, Kuiper produced the first high-resolution near-infrared spectrum of the giant star Betelgeuse,[13] and—in what would presage his future direction—he became the first person to see the near-infrared spectra of Jupiter and Saturn, the Galilean satellites of Jupiter, and the rings of Saturn.

Struve fully supported Kuiper's interest in planetary atmospheres, and regarded the discovery of the atmosphere of Titan as a significant breakthrough in the understanding of the origin of the Solar System. This in turn was closely related to the origin of binary-star systems, an area in which Struve and Kuiper had long collaborated. He also realized, in the words of historian Ronald E. Doel,

Interpreting the constitution and structure of planetary atmospheres at infrared wavelengths required a more extensive knowledge of molecular spectroscopy than that needed for studying the spectral features of most stars at visual wavelengths. Assessing the relative strength of particular bands seen against the heavy molecular absorptions characteristic of planetary atmospheres was rendered difficult, for example, by the apparent similarity of extremely weak and strong abundances; solving such problems meant spectroscopists had to investigate radiative transfer within planetary atmospheres. Few astrophysicists of [that] generation had the requisite spectroscopic skills to undertake such analyses, and it was for this reason that Struve had felt grateful that [the eminent spectroscopist Gerhard] Herzberg, with his extensive background in infrared spectroscopy, had joined the Yerkes-McDonald staff.[14]

By 1946–47, not only Kuiper and Herzberg but a full third of the Yerkes-McDonald staff were involved in some capacity in research on planetary atmospheres. A highlight of this exciting and productive era was a conference

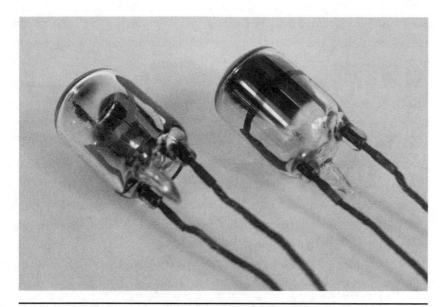

Figure 8.3 Original lead-sulfide infrared detectors used in Kuiper's early spectrometers. Photo by Dale Cruikshank.

on planetary atmospheres held at Yerkes Observatory in September 1947 in connection with the American Astronomical Society (AAS) meeting held at Northwestern University. Struve, who had grown weary of administrative duties, had just retired as director, with Kuiper succeeding him. Since it was the semicentennial of Yerkes Observatory's founding, both men wanted to have a conference celebrating the occasion; Kuiper wanted it to discuss planetary atmospheres, while Struve preferred his own special subject, stellar atmospheres. Though Kuiper was the director of the observatory, Struve had remained chairman of the astronomy department. At Yerkes, they had adjoining offices, and from the very beginning Struve believed that, though Kuiper was director, he was over him as chairman of the department. In addition, Struve had just been elected as president of the AAS. In the end they compromised—Struve did hold a symposium on stellar atmospheres at Northwestern, but on the last day of the meeting, everyone went to Yerkes to hear about planetary atmospheres.

The conference on planetary atmospheres was a brilliant success. Kuiper had invited such authorities as Harry Wexler of the US Weather Bureau and

Raymond J. Seeger, then with the Office of Naval Research, as well as leading researchers connected to the American rocket program (based on captured V-2s), including University of Iowa physicist James Van Allen. Many of the contacts forged at the conference would prove to be long and fruitful. The results of the conference—many of them tentative, since the participants were far from reaching a consensus—were published in a Kuiper-edited book, *The Atmospheres of the Earth and Planets*. The highlight was Kuiper's own classic paper, "Planetary Atmospheres and their Origin," in which he proposed classifying planetary atmospheres as either primary (in the case of the giant planets—Jupiter, Saturn, Uranus, and Neptune) or secondary (the terrestrial planets—Mercury, Venus, Earth, and Mars) and discussed the nature of their interactions with the solar wind (a term yet to be invented). The first edition of 1949 sold out almost immediately, and a new edition was called for in 1952. According to Doel, the book was more a "watershed" than a "point of departure."[15] The reason was that Kuiper's infrared spectrometer, even on the 82-inch McDonald telescope, had already reached the limits of its capabilities, and though Kuiper had hoped that Cashman would construct a second detector with a larger surface area and an analyzing slit, he never did. Another limitation of the original spectrometer was that it did not allow the lead-sulfide detector to be cooled with dry ice, an innovation that Kuiper estimated would improve the detector performance by a factor of three. With these constraints, the spectrometer worked well only on the brightest planets and stars; Uranus and Neptune were out of reach, Pluto out of the question. There would be no significant developments in this line until new detectors came out in the 1960s.

Meanwhile, Yerkes suffered the departure of some key staff—for example, Herzberg left at the end of 1947 to assume the directorship of a major national laboratory for fundamental spectroscopic research with the Canadian National Research Council in Ottawa—while the tensions between Struve and Kuiper came to a head, resulting in Struve's departure and Kuiper's losing the support of the staff and stepping down from the directorship. This was not necessarily a bad thing for Kuiper. Though Strömgren was not well suited to the role, he somehow managed to hang on until 1957. (Lured away to the Institute for Advanced Studies at Princeton, he took over Einstein's office.) Kuiper then became director of Yerkes Observatory for a second time.

There can be little doubt that Kuiper liked the idea of power—if only because it maximized his chances of getting the equipment he needed to carry out his own research programs as efficiently as possible and with the help he needed. But being relieved of the responsibilities of the directorship was probably a net plus to his productivity as a scientist, and the years between his two stints as Yerkes director were to rank among his most creative as an astronomer. Though some of his colleagues unquestionably saw Kuiper's transition from stellar astronomy to Solar System astronomy as abrupt—even as a kind of betrayal—the transition was actually gradual and inevitable.

Among the significant researches of this period must be mentioned his 1947–48 attempt to duplicate his discovery of the methane atmosphere around Titan for other outer satellites. In pursuit of this result, he obtained low-dispersion spectra covering the wavelength region between 330 and 880 nm of the four Galilean satellites of Jupiter, Saturn's satellite Rhea, Neptune's satellite Triton, and Pluto. However, he concluded, "No evidence of a CH_4 atmosphere is found from these data, or from better spectra obtained more recently."[16]

Cosmogony and Pluto

Kuiper's binary-star studies were also dovetailing with his interest in cosmogony—the problem of the origin of the Solar System, which, as he had intimated to Bok long before, had always been his great ambition. He had shown that binary stars are common; according to his statistics, they made up more than half of all stars. He also argued, quite plausibly, that the formation of planetary systems was a special case of the general process of binary-star formation. But if that were true, planetary systems—which most astronomers still believed were very rare—might be commonplace and even rampant throughout the Universe.

As with his early interest in infrared spectroscopy and rocket-borne studies of the upper atmosphere, Kuiper's return to cosmogony was indebted to German wartime development. In 1943, in German-occupied Strasbourg, a German physicist named Carl von Weizsäcker published a paper in the

German journal *Zeitschrift für Astrophysik*—since it was not available in the United States because of the war, its importance was not appreciated by American astronomers at the time—in which he imagined the Solar System forming from an eddying vortex of gas and dust in turbulent motion. Though similar ideas had been proposed before—the basic idea went back to the nebular hypothesis first suggested by philosopher Immanuel Kant in 1755 and developed by Laplace in the late eighteenth century—they had all struggled with what was called the "angular momentum problem": the fact that though the Sun had almost all the mass of the Solar System while the planets contained almost all the angular momentum. Seeing no way around this fundamental difficulty, in the first decades of the twentieth century most astronomers entertained the idea that the Solar System had been the product of a near brush of a passing star (perhaps a dark star) with the Sun, from which had streamed the material that would coalesce into the planets. In the 1930s, however, Princeton's Henry Norris Russell and the man who succeeded him as director of the Princeton University Observatory, Lyman Spitzer, showed that thermal diffusion of the hot gases generated in such processes would be incapable of coalescing into solid bodies like planets.

Von Weizsäcker envisaged a primordial solar nebula containing large vortices roughly equal to the Bode's-law distances between the planets. Dust and gas would be forced from the larger vortices into smaller whirlpools, and collisions between them would generate angular momentum even as the Sun lost angular momentum through the dissipation of large volumes of hydrogen and helium from the primeval solar nebula. The eddies were supposed to coalesce into planets.

Though Kuiper accepted von Weizsäcker's overall scheme (especially as modified by Chandrasekhar, who pointed out that a solar nebula would not form the large vortices but ones with a range of widths), he found it unworkable in some of the details, and so presented his own version in a series of papers published in the early 1950s.[17] Whereas von Weizsäcker had imagined a rather beautifully regular system of vortices with the planets forming at Bode's-law distances, Kuiper proposed that the distribution of eddies did not favor any preferential distances to the Sun. In his version, density fluctuations—gravitational instabilities—would arise in the solar

Figure 8.4 Kuiper lecturing on the structure of the Solar System and his views on the origin of Pluto. Circa 1955. Dale Cruikshank collection.

nebula. Under the right conditions, the initial density fluctuations would be amplified by gravity; the solar nebula would begin to break up and condense into protoplanets, the seeds of planets proper.

According to Kuiper, the gravitational instabilities in the solar nebula that produced the planets would also explain the general distribution of asteroids. Moreover, he speculated that such zones existed in the outer Solar System, even beyond Pluto, and that here comets were likely to form. Meanwhile, Jan Oort—who as a visiting astronomer at Yerkes had given a series of lectures on the structure of the Galaxy in late 1947, which surely influenced Kuiper's thought—was also working on comets. In 1950, Oort, building on previous suggestions by E. J. Öpik in 1932 and Adrianus J. J. Van Woerkom in 1948, published a paper that would largely settle the hitherto much-debated question of the source (if not the origin) of comets. Oort proposed the existence of a cloud of comets, now known as the Oort cloud, moving permanently with the Sun. He showed that the observed distribution of comet orbits having periods greater than about two hundred years could be explained by supposing this cloud to consist of hundreds of billions of comets surrounding the Sun at distances of 50,000–150,000 AU and affected from time to time by the perturbing effects of passing stars. These stars would act like bullets passing through a cloud of gnats, scattering a few of them inward toward the Sun where they could be seen from Earth. Kuiper's idea was different. Pluto, at 40 AU from the Sun, moved through what he calculated must have been the outer fringe of the solar nebula. As it passed, it could have gravitationally diverted comets toward the massive outer planets, which in turn would have thrown them outward into a distant comet cloud. Thus Pluto provided a plausible mechanism for supplying comets to a distant comet cloud, but only if it was sufficiently massive to play such a role. At the time, the estimates of Pluto's mass depended on the putative perturbations on Uranus derived from the questionable residuals of that planet.

Thus Pluto was potentially an important player in Kuiper's developing cosmogonic schemes, and as early as 1948–49, he attempted to get a handle on its size—and thus its mass—by measuring it with the 82-inch telescope at McDonald Observatory.

Kuiper employed a device called a disk-meter, which had been invented by Henri Camichel of the Pic du Midi Observatory in 1944 and used by him to measure with high accuracy the diameters of the Galilean satellites and Titan. So far, of course, astronomers had failed to make out a disk when viewing Pluto, so they had nothing to measure. Kuiper was hopeful, but these early attempts were disappointing. He wrote,

On repeated occasions, during 1948 and 1949, the writer has attempted to measure the diameter of Pluto with the 82-inch telescope. A disk meter, designed to produce a small artificial luminous disk of controllable brightness, color, and diameter, was placed at the Cassegrain focus. . . .

Owing to the faintness of Pluto no reliable results were obtained, in spite of very considerable efforts. The nearest approach to an actual measurement was made on November 4–5, 1949, when seeing was 9 on a scale of 10 for several hours and the mirror was essentially free from distortions. . . . But it was obvious that the measurement was on the very threshold of the capabilities of the telescope (if not beyond it), even with the exceptional conditions prevailing on that night. It was concluded that further efforts would be futile and that the observation could be successful only with the 200-inch telescope.[18]

The 200-inch telescope at Palomar Observatory was then very new. The first plates had been taken with it by Edwin Hubble only the previous January. At the meeting of the AAS in December 1949, Kuiper gave a presentation on his 82-inch results, and afterward consulted astronomer Walter Baade regarding the feasibility of adapting the disk-meter for use on the 200-inch. Baade had been one of the few professional astronomers to devote some of his time to the planets, having made important observations of Pluto in the 1930s.[19] He must have been encouraging, for after a Lowell Observatory–sponsored conference on planets in Pasadena the following March, Kuiper discussed the matter further with the director of the (combined) Mount Wilson and Palomar Observatories, Ira Sprague Bowen. Bowen granted him time on the 200-inch telescope for Pluto observations in March 1950. For the disk-meter used with the 82-inch to be fitted inside the prime focus cage of the 200-inch, the optics had to be modified. With the help of Hubble's longtime assistant Milton L. Humason, the disk-meter was fitted at the prime focus of the 200-inch on March 14. They left in the optical path a lens that Humason was using in his photographic program. Pluto Researcher Robert L. Marcialis writes, "On March 22, 04:00 UT, . . . Kuiper and Humason took turns measuring the diameters of Pluto and a nearby 11th magnitude star. The values they got were the same, and were 0.23"±0.01" for Pluto, and 0.11" for the star (presumably with comparable uncertainty)."[20] The derived diameter of Pluto was 6,000 km (3,700 miles), about 0.46 that of Earth. On the most reasonable assumptions, such as that Pluto's density was the same as (or at least no more than) Earth's, its mass ought to be only a tenth of Earth's. This was much less than the 0.94 Earth mass value derived from celestial mechanics using the Uranus residuals and the two prediscovery observations of Neptune by Lalande.[21]

Unless one were willing to accept the degenerate matter theory—the idea that Pluto was a white dwarf—the large mass simply had to be given up.

Kuiper—who knew a bit about white dwarfs after all—had been dismissive of the whole white-dwarf idea, pointing out that for a mass smaller than that of Earth, degeneracy leading to such a high density could not set in.

Several technical problems affected Kuiper's measurement that made it—as small as it was—still larger than it ought to have been.[22] One was the effect of Humason's extra lens in the optical path; another was that Kuiper compared Pluto's measured diameter with that of the star as if he were actually resolving the latter—a clear impossibility. Further, the image of Pluto's (then unknown) satellite Charon blended with that of Pluto, which meant that even though Charon was not actually resolved, it influenced the measurement.

It is not improbable that Kuiper made these mistakes unconsciously as he tried to get closer to the expected or desired result—in the same way that William Herschel, in the aftermath of his discovery of Uranus, had produced measures showing that it was getting larger because of his expectation that it was a comet approaching Earth, even though it was actually shrinking. If Kuiper's various mistakes are corrected, his measurement comes out much lower, at about 3,100 km (1,930 miles).

In addition to the disk-meter measurement, Kuiper determined Pluto's color by measuring its brightness through two optical filters, getting a result that he interpreted as evidence for a frost-covered surface with a reflectivity of about 17 percent:

> Such a body must have some atmosphere, though most of its original atmosphere will have frozen out owing to the low equilibrium temperature for Pluto, 40°–50°K. Both the atmosphere and the condensation products will prevent the albedo from being extremely low: nearly all snows are white, the crystals being small (H_2O, CO_2, CH_4, *etc.*). On the other hand, the albedo need not be that of freshly fallen snow, 0.7–0.8, because of several effects, including grit deposited by comets and meteors, which will darken snow over the ages. However, the rocky surface of Pluto would be expected to be invisible, which may explain why its color index is only slightly different from the Sun, quite contrary to the results for Mercury, Mars, and the Moon. For Pluto . . . is consistent with a gritty snow surface.[23]

According Marcialis, "This view of Pluto has proven to be way ahead of its time. Nearly three decades elapsed before the community as a whole embraced it, for the old, Earth-like mass estimates proved a difficult tenet to renounce."[24]

Kuiper never played only a single card. He knew of another way of resolving the mass question: find a satellite of Pluto. After using the 82-inch McDonald telescope to add two outer-planet satellites in the late 1940s (Miranda to Uranus's retinue and Nereid to Neptune's), Kuiper launched a search for a Pluto satellite. He obtained five plates with the 82-inch telescope in January 1950: three with thirty minutes of exposure and two with an hour. The telescope mirror was masked to 54 inches to improve the image sharpness. No satellites were found, despite a claimed limiting magnitude of 18 or 19. Humason, using the 200-inch Palomar reflector, carried out a similar search on April 15–16 that same year. The result was again negative.

For his theory of the origin of the Solar System, which was in progress during the attempts to measure Pluto's diameter, Kuiper desperately wanted a reasonable estimate of Pluto's mass. The mass was strongly dependent on the diameter since this quantity is raised to the third power in the mass calculation. Density, the other factor in the mass equation, could be that of ice, rock, or metal—a range of a factor of about six. The dependency of Pluto's mass on two essentially unknown quantities left the question open, as it would remain for several years to come. In a 1951 paper, Kuiper placed a mass-dependent burden on Pluto: "The planet Pluto, which sweeps through the whole zone from 30 to 50 astr. units, is held responsible for having started the scattering of the comets throughout the solar system,"[25] a process then amplified by Neptune and the larger planets and eventually resulting in the two basic comet dynamical families—Oort cloud comets and the short-period comets.

We now know that Pluto has a satellite, Charon, whose blue magnitude at the time would have been about 17.54 and whose orbital properties would reveal Pluto's mass. Kuiper and Humason failed to find it for the same reason that early exoplanet searchers would later fail to notice "hot Jupiters," the Jupiter-size planets orbiting very close to their stars: by analogy with what they knew, they expected something much farther out. In fact, Charon's disk would have been part of the saturated core of Pluto in all these photographic plates. When it was finally discovered, in 1978, it was on plates taken with a modest 61-inch telescope. The satellite was not resolved on these plates either, but it produced distortions in Pluto's image, and the discoverer, James W. Christy of the U.S. Naval Observatory, realized that these distortions were caused by a separate object. We will say more about Charon's discovery in a later chapter; suffice it to say that it involved Christy to see what was there rather than what he expected to see.

9

Planetary Science, New Technology, and the Discovery of Ice

The Birth of Planetary Science

THE QUESTIONS OF Pluto's size, mass, and composition went unresolved for many years. Meanwhile astronomers witnessed a dramatic split of their field into those who pursued the classical problems of astrophysics and a new breed calling themselves "planetary astronomers" or "planetary scientists." Astronomers of the first group largely concerned themselves with understanding stars, the gas and dust between them, clusters of stars, galaxies, and even galaxy clusters and superclusters. But as modern astronomy surged ahead and made huge strides in understanding the grand panorama of the Universe, scientists interested primarily in the planets and the workings of the Solar System also gathered momentum, becoming increasingly cross-disciplinary with the inclusion of not only astronomers in the usual sense, but also geologists, atmospheric scientists, chemists, and biologists into their fold. This cross-disciplinary enterprise has become the discipline of planetary science.

As the most prominent astronomer with an interest in the planets from both observational and theoretical points of view, Kuiper played a key role in the emergence of planetary science. However, while Kuiper was making telescopic observations enabled by his new infrared instruments and developing his theory of the origin of the Solar System, geologists and chemists brought fresh thinking to the study of the planets through a new understanding of the structure and thermal history of Earth and the compositions of meteorites. Kuiper's academic and scientific stature, great as it was, was exceeded by that of a relative latecomer to the study of the Solar System.

In the early 1950s, Harold Clayton Urey, who had won the Nobel Prize in Chemistry at the age of forty-one, turned to thinking about the origin of the planets from the point of view of chemistry, also known as cosmo-chemistry. Urey opened his 1952 book, *The Planets*, with the question his readers might pose—why is a chemist writing about the bodies in the Solar System—and then acknowledges his own astonishment that "I or anyone of similar training and experience should be able to say anything on the subject."[1] Yet Urey's book weaves a thorough study of the chemistry of Earth and the other planets derived from applying the principles of chemistry to the solar nebula, explaining the abundances of the elements, the heat budget of Earth, and the role of the Moon. Lunar scientist Charles A. Wood views Urey's book, which was the first to codify the physics and chemistry of the planets, with restoring credibility to the study of the Solar System and, with its geochemical perspective, laying the foundation for planetary science. The structure of the Moon's surface features played heavily in Urey's ideas for the formation and evolution of the terrestrial planets, yet Wood notes that Urey was "wrong about virtually everything he said about the Moon" while acknowledging the important role he played in convincing "NASA to under-take the Apollo program."[2]

Like Kuiper, Urey came from humble origins but rose rapidly. As the son of a pastor in a small Indiana town, for three years he taught in country schools. A BS in zoology led him to a two-year job as a chemist in indus-try, and then to a position as a chemistry instructor at his alma mater, the University of Montana. He entered the University of California in 1921 and received his PhD in chemistry in 1923. In 1929, he was appointed associ-ate professor in chemistry at Columbia University, where he pursued his research on atomic and molecular structure. In 1931, he devised a method allowing him to search for possible heavy isotopes of hydrogen, and this led to the discovery of deuterium, for which he was awarded the 1934 Nobel Prize in Chemistry.[3]

Urey's expertise in the separation of isotopes brought him into the Manhattan Project during World War II, where the efficient separation of the fissile isotopes of uranium and the extraction of plutonium from irradiated uranium were techniques vital to the development of nuclear bombs.[4] After the war, and from a faculty position at Kuiper's University of Chicago, his interests turned to questions of the origin of Earth and planets that might be answered from their compositions and thermal histories.

Initially embracing, like Kuiper, the von Weizsäcker concept of a turbulent solar nebula, Urey envisioned a scenario for the rapid formation of the terrestrial planets as relatively cold bodies, in contrast to the longer formation epoch and much higher temperatures in Kuiper's theory. Kuiper approached solar-system formation from his studies of binary-star systems, while Urey approached the subject through chemical considerations and the abundances of the elements as they were known at the time. The differences in how they attacked a common problem of such a large scale, but for which there was little direct evidence, led to a rancorous debate between Urey and Kuiper. The most specific point of disagreement between these two titans concerned the Moon. Urey posited a cold origin and development of the Moon, asserting that it never completely melted. Kuiper's view was diametrically opposed—the Moon had formed hot and subsequently melted throughout by the heat of the radiogenic elements it contained. Numerous publications from each combatant annunciated their differences, but while Urey addressed and named Kuiper directly in his point-by-point refutation of the discordant ideas, Kuiper's publications did not even cite Urey's, and his responses to the issues raised conveyed an aloofness that particularly irked Urey. The Urey-Kuiper sparring, which Carl Sagan later characterized as a "maddening counterproductive feud,"[5] went on for some fifteen years. The clamorous bickering between two scientists over somewhat arcane topics could have remained entirely invisible to the broader public had it not been for the emergence of the American program to send humans to the Moon. Both Urey and Kuiper had a stake in the program to send astronauts to the Moon and return them safely to Earth.

The key question about the Moon relevant to a safe landing (and departure) was the bearing strength of the surface—its ability to sustain the weight of the landing craft and the astronauts on board. Even the best photographs

Figure 9.1 Urey, Kuiper, von Weizsäcker, and fellow faculty member and geochemist Harrison S. Brown at the University of Chicago, circa 1948. Courtesy of Special Collections Research Center, University of Chicago Library.

of the Moon could resolve surface structures no smaller than about a kilometer, and visual observations could do only slightly better. Rocks and craters smaller than this could turn an attempted lunar landing into a disaster. Another unknown was the thickness and mechanical properties of the dust from lunar rock pulverized repeatedly by impacting bodies for more than a billion years. Some speculated that the dust layer could be a kilometer thick and that a landing craft and its occupants would sink out of sight. Urey leaned in this direction, as did the vocal astrophysicist Tommy Gold of Cornell University. Kuiper had concluded from his own visual observations with the largest available telescope (the McDonald 82-inch) and treks with his students across lava fields in Hawaii and the desert in the Southwestern United States that the volcanic lavas that he was confident covered the lunar maria would bear the weight of a lander and humans out for a stroll.

All these scientists offered their views on the safety of a lunar landing to NASA, the agency charged by President John F. Kennedy on September 12, 1962, to carry astronauts to the Moon and return them safely to Earth.[6] Considering that a trip to the Moon had vital geopolitical implications in the post-Sputnik world, as well as enormous technical challenges, NASA marshalled all the vocal and informed parties, and with their advice and council embarked on three nearly simultaneous pathways to establish to a reasonable certainty that a lunar landing would be safe. These were the Ranger, Surveyor, and Lunar Orbiter programs, conducted over a seven-year period.

NASA's Ranger program called for a volley of spacecraft intended to crash into the Moon while taking images of the last minutes of the flight to resolve surface structures, specifically rocks and craters about 10 m in diameter, in significantly more detail than photographs or visual observations from Earth could provide. Kuiper was the principal investigator for the Ranger missions, working with close colleagues, particularly British-born astronomer and lunar expert, Ewen A. Whitaker, to help select the intended impact sites. Urey, Gold, geologist Eugene Shoemaker, and astronomer Zdenek Kopal were all advisors to NASA on the Ranger program.

The program suffered several failures before Ranger 7 successfully imaged a region of Oceanus Procellarum in the final seconds before its crash on the surface on July 31, 1962. Rangers 8 (February 20, 1965) and 9 (March 24, 1965) were also successful. From our perspective, half a century later and after many successful probes deep into the Solar System, it is difficult to imagine how hard it was to hit the Moon with a rocket launched from Earth. Rangers 1 and 2 failed to leave Earth orbit, and Rangers 3 and 5 missed the Moon entirely. Since then, spacecraft navigation has been refined to an art. The highly accurate targeting of Pluto at the conclusion of a nearly decade-long flight spanning 4.5 billion km in July 2015 speaks to the progress made in celestial navigation.

At the press conference at NASA's Jet Propulsion Laboratory in Southern California, announcing the success of Ranger 7, Kuiper spoke to the achievement of an improvement by a factor of one thousand in resolution of lunar-surface features over that previously possible in photographs. Kuiper's conviction that the lunar surface could sustain a lunar landing attempt was strengthened by what he saw in the Ranger 7 images, but Urey was not convinced. Although NASA invited Urey to that first press conference and a later one, Urey declined to attend in a show of what was interpreted by Richard S. Lewis in an interview for the *Bulletin of the Atomic Scientists* as sour grapes.[7]

Figure 9.2 Kuiper at a press conference at NASA Headquarters reporting Ranger 7 results in 1964. NASA photo 64-H-2214.

NASA's Surveyor program successfully soft-landed five small robotic space-craft on the Moon between June 1966 and January 1968, proving that the surface was competent to hold a larger landing craft and humans. Kuiper and Shoemaker were the principal scientists on the Surveyor missions. The last pre-Apollo robotic studies of the Moon were made by five repurposed spy satellites dubbed Lunar Orbiters 1–5. Images from these spacecraft were key in selecting the landing sites for the six successful Apollo landings, begin-ning with Apollo 11 in July 1969.

While the intellectual foundations of planetary science were laid by spe-cialists in different fields—notably astronomy, chemistry, and geology, as described above—the study of hundreds of pounds of lunar rock and soil in the laboratory provided the tangible cement to give strength and perma-nence to the whole enterprise.

Since sustaining the birthing pains of the Kuiper-Urey affair and other, less rancorous events, planetary science as a field of study and research has become recognized by several major universities by designating departments of planetary science and granting advanced degrees in the subject. The six Apollo expeditions to the Moon and their precursor robotic missions in the 1960s were not the only factors leading to the definition and codification of planetary science. While preparations for the manned landings proceeded, the Soviet Union and United States sent a series of unmanned (robotic) space probes to the Moon and planets. Not only were the relatively nearby Venus and Mars explored by both countries, but Pioneers 10 (March 1972) and 11 (April 1973) were launched toward successful encounters with Jupiter and Saturn. However, geoscientists were clearly drawn by the availability of more than eight hundred pounds of lunar rocks and soil returned by the Apollo astronauts. Geochemists, mineralogists, and petrologists applied the most advanced analytical techniques to understanding the chemistry and structure of the rocks and soils gathered from six different regions on the Moon. It was the determination by geochemists of the ages of lunar-rock formations that established with certainty that the Moon formed at about the same time as Earth, and that the modification of its surface by volcanic processes had occurred very early in the Moon's history.

Geologists trained to decipher the stratigraphy and morphology of terrestrial geological features applied their skills to the features of the Moon and planets, while meteorologists investigated the phenomena of other worlds that possessed atmospheres. Despite significant setbacks to planetary exploration caused by occasional launch failures and spacecraft malfunction, great progress was made in the 1960s and '70s in unraveling the complexities of planet surfaces, interiors, and atmospheres. For a time, astronomers, even those specializing in Solar System studies, appeared to be under threat of being crowded out of the new field of planetary science. Astronomical observations have remained relevant, however, owing to outstanding technological improvements in telescopic instruments and a continuous supply of newly discovered asteroids, comets, planetary satellites, and an entire zone of the Solar System beyond Neptune and Pluto. Even more broadly defined, planetary science will eventually include the study of the planets (more than 3,500 at last count) currently being discovered by astronomers around other stars in our Galaxy. Embracing all the fields of astronomy, geology, atmospheric science, and more recently biology,

Solar System exploration and planetary science are flourishing well into the twenty-first century.

The almost-formal adoption of the Moon by geoscientists can be conveniently dated to January 5–8, 1970, when several hundred scientists who had been studying the Apollo lunar samples with every technique known at that time gathered in Houston for the first Lunar Science Conference. A series of annual events, always held in Houston, has continued to the present day, eventually widening in scope to include other Solar System bodies and becoming the Lunar and Planetary Science Conference. These meetings are indispensable venues for sharing new results and working out new ideas in the rapidly growing international enterprise that is planetary exploration.

Ices in the Coldest Regions of the Solar System

As we have forged increasingly farther from Earth in exploring the other planets, it has become increasingly clear that, in addition to rock, metals, and gases, ices are fundamental components of planets, rings and moons, and asteroids and comets. The realization of the importance of ices has been a comparatively recent development, largely because they have long been difficult for astronomers to detect and identify. Earth clearly has a vast amount of water ice in the caps at the poles, whose extents wax and wane with the seasons. Mars, too, has long been known to have polar caps that behave similarly. At the lowest temperatures encountered on Earth, the only constituent of its atmosphere that freezes is water. The other principal components of our atmosphere, such as nitrogen, oxygen, and carbon dioxide, remain gases.

On Mars, being farther from the Sun, winter temperatures in the polar regions are low enough to freeze not only water vapor but carbon dioxide, forming dry ice. Martian polar caps have been observed since the seventeenth century. At least since the time of William Herschel, astronomers have generally believed them to consist of water ice. Percival Lowell argued for a relatively mild Martian climate (about the same as that of the south of England) and invoked a seasonal Martian water cycle as underlying the canal network through which the local inhabitants pumped water from the poles to the thirsty temperate and equatorial latitudes to grow vegetation. At least a few scientists, including Alfred Russell Wallace (with Charles Darwin, the cofounder of the theory of evolution by natural selection), believed that Lowell's estimates of Martian temperatures were too high and instead envisaged a frigid Martian environment in which the caps were dry ice. For a long time the matter remained unresolved. Though few astronomers accepted Lowell's ideas about Mars as an abode of life (intelligent life, at any rate), no one could deny the observational fact, supported by high-quality photographics, that the bright polar caps advanced toward the equator and retreated toward the poles with the seasons. The presence of some kind of ice was clear. Finally, it was shown that both frozen carbon dioxide and water play a role, with ice of each kind coming and going in a complex cycle of Martian climate that includes not only seasonal effects but long-term effects from the changing tilt of Mars's axis and the shape of its orbit, which play out over many centuries.

Farther from the Sun, the temperatures are low enough to cause still other molecules that are experienced only as gases on Earth (they can, of course, be produced artificially in laboratories) to freeze on a planetary surface. These include not only water and carbon dioxide, but also nitrogen, oxygen, ammonia, and other molecules, which condense into ices on the surfaces and form crystals in the atmospheres of these cryogenic worlds. Naturally, we want to find out which molecules will predominate in these environments, and in answering this question, we begin with the observation that, as in the cosmos overall, the most abundant atoms in the Solar System are hydrogen (H, which by itself condenses only at exceedingly low temperature) and helium (He, which doesn't make molecules and can't exist as an ice except under extreme conditions). The next most abundant atoms occurring on planets are carbon (C), oxygen (O), and nitrogen (N), which are commonly linked with hydrogen to make molecules of water (H_2O), methane (CH_4), and ammonia (NH_3). They also join with one another to make carbon monoxide (CO), carbon dioxide (CO_2), and with themselves to make dinitrogen (N_2) and dioxygen (O_2), which are the two main components of Earth's atmosphere. Other, more complex molecules also occur, but the simple ones noted here are the most common.

Ices were early recognized as a key component of comets from studies of their behavior as they approach the Sun from great distances. As they near the Sun, they begin to emit gases, giving a long tail that might extend for tens of millions of kilometers. Most comets also have a second tail, made not of gases but of extremely tiny dust particles. The gas and dust tails have different colors and are often seen to diverge from one another as the gas molecules and dust particles react differently to the solar wind. The concept thus emerged of comets as "dirty snowballs" (a term commonly associated with Harvard astronomer Fred L. Whipple, who published an important paper on comets in 1950).[8] The "dirty snowball" is the nucleus, which consists of a mix of ices and dust. As the comet approaches the Sun, the ices, including water ice and ices formed of various carbon- and nitrogen-bearing molecules, warm and vaporize, unleashing the embedded dust particles. This is the homely stuff of which comets are made and the simple physics underlying these awe-inspiring spectacles of nature from which (as with eclipses) our superstitious forbears quaked with fear. Many an astronomer, at an impressionable early age, was excited to wonder by the appearance of a great comet, and determined at once to make the study of the night sky a lifelong career.

Ices could be predicted with a little elementary physics and chemistry, but they could not be detected with telescopes and their auxiliary instruments, such as cameras and spectroscopes, when Pluto was discovered, or for many years thereafter. The elusiveness of ices arises from their basic molecular properties, and the limited region of the spectrum that was accessible with photographic emulsions.

Searching for Ice with New Technology

Powerful as conventional photography is for taking long exposures to reveal faint objects, including stars, planets, satellites and asteroids, the wavelength span over which photographic emulsions are sensitive is only slightly wider than the visible spectrum, extending into the ultraviolet and the near-infrared to about 1 μm. Ices have their strongest diagnostic spectral features (absorption bands) at wavelengths longer than about 1 μm, so the spectrum of an icy object measured with a conventional photographic spectrograph— even if deployed on a very large telescope—registers nothing more than reflected sunlight and no spectral bands to reveal the presence of ice. Gases, on the other hand, have a wealth of spectral bands in the photographic region of the spectrum, and gases in the atmospheres of the Sun, some of the planets, and those emitted by comets during their harrowing passages close to the Sun had been seen, and in some cases identified with specific atoms or molecules since the late 1800s. In some cases the identifications were erroneous; spectral features attributed to the planets, especially Mars, were confused with those caused by Earth's own atmosphere.

It was only with lead-sulfide detectors in the 1940s, as described in the preceding chapter, that astronomers could probe wavelengths longer than the limit of photographic emulsions. As we have seen, Kuiper was the scientist who was most responsible for the initial breakthrough leading to infrared astronomy. His pioneering thrust forward with the just-declassified lead-sulfide sensor sufficiently sensitive to detect planets and stars sunk a deep root into the information-rich soil of the hitherto unavailable near-infrared. Between 1 and 2.5 μm, stars of all temperatures have distinct and diagnostic spectral lines and bands, and the gases of the giant planets display their molecular finery in great detail. And, contrary to the complications at infrared's longer wavelengths, not only is Earth's atmosphere effectively transparent in the near-infrared, the objects themselves are significantly brighter there.

Kuiper had near-infrared astronomical spectroscopy almost to himself for several years, using the spectrometer he had built at Yerkes Observatory and the lead-sulfide sensor provided by Cashman. At conferences in the United States and in Europe, and in a book he prepared from a seminal conference on planetary atmospheres he organized on Yerkes Observatory's fiftieth anniversary, he presented his early and groundbreaking observations. The infrared-friendly observing conditions at the McDonald Observatory and its new 82-inch telescope enabled the discoveries of the high-order hydrogen lines in the spectra of stars, the strong bands of methane and ammonia on Jupiter, bands of carbon dioxide in the atmosphere of Venus, and the first such spectra of Jupiter's moons and Saturn's rings. This was no accident, for it is precisely in this newly opened spectral region that many of the molecules in the atmospheres of the planets and stars, as well as the ices on cold planetary moons, could be detected.

In 1957, ten years after the original lead-sulfide spectrometer went into use, Kuiper presented a summary of his early results at a meeting of the AAS.

Among the most noteworthy observations were those of the four bright satellites of Jupiter (Io, Europa, Ganymede, and Callisto) and the rings of Saturn.[9] Two of the Jovian satellites, he noted in the published abstract of the paper he delivered, had the general characteristics of the spectrum of frozen water, as did the rings of Saturn. His spectra of these objects were suggestive, but not definitive. Also, they were registered in a purely analog fashion, long before digital data acquisition became possible, which made difficult the various laboratory steps in the processing and in the comparison to the spectrum of water ice. Additionally, his spectra had very low wavelength resolution, a consequence of the design of the spectrometer and the necessity to use it at its lowest resolution to be able to get the data on these faint objects. As it happens, the infrared spectrum of water ice has very broad spectral absorption features that don't require high resolution to discern; therefore—at least in hindsight—his conclusions were warranted. Fatefully, Kuiper did not publish a full description of the actual spectra from which he drew the conclusions outlined—with rather little detail—in his meeting abstract, and his conclusions about the first detection of ices beyond Mars and water ice beyond Earth drew little attention beyond his small cadre of students, including coauthor Cruikshank, some years later.

Partly because of conflicts with the other Yerkes astronomers, who remained chiefly interested in stellar astronomy and in the classical problems of astrophysics and did not sympathize with Kuiper's desire to allocate increasingly more of the observatory's limited resources to studies of the Solar System (including the Moon), Kuiper left the University of Chicago and Yerkes Observatory. He relocated to the University of Arizona in 1960, where he found a more supportive administration and established the Lunar and Planetary Laboratory. Drawing on local talent and the experience of scientists in the atmospheric physics group at the University of Arizona, he designed and built a new, more powerful spectrometer that used a diffraction grating instead of a prism. This spectrometer achieved higher spectral resolution without sacrificing sensitivity to weak sources. At the same time, lead-sulfide detectors had become commercially available and were much more affordable. Two young students, Dale Cruikshank and Alan Binder, who had both worked as summer assistants at Yerkes Observatory from 1958 to 1960, followed Kuiper to Arizona, where they began graduate studies and worked in Kuiper's spectroscopy lab. Their main tasks at the time were to maintain and improve the spectrometer and to accompany Kuiper to the telescope to operate it under his supervision. Over the next few years, Cruikshank and Binder built several new spectrometers and a photographic spectrograph to extend the observational programs underway in the lab and at the telescope.

The lead-sulfide detectors used in the Arizona version of the spectrometer were made in various configurations suitable for spectroscopy by Eastman Kodak Company, and cost only about $25 each. The procedure was to buy a dozen or so of each size desired, test each one in the spectrometer, keep the best two or three, and discard the rest. Testing consisted of attaching wires to the detector, mounting the little detector cup housing to the end of an eight-inch nylon rod, and then inserting the whole thing into a chamber containing a slurry of crushed dry ice and acetone. Lead-sulfide sensors will only work well if they are very cold, about −78° C (−108° F), and the slushy mixture of dry ice and acetone provides this degree of chill. A fresh batch of dry ice had to be crushed and packed into the cooling chamber every few hours to keep the temperature optimized; the acetone lasted for months at a time.

Once the detector had cooled sufficiently, tests were made to establish the level of the intrinsic electrical noise generated by the motion of the electrons in the lead-sulfide material. Only the detectors with the lowest electronic noise were kept because they would be the most sensitive for measurements at the telescope.

The new Arizona spectrometer with its improved lead-sulfide detectors was used frequently at the newly established Kitt Peak National Observatory and on two-week forays to McDonald Observatory, where Kuiper was granted observing time on the 82-inch telescope that he had helped conceive. With his faithful assistants in tow and the spectrometer in the back seat of the car, the McDonald expeditions crossed the arid landscape from Tucson across eastern Arizona and New Mexico to the observatory in west Texas about twice per year. Kuiper seemed to prefer the Christmas season, when academic duties were light and the long winter nights gave the most observing hours.

At that time, the McDonald telescope (now known as the Otto Struve telescope, after the Yerkes Observatory director during whose directorship it came to fruition) was the fourth largest both in the country and in the world. It had been conceived in the late 1930s by a group of prominent astronomers at Yerkes Observatory, including Struve, Kuiper, Bengt Stromgren, and Georges van Biesbroeck, all of whom saw the need for a large telescope in a location where observing conditions were favorable, as was no longer the case (if it had ever been) in southern Wisconsin. The Mount Wilson 100-inch telescope, located in the San Gabriel mountains above Pasadena, California, had been in operation since 1917, and remained the world's largest telescope until the Palomar 200-inch was finished in 1948. Another large telescope, the 120-inch Shane reflector, went into operation at Lick Observatory in 1959. Competition for time on these large telescopes was of course keen, and those who had access to them in general made the most (and most important) discoveries. The Yerkes astronomers naturally wanted their own large telescope, and thus the 82-inch reflector was born. The telescope was at first operated jointly by the University of Chicago and the University of Texas, and eventually by Texas alone. In the dark sky and dry air of west Texas, the site of the observatory on Mount Locke proved to be highly favorable for many kinds of astronomical observations, including those in the infrared.

Kuiper was not the only astronomer forging ahead into this new wavelength region of the spectrum at the frontier of technical capability. In the Soviet Union, Vasili Ivanovich Moroz, a young astronomer at Moscow State University, had also acquired a lead-sulfide sensor. He did so, however, through a rather curious route. Sometime in the early 1960s, an American-made heat-seeking missile using a lead-sulfide sensor found its way into the hands of the Chinese military. In a spirit of friendly international cooperation between the two Communist countries that would lose its luster shortly thereafter, the Chinese gave the sensor to Soviet military technologists, who reverse-engineered it and began to manufacture their own lead-sulfide detectors. Surprisingly, one of these made its way into the civilian science sector, where Moroz managed to acquire it and use it in an astronomical spectrometer. With the 50-inch reflector at the Astrophysical Observatory in Crimea, he observed many bright stars and planets, and pushed his instrument to the limits by taking spectra of Jupiter's four satellites. Unaware of Kuiper's earlier abstract, in 1964 he published in the Soviet *Astronomical*

Journal his results that two of the satellites (Europa and Ganymede) show the presence of water ice.[10]

The story of these early ice detections took another interesting twist when, in 1972, a PhD student at the Massachusetts Institute of Technology and his professor teamed with an astronomer at the Kitt Peak National Observatory to use a powerful new spectrometer to rediscover water ice on Europa and Ganymede. Their spectra were indeed excellent and the conclusions incontrovertible, but they had been unaware of Moroz's work and were either unaware of the earlier Kuiper abstract or chose to discount it because it showed no data. The discovery made news across the United States and appeared in a front-page article in the *New York Times*.[11] The lead author was student Carl Pilcher, who a few years later with his new PhD in hand joined the planetary group at the University of Hawaii's Institute for Astronomy, where he became a close colleague and friend of coauthor Cruikshank. Additionally, while the MIT group was getting spectra of Jupiter's moons, two young associates of Kuiper at the University of Arizona, Uwe Fink and Harold Larson, were observing with an entirely new infrared technique. They, too, "discovered" water ice on Europa and Ganymede.

Though the question of priority is usually of little importance, and in many cases several individuals or teams reach the same discovery independently and nearly simultaneously, a bitter exchange ensued between the MIT group and the University of Arizona group. Charges of plagiarism and other improprieties flew between the opposing parties, and the affair was even elevated to the chairperson of the Division for Planetary Sciences of the AAS for some sort of (nonbinding) judgment. Moroz's claim to the discovery, which was as good as any of the others', was not even considered, while Kuiper decided to dig out his old spectra—the subject of his abstract of 1957—and use them to buttress his own claim by publishing them in his own journal, the *Communications of the Lunar and Planetary Laboratory*, in 1973, the year of his death.[12] That publication was patterned after the annals of astronomical observatories that appeared irregularly and circulated on an exchange basis between observatories and institutions in Europe and the United States until 1973.[13] Thus it had very limited circulation. But at least Kuiper had managed to set the record straight with the presentation of the actual data from which he had drawn his original conclusions some sixteen years earlier. Regrettably, Moroz would never fully reconcile himself to the claim that Kuiper had seen the water ice first. He also felt that the MIT and Arizona groups had failed to acknowledge his own prior role.[14] If so, it wasn't the first instance of Americentrism in scientific research.

Leaving the priority issue aside, the discovery was of the first importance. It marked the first detection of water ice beyond Earth and set the stage for further discoveries of not only water ice, but also the ices of other molecular compositions on the small, cold bodies of the Solar System that range far beyond the orbit of Pluto, as described below. It also illustrates the way scientific discoveries often come about—early data of limited quality indicative of a new result are presented, but are not universally accepted or are accepted provisionally or with reservations even by colleagues working (often competitively) in the same or allied fields. As new techniques are

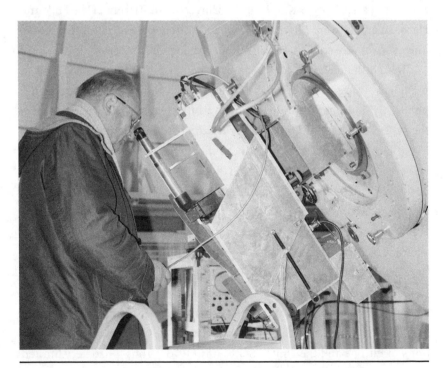

Figure 9.3 Kuiper observing with the Fourier-transform infrared spectrometer at the 61-inch telescope, Catalina Observatory, 1970. Photo by Dale Cruikshank.

developed—in astronomy, at any rate, this usually means the development of more sensitive detection equipment deployed on larger telescopes—more and better data are acquired until the discovery is established to the agreement (if not the joy) of all. Sometimes, of course, better data refute the original claims made from the initial observations, and the whole matter goes off sideways. It is this circuitous pathway to the truth, often with branches that turn out to be dead ends, that characterizes how science inches ahead toward a more complete understanding of the natural world. In this way science is distinguished from other—less empirical—provinces of human intellectual endeavor, where asserting a thing is sufficient to believe it.

Ice and the Formation of the Planets

To see where ices and rocks fit into the compositions of planets and smaller bodies in the Solar System, it seems worthwhile to summarize our understanding of how the Sun and planets came to be, noting the roles of the various kinds of materials of which each is composed.

The Sun began as a local condensation of hydrogen, helium, other gases, and dust in our Galaxy, the Milky Way, which was pulled together by gravity. Hydrogen is by far the most abundant substance in existence, its single proton and solitary electron having combined soon after the big bang (13.8 billion years ago, according to the latest estimate), from whence sprung the totality of our universe of energy and atoms (and ourselves). The giant molecular cloud in which the Sun formed was also the birthplace of several other stars of various sizes that formed at about the same time. The fragment of the solar nebula, the cloud surrounding the nascent Sun, began to rotate and flatten into a disk while gravity pulled the heavier materials—metals and rock-forming silicates—to the inner region of the disk, leaving the lighter elements largely concentrated in the outer regions. Under the inexorable and growing strength of the gravity field, the innermost zone became very dense and hot, eventually igniting the nuclear chain reaction that powers the Sun—the fusion of hydrogen into helium, which generates prodigious amounts of energy.

The cloudy disk of gas and dust surrounding the newborn Sun fragmented into smaller condensed units that began to cool and form the first planetesimals—the feedstock of the planets yet to come—the proto-asteroids, and myriad other bodies of the early Solar System. Close to the Sun, as the hot gases cooled, solid dust condensed. The first condensates were the refractory oxides that can be studied directly in the carbonaceous meteorites that preserve them to the present day. These are calcium-aluminum-rich inclusions, which appear as light-colored lumps and smears in the otherwise black and gray carbon-rich carbonaceous chondrite meteorites. Further cooling of the near-Sun gas and dust allowed the condensation of silicate minerals and metals, plus the still-mysterious pea-size chondrules comprising most common meteorites—the ordinary chondrites.

Much farther from the Sun, cooling progressed rapidly, eventually reaching the temperatures at which water vapor condensed into ice. This is generally thought to have defined a "frost line" at about 5 AU from the Sun, where Jupiter is now located (though not necessarily where it formed; as we shall see, one of the unexpected developments of the past few decades is the realization that the giant planets migrated early in the history of the Solar System, with far flung consequences). At greater distances still, other more volatile molecules in the cooling solar nebula condensed, yielding icy rims of methane, ammonia, and carbon dioxide on grains of silicate dust.[15] At the greatest distances from the fledgling Sun, where the temperature was only a few tens of degrees above absolute zero, the most volatile ices formed—nitrogen and carbon monoxide. During the Solar System's birth in its cloud of dust and ice, the temperature of the Sun changed, the density and the temperature distribution in the solar nebula changed, and the frost lines shifted inward and outward.[16] The raw material that coagulated and became each large planet and smaller planetesimal depended on the local composition and other conditions in its region within the solar nebula as the nebula itself evolved.

When described in these terms, the formation of the Solar System appears relatively straightforward. In fact, it was a dynamic and complex process about which the details are even now somewhat uncertain. For example, the discoveries of planets orbiting other stars that are reported on an almost daily basis are presenting a serious challenge to the simplistic picture we just described.[17] But regardless of all its complexities and nonintuitive twists and turns, the origins of stars and planetary systems are events that have occurred billions of times in our Galaxy alone. With our telescopes on the ground and in space, we can watch some stages of this process taking place. But the formation process does not occupy more than an eyeblink on the cosmic timescale—it takes just a few million years, and is then generally blocked or at least obscured behind a cloak of dust. That dust is, of course, part of the cocoon of material from which the enclosed star and planets condense. Only when the star has become hot and energetic enough to blow away the remaining dust with the pressure exerted by its heat and light does the nebula thin out sufficiently to permit a clear optical view. Until that clearing occurs, by which time the planets have already largely formed, we are able to discern what's going on inside the dusty cocoon only by the heat radiated therein, to which the dust is transparent at long infrared wavelengths. That is to say, only now—with modern sensitive infrared-optimized telescopes—can we observe newly forming planetary systems around young stars within their dusty envelopes, and sometimes with startling clarity, whereas before they were subject only of obscure speculation. These new systems are currently forming all through our Galaxy and must be in other galaxies throughout the Universe. At least some of them are near enough to allow us a ringside view.

Just as Kuiper's spectrometer and its successors have revealed more about the Solar System's objects in the present day, so are they now peering into the hearts of newly forming planetary systems elsewhere in the Galaxy, informing us of the conditions and events likely to mirror those of the solar nebula when the Solar System formed more than 4.6 billion years ago.

We return from these distant vistas to the original discovery and verification of water ice on at least two of Jupiter's large satellites, as well as its inferred importance in comets. These observations were the first of many that have led to the emergence of a compelling and convincing confirmation of the fundamental concept of the sequence in which the bodies in the Solar System solidified. Theory and observations have been brought into accord; predictions leading to discoveries, with the discoveries improving the theory. Science is, in that sense, an exquisitely crafted and well-tuned musical instrument. As a scientist who theorized about the formation of the Solar System, and did so years before these discoveries on Jupiter's satellites came within reach of astronomical technology,[18] Kuiper saw what was coming and knew what to expect far beyond Jupiter. Although limited by the available technology, even in 1950 he was pushing to the edge of the known Solar System with mere visual measurements of Pluto's size with the Palomar 200-inch telescope while speculating about the composition and nature of the planet's surface and atmosphere, as we saw in the previous chapter.[19]

10

Whence the Ices?

Chemistry in the Solar System

SCIENTISTS OF VARIOUS DISCIPLINES such as astronomy, geology, and biology look at Earth and its neighboring planets and wonder when and how it all began. The "when" is well established by the evidence presented by the numerous meteorites that have fallen to Earth, which we can collect and analyze in detail. The origin of the Solar System has been dated to the time when the first solid materials in meteorites condensed from a hot vapor of rocky and metallic elements into solid components. By careful analysis of tiny amounts of radioactive elements, such as potassium-40, thorium-232, and uranium-238, which undergo natural radioactive decay into argon-40, lead-208, and lead-206, respectively, the age of the earliest condensed components of meteorites has been determined to be 4.567 billion years.[1] The age of the oldest rocks on Earth so far known is of the same order but just a bit less, at about 4.4 billion years, as determined from a recent discovery in the Jack Hills of Australia.[2]

So much for the when. The "how" is somewhat more complicated, but astronomical observations of planetary systems at various stages of formation around other stars have given fresh insights into just what goes on. Without reviewing the entire subject, we note that one of the key components of any scenario for the formation of our own Solar System is the composition of the initial material from which the Sun and the planets formed.

The starting composition is critical, but so are the chemical reactions that took place in the solar nebula, that portion of the molecular cloud of ice, gas, and dust from which derived everything the Solar System now contains.

The specific part of the Solar System's formation that interests us here involves the condensation of ices in the outer part of the nebula. In 1972–73, John S. Lewis of MIT, a former PhD student of Harold Urey and then a young member of a small and rather exclusive clan of cosmochemists, first laid out the sequence of the condensation, or solidification, of the materials thought to constitute the solar nebula.[3] He knew the compositional makeup of the present-day Solar System, as well as the contents of the giant molecular clouds strewn about the Galaxy that resemble the one that gave birth to our Sun. Lewis applied well-known chemical principles explaining the formation of atoms into molecules, and molecules into grains of dust and ice, to set out the basic chemical themes underlying the formation of the Solar System as well as any planetary systems that might exist around other stars resembling our Sun in mass and composition. At that time, no exoplanet systems were known. Now, more than two thousand exoplanets have been identified and confirmed (mostly from observations with NASA's Kepler space telescope), and many of these orbit solar-type stars.[4] Lewis's study of the Solar System provides a template for interpreting the compositions of those planets. Of course, each planet in each system has its own unique story to tell, just as do those circling the Sun, and the accumulating evidence shows that a great many exoplanets are utterly unlike any in the Solar System.

As noted in the previous chapter, the role of water ice in the colder regions of the Solar System was already being suggested by the infrared observations of the satellites of Jupiter and Saturn's rings by the time Lewis published his early work. However, no direct evidence was available for the other ices presumed to form in yet colder zones, where ices of methane, ammonia, carbon monoxide, nitrogen, and other molecules were expected to exist. At that time, these molecules were found only as gases in the atmosphere of the giant planets. Kuiper and others had developed theoretical views of the origin of the Solar System in the mid-1940s through the 1960s that predicted in broad terms the presence of various kinds of ice in the colder outer realm of the giant planets and comets. In 1950, Kuiper had envisioned surface frosts of methane, water, carbon dioxide, and possibly others, on Pluto and by extension to other distant small bodies.[5] Though some of these earlier ideas were on the right track, it was up to John Lewis to bring to the field a quantitative perspective based on detailed chemical calculations and keen insights.

New ideas or theoretical concepts on the chemistry of the outer Solar System, like those offered by Lewis, are usually met by unsentimental, practical-minded observers in two major ways. Most will admit that the ideas or concepts make sense and go blissfully and unconcernedly ahead with their own programs, in the recognition that there is, at least at present, no possible experimental means of testing it. A few will be less complacent; for them, the idea will stick, incubate, and resurface when better technology allows observations that can either verify or refute the theory. In the best-case scenario, observations will be obtained that not only verify or refute, but push the concept forward by revealing details not originally predicted.

Preliminary theories of the origin of the Sun and planets, which combined with the chemical calculations of Lewis and others to provide increasingly detailed understanding of the gas, dust, and ice elsewhere in the Galaxy, were bolstered by the early infrared observations of ice on Jupiter's moons and Saturn's rings. The stage was thus being set for the growing recognition that ices of several kinds are a fundamental component of all the bodies in the cold, outer regions of the Solar System. However, details still needed to be filled in, and those awaited improvements in the observational capabilities that would allow infrared astronomers to detect and measure smaller, fainter objects than Jupiter's four large moons or Saturn's rings.

Going back at least to Tycho Brahe's refined instruments for measuring the positions of the planets and Galileo's telescopes that revealed a world in the Moon and satellites orbiting Jupiter in the image of a small Copernican system, discoveries in astronomy have always depended critically on the development of new technology and new techniques. Larger and better telescopes certainly help, but the development of more capable sensors (detectors) has arguably been as, or even more, important. For almost two centuries after the invention of the telescope, planetary astronomy depended on the only sensor available—the human eye. Marvelous as it is for everyday purposes, the eye is seriously wanting as an astronomical detector—not least because, as we have seen, it is unable to register the infrared part of the electromagnetic spectrum. Photography represented an advance over the eye for some purposes because of its ability to build up an image over time, particularly useful for the study of faint objects. Thus Pluto, which at stellar magnitude 14 is always at least one hundred times fainter than Neptune, would never have been discovered had Lowell Observatory astronomers had to rely on the same equipment and methodology that Challis or Galle had used eighty-four years earlier. As we have discussed earlier, for understanding the nature of ices in the outer Solar System, even standard photography offered little.

The introduction of lead-sulfide sensors for detection of infrared radiation and, more recently, of charge-coupled devices (CCDs), which are highly efficient sensors for registering photons, has completely revolutionized ground-based astronomy. Most spectacularly, rocket technology—developed in World War II while lead-sulfide sensors were first being made—has allowed us to put instruments not only above the confines of Earth's turbulent ocean of air but also in the vicinity of the planets themselves, and in some cases on their actual surfaces.

In terms of infrared astronomy and the study of ices in the outer Solar System, we can put all this into perspective: from the time of the initial discovery of water ice on the Jovian moons to the present day, telescopes have increased in size from about 3 m to 10 m diameter, a gain of $(10/3)^2$, or about a factor of 10 in the faintness of stars that can be detected. That is significant, certainly. However, over the same period the sensitivity of infrared detectors has been much more dramatic. Consider that whereas Kuiper and Moroz and other pioneering infrared observers used a single detector (one pixel in modern terms), astronomers now use detector arrays with 16 million pixels, and each of those pixels is much more sensitive than Kuiper's. This new technology used on a 10-meter telescope has 1.4 trillion times the capability of the single-pixel sensor on a telescope considered large in the 1970s. And, of course, by combining multiple new technologies, the increase of power is synergistic. With larger telescopes and more capable detectors, together with advanced optical techniques at superior observatory sites on the ground and the ability to observe in space, astronomers can now study bodies that are twenty stellar magnitudes—about 100 million times fainter—than the four large Jovian moons.

While progress in astronomy awaited developments in technology, without persistent attempts to push the observations to the limit, technological breakthroughs would never have been made. In any case, astronomers—being impatient, ambitious, and inquisitive sorts—weren't willing to bide their time from 1958 to today waiting for engineers and physicists to give them the technology they needed. The possibility of discovery is often found at the margins of what is achievable at a given time, and astronomers have worked diligently to eke out every bit of information allowed by the capabilities on hand. This is as it has always been, and will continue to be so into the far foreseeable future.

As planetary scientists at the University of Arizona and MIT steadily improved their equipment, they made ever more detailed studies of the water ice on Jupiter's and Saturn's satellites. Though water ice had been convincingly detected on Europa and Ganymede, the satellites of Saturn were, by virtue of their faintness, bigger challenges. The first spectra clearly showing water ice on the four brightest were presented by Uwe Fink and his colleagues at the University of Arizona's Lunar and Planetary Laboratory.[6] Their measurements were made with an entirely new kind of astronomical spectrometer, the Fourier-transform interferometer. In the mid-1970s, as increasingly fainter objects were coming within range of advancing detector capabilities and spectrometer technology, coauthor Cruikshank and his colleagues Carl Pilcher and David Morrison in Hawaii were extending the search even farther from the Sun for clues about satellite compositions. Over the course of a few years the moons of Uranus and Neptune, and even Pluto itself, were coming within reach of infrared detectors. The brightest and most accessible candidate was Neptune's large moon Triton, while Pluto was a rather harder target. Unfortunately, the faintest of these objects remained well beyond the reach of the traditional near-infrared techniques using spectrometers, so Cruikshank, Pilcher, and Morrison decided to try a new approach. Their idea was to test for the presence of ice by using a set of filters transmitting specific wavelengths characteristic of ice composition, as the strength of the signal through these filters would be greater than could be achieved with a spectrometer. It was no longer spectroscopy but three-filter photometry, and while it wouldn't produce a spectrum, it would allow the observation of the much fainter objects, like Triton and Pluto, in which they were interested.

What kind of ice might be expected? Lewis's calculations had shown that in addition to water, other plausible candidates included methane, ammonia, carbon monoxide, and carbon dioxide, all consistent with Kuiper's 1950 insight. The infrared spectra of ices of these compositions had all been measured in the laboratory by a few scientists, notably Hugh H. Kieffer at the California Institute of Technology (Caltech). The lab spectra showed a series of absorption bands in which methane ice absorbed preferentially in narrow spectral regions, which were different from those of ammonia or water ice. Water ice, in particular, had its own distinctive absorption-band pattern. Rocky minerals, on the other hand, behaved differently from all the ices. The three-filter technique could distinguish among four possibilities: water ice, ammonia ice, methane ice, and rock. The latter could not be ruled out, and indeed, in the mid-1970s it seemed entirely possible that some outer Solar System bodies, particularly Pluto, had rocky surfaces. Measured through two specially designed filters and one standard filter in a sensitive infrared photometer, each of these four materials would show a pattern that could be related to their spectral properties as seen in the lab.

The three-filter technique seemed very promising at the time, but the astronomers were faced with a dilemma regarding where to conduct their tests. Since 1970, their institution, the University of Hawaii's Institute for Astronomy, had been operating a 2.2-meter telescope on the extremely dry summit of Mauna Kea, at 4,205 m (13,796 ft) the highest island mountain in the world.[7]

Unfortunately, when Cruikshank, Pilcher, and Morrison set out to begin their work on outer–Solar System ices, though it proffered the excellent 2.2-meter telescope, it did not yet have all the ancillary equipment needed for such observations. Despite their University of Hawaii affiliation, and access to the world's premier location for infrared astronomy, they had to carry out their observations elsewhere. The Kitt Peak National Observatory, in the Quinlan Mountains south of Tucson, Arizona, had a state-of-the-art infrared detector system, at the heart of which was not a lead-sulfide detector but a much-improved detector made of indium antimonide. This detector and its chilled containing device, cooled with liquid nitrogen at $-196°$ C ($-321°$ F) instead of dry ice, had been built by Kitt Peak astronomer Richard Joyce, who named his instrument the "Blue Toad" because its aluminum outer shell was made of aluminum anodized with a bright blue color. Cruikshank, Pilcher, and Morrison successfully applied for observing time on the 4-meter Mayall reflecting telescope at Kitt Peak, then the second-largest telescope in the world after only the venerable Palomar reflector, and Joyce agreed to mount their special filters, only 0.375 inches in diameter, in the light path inside the Blue Toad.

In March 1976, the filters had been mounted in the photometer and the beautiful 4-meter telescope was ready for the Pluto observations The Hawaii astronomers arrived at Kitt Peak ready to bring Pluto into view. The Blue Toad was a roughly cubical device, about 12 inches on a side, attached at the bottom of the telescope's tube structure—a tiny projection through a jumble of cables, hoses, and electronics-filled boxes just behind the big 4-m diameter mirror, where the light is brought to a focus. Operating the Blue Toad required one of the observers to ride in a mesh cage built around the bottom of the telescope as it moved across the sky to acquire and track the target object for the length of time necessary to acquire the observational data. They would watch the target through a guide eyepiece, making occasional tiny corrections to the telescope's position to ensure that the target would remain in exactly the right place for the light to go through the filters and onto the sensor. Only one astronomer could ride in the cage at a time, so the observers took turns during the cold and dark of the night guiding the telescope while the other two watched in a warm control room as the data slowly came in to the recording computer and the viewing screen.

Although Pluto is vastly fainter than the faintest star visible to the naked eye, a 4-meter telescope has 100,000 times the eye's light-gathering power, so Pluto was easily visible through the eyepiece. Detecting it with the special filters in the light path was more difficult, but after a few trial measurements, the Blue Toad had performed hoppingly. It appeared that the job could be done according to expectation.

Visually guiding a giant telescope on Pluto for an hour or more at a time has aesthetic benefits that transcend the minor physical discomfort of the cold open-air exposure of the telescope and observer—for example, the instantaneous thrill of moving with the push of a button a multiton optical-mechanical monster machine precisely and smoothly by an utterly tiny angle on the sky to match Pluto's position minute-by-minute, all while riding along in the cage with it. Over fifteen to twenty minutes, the orbital path of Pluto, nearly 3 billion miles away, is readily seen as the planet slowly changes position relative to its field of background stars. It is both exciting and humbling to peer with one's own eye through the darkness of space to the edge of the Solar System and ponder the scale of it all, knowing that its most distant known planet, since its discovery in 1930, has moved only a fraction of its 248-year orbit around the Sun.

Modern giant telescopes now use measurement instruments, including spectrometers, of such size and complexity that the astronomer collects the data in a warm control room at the observatory—or quite often from the convenience of an office in their home institution on the other side of a continent, or even of the world—observing remotely with the on-site assistance of an operator in the telescope control room. This arrangement affords greater physical comfort for all concerned and usually makes for greater overall efficiency in the operation, but it comes not without loss, for it takes

Figure 10.1 Pilcher, Morrison, and Cruikshank in the instrument cage of the 4-meter Mayall telescope at Kitt Peak National Observatory in 1976, when they found the evidence for frozen methane on Pluto. Dale Cruikshank collection.

the astronomer away from direct contact with the sky. Astronomers of earlier generations and through most of the twentieth century can only reflect with nostalgia on the days of deep and compelling intimacy with the sky.

The first of the three nights of precious telescope time awarded to this project was lost to clouds over the observatory, and the telescope sat idle. There were no rides in the cage that night. But on the second night Pluto could be seen through some thin cirrus clouds. The initial measurements suggested that Pluto's surface might indeed be covered with frozen water. However, since cirrus clouds consist of tiny ice crystals in our own atmosphere, these measurements were regarded as unreliable. But the third time was the charm: the last of the assigned nights was free of clouds, and once again the giant telescope was turned to the little planet. The first and subsequent measurements showed clearly that Pluto's surface has not water ice but frozen methane—the first detection of any ice other than water in the Solar System beyond Mars.

As these new and definitive measurements came in on the recording instruments, howls of excitement reverberated through the big Mayall telescope dome. Other astronomers working on the mountain that night must have heard them from their own open domes. The discovery was a first, and full of implications for understanding the composition of outer–Solar System bodies. The clear and immediate cognizance of an important new result is another deep aesthetic pleasure of the scientific endeavor, and one that occurs but rarely. These moments are savored by those fortunate enough to have them, and conveyed with enthusiasm to young and aspiring scientists as the brass ring that they too may be privileged to grasp from nature's offering hand. There must be few greater rewards.

With the detection of methane on Pluto, the chemical calculations by Lewis and the earlier speculations by Kuiper had been underpinned with the beginnings of a solid observational framework. This discovery immediately led to a new view of the surface of Pluto, and would align the path toward future discoveries of additional chemically favored ices in the pervading coldness of the trans-Jovian Solar System.

11

Icy Earth and Beyond

The ice was here, the ice was there,
The ice was all around.
It cracked and growled, and roared and howled,
Like noises in a swound!
"THE RIME OF THE ANCIENT MARINER,"
SAMUEL TAYLOR COLERIDGE

Earth's Story in Ice

ON A WARM SUMMER DAY, and almost any time of year in the tropics, it is hard to imagine that Earth is an icy planet. However, on a bitter cold and frozen day in the winter, when snow covers the ground and warmth seems to threaten perennial absence, it is less difficult to imagine. The snow and ice come and go, not only with the changing seasons, but also on much grander time scales, and have done so for much of Earth's lifetime. The icy eras alternate with extended times of high temperature and no ice at all. A thorough freeze-over was in place some 2.2 billion years ago, when "snowball earth" may have had no uncovered land or exposed sea water whatsoever. This was followed by a billion years of more temperate climate, during which great changes occurred in the atmosphere, ocean, and most profoundly the biosphere.

Earth—the third planet from the Sun, and in striking contrast to the two planets whose orbits lie inside it—is, in fact, an icy planet. There is no ice on the surface of Venus, while on Mercury there is none except in perpetually shadowed craters near its poles. Today Earth has two polar caps of solid ice and glaciers in mountainous regions at the equator. At both the south and north poles, accumulating snow compressed into ice covers thousands of square miles of land (Antarctica) and sea (the northern Arctic). Antarctic ice is rooted in a continental land mass that has been pushed downward into Earth's slightly soft, rocky mantle by the weight of the ice itself, but the Arctic has no supporting land, and the ice sheet floats on the Arctic Ocean. The one exception in the north is Greenland. In Greenland, a 3-km thick layer of ice more than twice the size of Texas lies in a shallow elongated bowl pressed by the weight of the ice itself into Earth's crust.

In the early twenty-first century, we are still in an epoch of ice that began some 40 million years ago and has alternately presented long periods of intense cold (when glaciers grew and sea levels fell) followed by warmer interglacial times. The current interglacial has lasted about ten thousand years, during which all of what we think of as civilization has developed. The current balmy times were preceded by several glacial periods.

As Earth's climate continues its present trend of warming, the ice in both polar regions is melting. The effect of this melting is to raise sea levels worldwide and to expose some of the land formerly completely covered by ice. Though some of this likely would have happened gradually because of natural cycles, human activity has superimposed a dramatic—and historically quite unprecedented—increase in warming. Beginning during the Industrial Revolution and continued unabated ever since, humans have been pumping increasingly more greenhouse gases into the atmosphere. These gases, primarily carbon dioxide from the combustion of fossil fuels, enhance the air's ability to trap heat from the Sun, thus blanketing us in an increasingly effective foil to the natural balance of heat and cold that would keep the worldwide climate stable over long periods. The physics, which was already well understood by physicists in the nineteenth century, is not in doubt. Scientists are also in agreement, at least in broad terms, about the effect. Many of these effects would not be known if not for planetary scientists, who have investigated since the early 1960s how Venus, a planet almost identical to Earth in size and mass but closer to the Sun and lacking watery oceans, has followed such a different evolutionary track from Earth. The average temperature on the surface of Venus is 462° C (864° F), hot enough to melt lead. The reason for this high temperature is Venus's massive atmosphere of carbon dioxide. A runaway greenhouse effect—essentially an overwhelming positive feedback loop that acts as an extremely efficient heat trap for solar heat—has brought it to its present state. The same physics is operative on Earth, where stronger forces have thus far offset the positive feedback loop and where the greenhouse effect is not yet out of control, though clearly Venus marks the extreme point on a continuum as well as a cautionary tale as to how the climate of Earth is changing.

Of course, Earth's climate is extremely complicated, involving subtle combinations of astronomical, physical, and chemical effects. The tilt of Earth's axis changes over time, as does its distance from the Sun. Continents move around, and when large continental land masses are over the poles, large areas become covered with ice (like Antarctica now) and reflect much of the sunlight back into space. The amount of cloud cover has similar effects. Carbon dioxide levels have varied over time because of natural processes such as forest fires ignited by lightning strikes and by volcanic eruptions. Living organisms affecting climate did not begin with the nearly 8 billion people currently living on the planet. Huge meteor impacts have led to mass extinctions, and the Precambrian era saw the rise of photosynthesizing organisms like blue-green algae, which changed the composition of the atmosphere—producing what has been called the "oxygen holocaust," which was detrimental to earlier organisms not equipped with defenses to deal with the extremely toxic effects of oxygen. Our present-day Earth—with oxygen to breathe, and dressed in greenery—took shape only after about 2 billion years of planetary evolution, supplanting earlier, wilder scenes of vast billowing volcanoes poking out of huge oceans, and before that, of an embryonic planet being buffeted by huge planetesimals thrown sunward as the giant planets migrated around the Solar System. These craters have been largely rubbed and worn away by the erosive effects of tectonic forces, wind, and water, but they are still evident in the heavily cratered highlands of the Moon.

Even if we confine ourselves to the most recent 0.1 percent of Earth history, we see evidence of the shaping effects of natural cold and warm cycles. We can decipher the coming and going of several ice ages from the traces they left behind in the geologic record. Aerial views of the Rocky Mountains, the Sierra Nevada, the Alps, the Hindu Kush and the Andes clearly show the work that ice has done in past millennia. Even near the equator, on Mount Kilimanjaro and the Rwenzori Mountains in Africa and Mauna Kea in Hawaii, the evidence of present and past ice sheets is clear. Glaciers have an extraordinary ability to carve rocky landscapes into deep mountain valleys with characteristic shapes, so that even long after the ice has melted, the effects of scouring and grinding can be seen. The pulverized and broken soil and rocks carried by the moving ice sheets are often deposited in great heaps (moraines) along the sides of the glaciers and where they reach level land or sea and cease their relentless downward movement.

We do not have to look at mountains to see the work that ice does. The smooth landscapes of much of Canada and the Northern United States owe much to the grinding of great ice sheets of recent ice ages and the resulting piles of glacial drift left behind. As the Swiss scientist Louis Agassiz noted, "The glacier was God's great plough set at work ages ago to grind, furrow, and knead over, as it were, the surface of the earth." Even Hudson Bay—which looks superficially like a large impact basin—actually represents a depressed area of the crust that has not rebounded from the ice sheet that melted at the end of the last ice age. Across a large swath of North America, shallow lakes and thick deposits of fertile soils testify to the repeated incursions of moving ice from the northern polar regions and the deposits of the finely ground rocks left behind during their retreats as the climate warmed. Simple plants grew in the ponds and on the pulverized rock, taking minerals from those rocks and then leaving a film of organic matter when they died. Later plants could then take root in a layer of more nutritious newly developing soil.

The Huronian glaciation, the first ice age identified from the geologic record, occurred about 2.4 to 2.1 billion years ago, roughly the halfway mark of Earth's 4.6-billion-year history. So widespread was the ice during this period that Earth was a virtual snowball, with most, or perhaps all, of the solid surface covered with solid ice and the oceans choked with slush. The only life was microbial, and it was confined to the seas. The second identified great ice age was the Cryogenian from 850 to 635 million years ago. Life at that time was still primitive, though at the end of this ice age—perhaps even as a result—the first multicellular organisms appeared. Many evolutionary biologists regard the leap from unicellular to multicellular organisms to be much greater than from organic matter to single-celled organisms. Whether causally related or not—something that is still debated—this cold period was undeniably followed by the rapid diversification of complex aquatic life forms that defines the event known as the "Cambrian explosion."

Since a single glaciation on land can obliterate all the evidence left by previous events, sediments in the oceans provide a more reliable record of past ice ages than do landscape features. In some ocean regions, sediments have been accumulating for millions of years, allowing the effects of glacial fluctuations to be clearly distinguished, rather like the climatic cycles identifiable from tree rings. During just the last half million years, five cycles of glacial advance and retreat can be identified. These cycles belong to the Quaternary glaciation,[1] which began about 2.6 million years ago and, in fact, continues today—a concept perhaps difficult to reconcile with the present-day warming of the planet and the recording of record-high temperatures worldwide with each passing year. However, scientists, unlike the general public and weather forecasters, are obliged to take the long view, and do not fail to distinguish long-term trends from short-term fluctuations.

Surprisingly—and a fact only recently recognized—the cold temperatures in the air and water that spawn the great glaciers and ice sheets also affect what is happening on the ocean floor. The sideways motions of the continents we call continental drift are driven by hot currents in the mantle of Earth. Convection currents within near-molten rock in the outermost 45 percent of the planet drag along the stiffer and much-thinner surface plates, creating new ocean-bottom material along cracks where molten lava emerges into the sea water to form ridges. The continuous formation of new sea floor along these ridges causes the earlier lavas to be pushed away toward the continents, so that the record piles up rather like the sequence of pages of a book. The youngest rocks at the ridge and the oldest are pressed up against and squeezed below the thicker continents. When much of Earth's water is frozen in ice sheets on land, sea level is reduced by as much as 200 m, relieving some of the overhead pressure of the sea-floor water column. Lavas from the mantle therefore find it easier to reach the surface and form thicker deposits of sea-floor rock. In contrast, when warmer periods have reduced the continental ice load and much of the water is returned to the sea, the pressure on the ocean bottom is increased and the volume of lavas that erupts along the ridges diminishes, creating thinner layers. The great ice ages manifest on the sea surface and on the continents are thus recorded on the topography of the ocean floor.

As ice goes about its business of molding continental landscapes, creating our soils, and shaping the ocean floor, it also preserves vital records of Earth's history. In Antarctica and Greenland, where the ice has formed from the accumulated and uninterrupted snowfall of thousands of years, tiny bubbles of air, ash from volcanic eruptions, dust from large-scale dust storms, and even pollen grains and bacteria have been quietly and neatly packed away in cold storage. By drilling vertically into the ice, scientists extract cylindrical core samples that are true time capsules of the planet's past. These samples reveal the changing chemistry of the atmosphere and other information in a time sequence in which more distant times are chronicled at increasing depths from the surface. EPICA, a very long core sample extracted in an Antarctic location called Dome C, reached a depth of 3,270 m (10,729 ft), extending the record back nearly 800,000 years from the present.[2] For a time this sample held the record for a probe into Earth's icy past, but a new, deeper core from the Allan Hills in Antarctica reported in 2017 topped it, reaching to a depth corresponding to an age of 2.7 million years.[3] Ice cores from other locations in Antarctica go back shorter time spans, depending on the thickness of the ice sheet. This is similar to what is done with tree rings, where the time sequences of multiple samples can be correlated with one another using, among other characteristics, layers of dust and volcanic ash simultaneously preserved in all of them.

The ancient ice at the base of the Antarctic pack is not the only old ice on Earth, for in the Kunlun Mountains on the high Tibetan Plateau, an international team of glaciologists has begun drilling into the Guliya ice cap in search of ice deposits recording Earth's frozen past. The Tibetan Plateau is called Earth's Third Pole because its 46,000 glaciers make up the largest ice deposit outside Antarctica and the Arctic. This ice pack is critical to the weather patterns and water resources in the highly populated areas of India, China, and Southeast Asia. Thus the history it records about the climate in the midlatitudes is of compelling interest both to understanding long-term trends and in projecting short-term consequences of human-induced climate change. It would be a major step forward in our understanding of the history of Earth and the evolution of life (including the emergence of multicellular organisms as noted above) if, for instance, billion-year-old ice samples in Tibet were to push the beginning of the Cryogenian ice age from 850 million years ago to some earlier date.[4]

The long cores of ice pulled from the drills are usually stored in sections in the ambient low temperatures of the locations from which they are extracted. Hundreds of them, resembling glassy, translucent noodles, are carefully arranged and cataloged in the sequence they were extracted and identified with the depths from which they came. Because ice deposition is not continuous, determining the absolute times for every level of core is a challenging business that depends in part on correlation with other information, such as large-scale volcanic activity around the world as measured from written historical sources or the evidence in the rock record on the continents. Nevertheless, it is not impossible.

As noted above, the variegated contents of the ice cores include tiny grains of pollen. Pollen is found in seed plants, which first appeared at the end of the Devonian period, about 360 million years ago. Today, seed plants include some of the most important plants on Earth. With their great variety of extraordinary shapes revealed under the microscope, pollen grains are highly durable little messengers of their parent plants. Wafted and stirred by local and global winds, pollen grains find their way to the frigid regions of Earth where falling snowflakes trap them from the air and deposit them on the surface, allowing them eventually to become part of the ice column. Preserved in this way, pollen is an important indicator of past climates and flora of Earth. The pollen record for the last two millennia is dominated by plants cultivated for food—cereal grains and other crops—but pollen found in older ice cores from glaciers around the world and in the Arctic and Antarctic give reliable information on the abundance and extent of seed plants in numerous terrestrial environments. This information, taken with that derived from pollen extracted from bogs and sediments, opens an important window on the past climate of Earth.

Falling snow on glaciers and polar ice sheets also traps gases from the atmosphere, allowing, for example, the atmospheric concentration of carbon dioxide, the greenhouse gas most implicated in the current dramatic episode of global warming, to be determined for past epochs. It is the carbon-dioxide concentration measured in tiny samples preserved as air bubbles in the ancient ices that give us a baseline to compare with atmospheric concentrations measured at monitoring stations such as that at Hawaii's Mauna Loa and Tasmania's Cape Grim. This kind of data has revealed that, though carbon dioxide in the atmosphere has varied significantly over many thousands of years from natural causes (volcanic gas emissions, for example), the accelerating upward trend in concentration that began at about the time of the Industrial Revolution (roughly in the year 1800) cannot plausibly be attributed to anything but the progressively increased use of fossil fuels.

While the palynologists (pollen specialists) and atmospheric chemists are interested in deriving a consistent and clear picture of the past climate conditions on our planet—a kind of climate biography of Earth—our points here (in a book on Pluto) are not quite that Earth-shattering. We emphasize here only that ice can preserve a unique and indispensable record of a dynamic environment in which various kinds of change and activity occur over very long timescales. The richness of this record on Earth prompts a hope—we shall see to what extent justified in the following chapters—that the ices of other planets and moons will be similarly informative and help us decipher the interesting biographies of other (icy) worlds.

The static quality of the ancient record of life encapsulated in the old ices of the polar worlds is punctuated by the dynamic quality of an amazing organism alive today and busily making a living in glacial ices around the world. Several species belonging to the genus *Mesenchytraeus* are annelids (ringed or segmented worms) that live entirely on and in ice. In fact, they are so dependent on the low temperature of the ice that at temperatures just slightly above freezing the membranes holding their bodies together disintegrate and they "liquefy." At propitious times of the night, these ice worms scavenge the snow algae on or near the surface, then retreat by means not entirely understood into the solid ice.

Thus, the ices of Earth, entirely made of frozen water, constitute an ecosystem where a highly specialized life form thrives. This terrestrial fact may constitute a paradigm that can guide our imaginations and researches as we explore ices in "hostile" environments beyond Earth that may be hiding biological ecosystems where they are least expected. In this spirit, later in our story we will indulge in brief dalliances with *Sulfolobus solfataricus* and *Hesiocaeca methanicola*. The reader for whom that invitation sounds appealing is encouraged to read on.

Before we leave Earth, it is worth noting that in a geological sense ice is actually a rock—harder than some but softer than other "classical" rocks. It is formed of crystallized water as other rocks are formed of mineral crystals such as quartz, feldspar, mica, etc. Note that meteorites are also rocks— albeit rocks from space. They contain metals such as the iron- and nickel-forming crystals kamacite and taenite. Ice as a rock calls to mind a story. In his course on glaciology at the University of Arizona, polar explorer Laurence M. Gould sometimes regaled his students with stories of the Arctic and Antarctica. Gould was the chief scientist and second-in-command to Admiral Richard E. Byrd on Byrd's first expedition to Antarctica and the South Pole, 1928–30. During his lengthy stay there, Gould conducted the first geological survey of the continent. In one of his glaciology classes in 1966, Gould told of a gathering in New York at the Explorers Club, an exclusive social club for explorers whose members have included the likes of Robert E. Peary, Matthew Henson, Roald Amundsen, Charles Lindbergh, Sir Edmund Hillary, Tenzing Norgay, and astronauts Neil Armstrong, Buzz Aldrin, and Michael Collins. For this party, chunks of glacial ice had been flown in from Alaska to chill the drinks. Thus, Gould impishly pointed out, scotch "on the rocks" was served, signifying that glacial ice is indeed rock.[5]

Ices Beyond Earth

The discovery of ices of various kinds in the Solar System's planetary bodies is traced out in a curious chronology. First, we look at the reasoning behind thinking that there might be ice elsewhere beyond Earth.

Ancient astronomers knew about the five planets in addition to Earth: Mercury, Venus, Mars, Jupiter, and Saturn. While there was some controversy over the placement of the planets relative to the Sun for a long time, Copernicus proposed in 1543 that the Sun is the center of the Solar System and the planets revolve around it. It eventually became clear that the slower-moving planets (Mars, Jupiter, and Saturn) were more distant from the Sun than are Mercury, Venus, and Earth. While Kepler worked out the proportions of the planets' orbits in the correct ratio as a part of his derivation of the famous three laws of planetary motion (1600–1619) the true distances between the planets were not known until the absolute distance of Earth from the Sun was determined through observations of a transit of Venus across the Sun's disk in 1769.[6] While it was easy to speculate that the planets more distant from the Sun than Earth would be colder, one more critical piece of information was needed. Newton determined the law by which the light from the Sun decreases with increasing distance,[7] showing that the light (or heat) intensity decreases with the square of the distance. This means that a planet five times farther from the Sun than Earth (specifically Jupiter) receives 5^2, or 25, times *less* light and heat than our planet does. Saturn, some nine times farther from the Sun than Earth, receives 9^2, or 81, times less light and heat. From this simple relationship, our intuition that objects more distant from the Sun will be colder is validated, and the degree of coldness can be calculated.

This reasoning is good enough for a rough estimate of planetary temperatures. The exact temperature depends on the presence or absence of an atmosphere, the reflectivity of the surface, its rotation period and axis tilt, plus a few other factors, but calculating the temperature of a planetary surface to within 10 or 20° is straightforward.

Water is so abundant on Earth that we might expect it to be widespread throughout the Solar System. It is. Water is composed of two of the most abundant elements—hydrogen and oxygen—and these elements combined in the proportion two hydrogen atoms with one oxygen atom make H_2O, a molecule that is stable under a very wide range of temperatures and pressures. Once made, a water molecule is not easily taken apart. We know from experience quite a lot about its properties as a gas, a liquid, and a solid; we also know how crucially important it is to life here on Earth, where water conveniently and simultaneously exists in all three forms.

Mars

By analogy with the abundance and behavior of water on Earth, it should exist elsewhere in the Solar System and lie frozen as ice on planetary bodies more distant from the Sun than Earth. Mars is sufficiently cold that any water there should occur as ice, possibly with a tiny amount in the gas phase—liquid seems out of the question because the atmospheric pressure on Mars is too low, and liquid water depends on both the right temperature and pressure ranges. These considerations about water on Mars were central to the speculation by Percival Lowell and others about the habitability of the planet and the plight of its putative inhabitants in their epic quest to survive in the predicament of their drying planet. While most of Mars's surface presented a desert-like landscape to visual telescopic observers in the late 1800s and early 1900s, the planet was clearly seen to have bright white polar caps that changed in size and prominence with the planet's seasonal cycle. This naturally suggested the regular coming and going of water ice that would condense from the atmosphere in the winter hemisphere and then evaporate or liquefy as spring came and the polar cap nearly disappeared.

The problem with this dynamic scenario was that there was no way to test the water-cycle hypothesis with hard data because telescopes had no suitable instruments to detect water in any form. Further exasperating astronomers was the difficulty in measuring the pressure of Mars's atmosphere. It was not until the late 1950s and early 60s that spectrographs were sufficiently sensitive to provide the kind of data needed to determine these critical physical conditions. While the quest for water in any form on Mars unfolded, Walter Adams and Charles E. St. John photographed Mars's spectrum with the Mount Wilson 100-inch telescope in 1925. They reported that small amounts of water vapor and oxygen exist in Mars's atmosphere, which must be much thinner than Earth's atmosphere. In 1947, Kuiper applied his new infrared spectrometer, attached to the McDonald Observatory 82-inch telescope, and found carbon dioxide in Mars's atmosphere—this is now known to be the most abundant gas in the planet's atmosphere, about 95 percent of the total. Small amounts of argon and water vapor make up most of the remaining 5 percent.

With the discovery that Mars's atmosphere is primarily carbon dioxide, the seasonal appearance and disappearance of the polar caps became clear—carbon dioxide condensing from the atmosphere in winter causes the cap to extend in size, whereupon in spring the thin layer of carbon-dioxide frost begins to sublimate back into the atmosphere,[8] exposing the rocky and dusty surface below. Long-lasting residual ice caps that contain both carbon dioxide and frozen water exist at each pole, but the dynamical exchange of the frozen solid on the ground with the gas in the atmosphere is entirely a manifestation of the annual carbon dioxide cycle.

Carbon dioxide on Mars was therefore the first ice found beyond Earth, and its discovery was quite recent, as we have shown. This discovery resulted from the application of a new technology, infrared spectroscopy, to the problem. Conventional spectroscopic techniques working in the visual spectrum were simply not sufficiently sensitive to measure the small amount of gas in Mars's atmosphere or the solid carbon dioxide in its polar caps.

The Galilean Moons of Jupiter

On the 7th day of January in the present year, 1610, in the first hour of the following night, when I was viewing the constellations of the heavens through a telescope, the planet Jupiter presented itself to my view, and as I had prepared for myself a very excellent instrument, I noticed a circumstance which I had never been able to notice before, . . . namely that three little stars, small but very bright, were near the planet; and although I believed them to belong to the number of the fixed stars, yet they made me somewhat wonder, because they seemed to be arranged exactly in a straight line, parallel to the ecliptic, and to be brighter than the rest of the stars, equal to them in magnitude. . . . When on January 8th, led by some fatality, I turned again to look at the same part of the heavens, I found a very different state of things, for there were three little stars all west of Jupiter, and nearer together than on the previous night.

GALILEO GALILEI, *SIDEREUS NUNCIUS*, MARCH 1610[9]

Astronomers are by nature compelled to push the limits of their capabilities—particularly their telescopes and associated light-sensing instruments—to wrest the secrets from increasingly fainter objects. Galileo in 1610 pushed the limits of his tiny spyglass telescope, and with the discovery of Jupiter's moons and other wonders in the sky revolutionized astronomy for all time. In the modern era, building larger telescopes rapidly increases the number of visible celestial bodies, so building a large telescope with a more sensitive light-detection instrument opens vistas on a vastly greater number of stars and galaxies than previously could be observed. The same is true for Solar System bodies—increased observational capability invariably leads to the discovery of new asteroids, comets, and planetary moons while providing more detail about their characteristics, similarities, and differences.

Although the four bright moons of Jupiter have been known since their discovery by Galileo, for three and a half centuries they did not yield much information to astronomers trying to learn even their most basic physical characteristics. All four are roughly as bright as the faintest stars visible to the naked eye, and while they were within reach of telescopes with spectrographs starting in the early 1900s, nothing much about their compositions and other properties emerged from the relatively few studies that were undertaken. In short, as physical entities they did not command much attention. However, their orbital motions around Jupiter, crossing over it's banded, cloudy face and periodically dipping into its shadow, provided a means for determining the longitudes of locations on Earth.

Galileo himself recognized the utility of determining longitude from the movements of the moons he discovered, and sought to develop the technique for finding longitude at sea, a critical and vexing problem for navigators in both commerce and war. Two things were critical. First, an accurate table must be calculated of the times of the satellite eclipses and transits as observed from, say, Florence, where Galileo was working. These predictive tables had to be based on the orbits of the satellites as estimated from the observations by Galileo and others with telescopes, and they had to be accurate to within about two minutes of time. Second, a navigator at sea had to be able to observe the predicted eclipses or other configurations of the satellites with one of the small telescopes of the day from a rocking and pitching ship, and he had to know the local time. Both requirements were difficult to meet, but when they could be achieved, the longitude of the observer at sea relative to that of Florence was simply $L = 15 (t_2 - t_1)$, with the times measured in hours. This applied to measurements made at about the same latitude of Florence; otherwise an additional correction was needed.[10]

Returning to the question of ice on Jupiter's moons, we saw in chapter 9 that these objects held their secrets well until the emergence of infrared-sensitive detectors during World War II. Kuiper's application of the new technology to one of the largest telescopes of the day finally began to reveal information on the compositions of these four Jovian satellites.

Even before the ices of the Galilean satellites were discovered, their densities provided hints of their bulk compositions. Density is mass per unit volume, and is usually thought of as the number of grams of a substance in a cubic centimeter (g/cm³), although it is often defined as kilograms per cubic meter. Liquid water has density of 1.0 g/cm³, while frozen water has density 0.934 g/cm³—which is why ice floats on liquid water. The density of ordinary silicate minerals is about 3.0 g/cm³, while iron, the most common metal, has density 7.9 g/cm³. A completely icy planetary satellite could have a density of about 1 g/cm³, while one composed of a combination of rock, metal, and ice could have density in the range of about 2 to 5 g/cm³. Thus, if we can determine the density of a planetary body by measuring its size and its gravity field, we can get a good idea of its internal composition—ice, rock, metal, or more likely some combination of those three components.

Once reliable measurements were available of the satellites' diameters, made by visual means at the end of the 1800s by such notable astronomers as E. E. Barnard (who also discovered Jupiter's fifth satellite in 1892—the first since Galileo) at Lick Observatory, as well as reasonably accurate estimates of their masses from the gravitational effects they have on one another, it became possible to calculate their densities with a useful degree of accuracy. Of course, subsequent measurements have improved knowledge of both size and mass for each of these moons, but the initial values gave good hints to the differences between them and to the probable compositions of their interiors.

The four Galilean satellites of Jupiter have densities indicating a significant amount of rocky material in their interiors, despite surface evidence for water ice on three of them and sulfurous amalgam on volcano-ridden Io.

Io has density 3.53 g/cm³ and is hot, suggesting little or no water in its interior; it is primarily composed of some combination of (mostly) rock and metal, in addition to the copious sulfur that erupts from its active volcanoes. The next satellite outward, Europa, clearly has an icy crust, but its density of 3.01 g/cm³ demonstrates that it contains a sizable amount of rock in the interior—about 90 percent—and the remaining 10 percent is water. Still farther out is Ganymede, just a bit larger than our own Moon. With a surface

rich in frozen water, it has a density 1.94 g/cm³, indicating that the proportions of rocky material in its interior is smaller than in either Io or Europa—
about half rock and half ice. Ganymede appears to retain most of the water
that is thought to have been the most abundant component of the cloud in
which the four large moons solidified (that cloud is called the Jovian subnebula to distinguish it from the much more massive, and hydrogen-rich, nebula
from which Jupiter itself took form). The large satellite that aggregated still
farther out in the subnebula is Callisto, which has a density of 1.83 g/cm³.
While Callisto has relatively little frozen water on its surface, its low density
indicates that its interior must be roughly equal parts frozen water and rock.

This pattern of decreasing density with distance of the four Galilean moons
is, in fact, the key point in a theory that all four coagulated in the Jovian
subnebula after the planet itself had largely come together, leaving relatively
little material—a combination of rocky material and ices—behind. James B.
Pollack, a student of Carl Sagan and NASA planetary scientist, authored
with his colleague Ray T. Reynolds a compelling theoretical study of Jupiter's
formation and the birth of the four large moons in the subnebula.[11] Jupiter
was born hot, raised to high temperature by the energy expended by the
countless bits and pieces pulled and stuck together by the gravity field. This
heat cooked much of the icy component of the subnebula out of the inner
zone, but decreasingly so at the distances of Ganymede and Callisto. Any
modest amount of ice that Io may have originally contained has apparently
been boiled away by the volcanic heat that still exists in its interior. The
heat within Io in the modern era is generated by the combined gravitational
pulls exerted by Jupiter and the other three moons, particularly Europa and
Ganymede, and shows no sign of diminishing even after some 4.5 billion
years circling the planet.

But this picture runs into difficulty when we look at tiny Amalthea, the
moon that Barnard discovered in 1892 with the great refracting telescope at
the Lick Observatory. Amalthea is only about one-third as far from Jupiter
as Io, yet its density is less than that of water, and it is thought to consist of
ice with a sponge-like structure with many void spaces. Some alternative
explanation is needed for the overall compositions of the Galilean satellites
and Amalthea. One possibility is that they all started with an abundance of
water, but incoming asteroids and space debris preferentially hit the large

inner moons, depleting them of that original water. This and other explanations all fall short in one way or another, and it is safe to say that the puzzle of the moons of Jupiter concerning their water or ice content is not fully solved.

We saw in a previous section how difficult (and contentious) the detection of water ice on some of the Galilean satellites had been, and how it depended on observations with newly invented instruments. Here we pick up that thread and give a few more details about the ices of the four large Jovian moons.

The measurements of the infrared spectra of the four satellites were not entirely straightforward in the beginning; Europa showed a mostly clear indication of water ice, as did Ganymede. But Callisto, the most distant from Jupiter, exhibited only a hint of the spectral bands diagnostic of ice. Io, the closest to Jupiter, was an entirely different matter, with its unusual yellow color and no indication whatsoever of water ice. Observational capabilities improved during the 1960s and 70s, eventually allowing astronomers to establish that water ice also exists on Callisto, although it is largely masked by the rocky material that covers its old, cratered surface.

A Close-Range Look

Although the identification of water ice on Europa and Ganymede was secure, a close-range look at all the major planets and all their satellites was a high priority in planetary exploration by robotic spacecraft. Accordingly, the United States launched two identical spacecraft toward Jupiter in 1977, Voyager 2 on August 20 and Voyager 1 on September 5. Voyager 1 flew by Jupiter in March 1979 on its way for an encounter with Saturn, while Voyager 2 passed Jupiter in July of that same year, also on the way to Saturn (and eventually Uranus and Neptune).[12] Both Voyagers sent back images of the Galilean satellites, showing the extraordinary and highly varied geologic terrains of each. Europa is entirely covered by a fresh and cracked ice crust with very few impact craters—later evidence emerged that the icy crust hides a global ocean of liquid water at some unknown but shallow depth. Ganymede's surface is heavily cratered, indicating that it is much older than that of Europa, and it displays a fantastic array of ice geological structures. Callisto's surface was found to be extremely old as well, but retains some large-scale geological features indicative of a history of liquid water in its early years.

One of the most extraordinary discoveries of the two Voyager spacecraft at Jupiter was the extreme volcanic activity on Io. While observations had been made from Earth that Io's spectrum is anomalous (with absorption bands that could not at the time be identified) and that the whole satellite displayed temporary outbursts of something very hot on its surface, the full picture did not come together until the active volcanic fountains and lava flows were seen in the spacecraft images. The new Voyager images, heat measurements made from another instrument aboard the spacecraft (an infrared radiometer spectrometer), the Earth-based observations, and a set of spectral measurements in the laboratory all supported the view that Io is the most active planetary body in the Solar System and that its surface is covered by lavas of various compositions as well as a layer of frozen sulfur dioxide and probably hydrogen sulfide. Sulfur dioxide and hydrogen sulfide are common volcanic gases in terrestrial volcanoes, but on Io, sulfur dioxide dominates the volcanic effluence. When a volcano injects it like a jet into the cold near vacuum above Io, some of it freezes into snow or ice particles and falls to the surface in beautifully symmetric and graceful umbrella-shaped plumes to form a frosty, icy covering. Although Io sports more than seventy volcanoes, somewhere between five and fifteen are active at any one time in a pattern that is constantly changing. Io's volcanoes are monitored routinely from Earth-based telescopes and from an occasional passing spacecraft— Cassini-Huygens (hereafter "Cassini," as it is more commonly known) on its way to Saturn in December 2000 and New Horizons en route to Pluto in February 2007.

The discovery of sulfur dioxide on Io adds another ice to the list of discoveries, joining water and carbon dioxide. Direct evidence for hydrogen sulfide is sketchy, and it is presently unconfirmed, as are a few other sulfur-bearing gases and ices.

Before moving on, let us ponder the possible implications of the sulfurous, hellish Ionian environment, and what great discoveries remain to unfold there. Speculation about the possibility of life beyond Earth has been a cottage industry for writers of both science and science fiction for centuries. But in the last fifty years, as the frontier of exploration has swept through the Solar System, engulfing all the planets and a vast number of asteroids, comets, and moons, the speculation has become very well informed by the uncovered facts about planetary environments. The information gleaned by spacecraft and Earth-bound telescopes has bolstered the view that life might occur in other places in the Solar System. This inspired speculation focuses on life-forms familiar to us, meaning organisms dependent on water, usually liquid in form. At the same time, "life as we know it" has taken on expanded dimensions with the discoveries of extremophiles—organisms thriving in environments of enormously high temperature, total darkness, high acidity, and even normally deadly emissions from radioactive elements.

Present-day Io has not given us any evidence for the presence of water, either as ice, liquid, or vapor, but water must have been present when the moon aggregated in the gaseous envelope surrounding the young Jupiter. Perhaps it has been cooked entirely out by the heating that gives rise to the volcanism we see now, but perhaps some water has been retained. Could organisms like *Solfataricus* have spawned in a sulfurous, aqueous enclave inside Io, and now await discovery by some future emissary from Earth on a voyage to explore the most intensely volcanic world in the Solar System?

Saturn's Moons

For ices, colder is always better, and as we look farther from the Sun than Jupiter, the next planet bearing moons is Saturn, nearly twice as distant. Saturn offers an additional icy bonus—the rings that make it the favorite planet of every cartoonist and probably everyone else who bothers to let their thoughts drift beyond Earth.

Saturn's family of moons just seems to keep growing. As this book goes to press, the count was sixty-two confirmed objects.[13] The discoveries began with Christiaan Huygens's detection of Titan, the largest satellite, in 1655 with a long, cumbersome telescope suspended from a pole with wires and ropes. Later telescopic observations with improved equipment revealed a few more, until the last satellite found by visual means was discovered in 1848 independently by William Cranch Bond in the United States and William Lassell in England, and named Hyperion. When photography, and later electronic imagery, was used to survey the region of the Solar System around Saturn, Phoebe was found first, and then tiny Janus. Voyagers 1 and 2 flew past Saturn in 1980 and 1981, respectively, revealing three more tiny moons, mostly close in to the planet, and some even sharing the same orbit with one another. But the champion satellite finder is Scott Sheppard, now of the Carnegie Institution in Washington, DC, but formerly a graduate student at the University of Hawaii. Using the telescopes at Mauna Kea Observatory on the Island of Hawaii and large charge-coupled devices (CCDs), Sheppard has found an additional twenty-five satellites orbiting Saturn, most at great distance from the planet, traveling in the normal prograde orbit, but several traveling backward in retrograde. Prograde means that the direction of motion around Saturn is the same as the direction of Saturn's orbital motion (and those of all the planets) around the Sun. (Looking down on the Solar System from the north, the direction of motion around the Sun is counterclockwise.)

Saturn's family of satellites range widely in size from the largest (Titan) to several bodies only a few kilometers across. After fleeting glimpses of the satellites as Voyager 1 and 2 flew past the planet, the entire Saturn system has been scrutinized in great detail by the Cassini spacecraft that entered orbit around the planet on June 30, 2004. Apart from the planet and its rings, the largest satellites naturally have been studied most closely in terms of their compositions and geological structures. The compositions, which were previously determined from spectroscopic measurements made with telescopes on Earth, are of special interest, as we will see below. We now have so much detailed information on these moons that more than just a few words are needed to describe them. The Cassini mission ended on September 15, 2017.

Mimas

Moving outward from Saturn, the first major satellite is Mimas, known from Voyager images and more recent pictures from Cassini for its heavily cratered surface and one very large crater named Herschel after the satellite's discoverer. The impact that formed the 130-km diameter Herschel was nearly large enough to destroy the satellite altogether. A few other moons in the Solar System have similarly large craters resulting from nearly planet-busting collisions. The surface of Mimas is made of water ice, and its density of 1.15 g/cm^3 shows that its interior is also ice.

Enceladus

The next icy moon of Saturn is Enceladus, small at diameter of 504 km, but remarkable in so many ways. Water ice covers all of Enceladus's surface, but at the south pole, from a series of nearly parallel cracks in the icy crust, plumes of ice and gas jet into space from an underground ocean of liquid water mixed with a variety of other chemicals. The spray of icy particles from the polar cracks, nicknamed the tiger stripes, spreads out in the surround-ing space in a disk that extends to other nearby moons and creates Saturn's E-ring. This tenuous ring is far outside the planet's main ring system and can barely be detected from Earth. But perhaps more remarkably, the Cas-sini spacecraft has flown close to Enceladus's surface and directly through the plumes, collecting and analyzing the gases and particles it encountered. Among the gases in the plumes is molecular hydrogen, a key ingredient in geothermal processes taking place in the interior of the satellite.[14] The analyses show that the ocean is salty and alkaline, possibly from the reac-tion of liquid water with iron- and magnesium-rich minerals in the moon's interior.[15] These chemical reactions produce a new mineral (serpentine) and hydrogen that can drive the formation of organic molecules like amino acids, which appear to be essential to the origin of life. This chemical process of serpentinization may forge a link between biology and geology. This extraor-dinary discovery is only one of many made by Cassini, and it demonstrates how a small, icy body that had previously received little attention is cata-pulted to the forefront of scientific inquiry from a few measurements and observations made with a nearby capable robotic spacecraft.

Tethys

Tethys, twice the diameter of Enceladus, is the next icy satellite. It also has a solid ice surface that bears the cratered witness of more than a billion years of impacts from objects drawn by gravity toward Saturn like moths to a flame, but encountering this obstacle on the way. A great icy chasm cuts across

Tethys's surface, the result of tectonic forces induced by some phenomenon net yet understood. The density of Tethys is only slightly more than that of water ice, and some rocky material may exist inside, deep below the icy mantle. The surface is stained yellow in some places where atomic particles trapped in Saturn's magnetic field strike the icy surface and cause chemical changes, also not yet understood.

Dione

Dione is next large satellite outward, but its density is about 1.48 g/cm^3, clearly indicating that there is a rocky core inside. The surface is nearly pure water ice, but in addition to thousands of impact craters, it is heavily cracked by tectonic forces. Like Tethys, it bears the yellowish stain from its exposure to the energetic particles in Saturn's magnetic field that blast its unprotected surface continuously.

Rhea

Saturn's second-largest satellite, 1,527 km in diameter, is Rhea, a moon with an icy surface like the others, but with a density of 1.24 g/cm^3, indicating that it is made of about 25 percent rock and 75 percent water ice. Its ancient surface is heavily cratered and fractured, although some of the large craters are clearly younger than others. A body this large and with so much rocky material inside may have been able to maintain an internal liquid water ocean by the heat generated from radioactive minerals.

Titan

Huygens's discovery in 1655 of Saturn's largest moon, Titan, continues to bear fruit from both Earth-based studies and the close-range scrutiny by Cassini and its many sensing instruments. Not only has Titan been studied in minute detail from close passes, an automated probe was dropped from Cassini into Titan's atmosphere for a two-and-a-half-hour thrill ride by parachute through the clouds and haze and down to the surface. Named Huygens after Titan's discoverer, the probe successfully deployed three parachutes in succession to descend several hundred kilometers from the upper edge of the atmosphere to the surface for a soft landing, making measurements of the surface and atmosphere all the way. This landing in January 2005 was an extraordinary engineering achievement and provided a scientific perspective on Titan that could not be achieved any other way.

Huygens landed in a pebble-covered, lake-bed-like flat terrain tinted yellow orange by particles of organic smog that form in the nitrogen-methane atmosphere and settle on the surface. Much of the surface and pebbles are water ice, but their icy characteristics are largely masked by the colored coating.

With a diameter of 5,152 km, Titan is nearly half again as large as Earth's Moon, and is almost as large as Jupiter's Ganymede. Titan, however, is the only moon in the Solar System with a dense atmosphere. Kuiper discovered the atmosphere in 1944 when he used his spectrograph to find methane, but it was later found to be mostly nitrogen; methane is only about 10 percent. Sunlight acting on this mix of nitrogen and methane induces many chemical changes that result in complex organic and inorganic molecules being formed, including Titan's organic colored smog, which is similar to that forming over cities with significant air pollution.

Titan's density of 1.88 g/cm^3 demonstrates that this body contains a large fraction of ice, but also a significant amount of rocky material. Heat released by the decay of radioactive elements in those rocks has probably melted the interior of Titan, but whether any liquid water remains is unclear. By the time Cassini finished its thirteen-year tour of the Saturn system, it had made 293 orbits of Saturn and 126 close passes by Titan, gathering an abundance of information about the surface and atmosphere. Still, many mysteries remain, and planetary scientists are making detailed plans for a future Titan lander that will explore some of the diverse landforms and liquid hydrocarbon lakes revealed by Cassini.

Hyperion, Iapetus, and Phoebe

Outside Titan's orbit, Saturn has three more icy moons of note. Hyperion is an irregularly shaped lump of very frothy ice that appears to have the structure of a sponge with many voids throughout its interior. Even the surface has spongelike characteristics. While Hyperion has many other unique properties, here we simply accept it into the icy fold presented by Saturn. Iapetus, with a density of 1.09 g/cm^3, is almost entirely made of ice, but its cratered surface tells a complex story. The other icy Saturn moons are highly reflective to varying degrees, as would be expected for frozen water, but while one half of Iapetus is bright white, the other side is a dark, dull, reddish color. The dim hemisphere is centered on the direction of Iapetus's orbital motion. Iapetus and all the other satellites of Saturn we have described (except Hyperion) are in locked, synchronous rotation around Saturn. That is, like Earth's Moon, they keep the same side toward Saturn throughout their orbital periods. This means that they have a front side and a back side in terms of the motion around Saturn, and Iapetus's front side is dull, as if it has swept up a patina of dark dust that lay in its orbital path. This is exactly what is generally thought to have happened, and it is probably ongoing.

The dark dust coating one entire hemisphere of Iapetus carries complex organic molecules, perhaps some minerals, and possibly some charcoal-like carbon. The thickness of the coating, revealed by the radar instrument aboard Cassini, is about 20 cm (8 in) thick. The coating is known to be young because it covers the entire landscape, with the only exceptions being some tiny impact craters that have cut through the thin layer and exposed bright water ice below. The size and number of these craters show that they are very recent, within a million years.

The search for the source of the dust on Iapetus takes us to the last Saturnian satellite we will consider. Phoebe is far from Saturn, and its orbit is retrograde and highly inclined relative to the orbits of the other satellites, the rings, and the equator of Saturn itself. Phoebe clearly has a different origin than the other satellites we have considered here, and to our best understanding, it was captured by Saturn's gravity long ago. Phoebe likely originated in the outermost Solar System in the region known as the Kuiper Belt, which we will discuss below. If so, it accreted in the region beyond Neptune from the ices and dust far from the Sun called the solar nebula. This occurred while the major planets were forming at various distances from the new, young Sun.

Phoebe's density is 1.64 g/cm^3, showing that it is a mix of ices and rocky material. Its shape, viewed closely by Cassini, is irregular, and its surface is heavily scarred by impacts. An astonishing discovery in 2009 revealed a gigantic, tenuous ring surrounding Saturn in which Phoebe is embedded.[16] This dust ring appears to have been formed when a large comet or asteroid-like body collided with Phoebe and liberated a huge amount of dust that slowly spread completely around Phoebe's orbit. Studies of the fate of individual tiny particles that form the ring show that particles of various sizes slowly spiral inward toward Saturn, and on their way, they encounter Iapetus, which sweeps them up to form the dirty coating on its leading hemisphere. Some particles get past Iapetus and are swept up by Hyperion, the sponge-like lump of ice that is unusual in so many ways.

Because Saturn is nearly twice as far from the Sun than Jupiter, the temperatures of the planet, its rings, and its moons are generally lower. The temperatures of Saturn's rings and moons range widely from about 85 to 170° K (roughly –190 to –103° C).[17] The lower of these temperatures enables ices more volatile (subject to evaporation) than water to be stable over astronomical time scales (e.g., the age of the Solar System). In fact, at 85° K, frozen water is essentially rock-like. It is hard and brittle, and its evaporation rate is exceedingly slow. Carbon-dioxide ice is less stable (more volatile), and at 100° K, it is likely to evaporate from the sunlit part of a Saturn satellite in just a few years. Shadowing places on the cratered surfaces of Saturn's satellites are much colder than the sunny spots. In regions of large topographic relief in the latitudes far from the equator, and certainly in both the north and south polar regions, the temperatures are very low, giving carbon dioxide molecules hiding places where they can persist for potentially billions of years. If a patch of frozen carbon dioxide, or even just a few molecules, forms in a warm part of a satellite's surface, the sunlight evaporates it and a sizeable fraction of those molecules can migrate to a colder place where they freeze back. Some molecules evaporated from a surface will be ejected into space and lost, but calculations of the migration of frosts on satellite surfaces show that the cold polar regions and other places that remain in long-term shadow are repositories for molecules such as carbon dioxide and other materials that are volatile in warmer regions.

Carbon dioxide also can survive longer if the individual molecules are trapped inside a cage of water molecules. This arrangement is called a clathrate, and is better known in some environments on Earth where methane molecules are trapped in water cages, making a methane clathrate. The spectroscopic data for carbon dioxide on Hyperion, Iapetus, and Dione show a wavelength shift of the main absorption band, suggesting that the carbon dioxide molecules are held in icy cages formed by crystalline water structures.[18] Frozen water can form various types of cages in combinations of 34, 46, or 136 water molecules, each capable in the right conditions of trapping carbon dioxide molecules.

The coexistence of water and carbon dioxide ices, and the special ways they seem to be combined on Saturn's satellites, tells an important story about the way the ices interact at a molecular level. Those details suggest that some of the carbon dioxide might be made locally, molecule by molecule, from oxygen atoms extracted from the water that combines with carbon atoms falling on the surface of the moon from space as micrometeorites. Meanwhile, a reservoir of "original" carbon dioxide, which was part of the great cloud of ice lumps and grains from which the satellites formed early in Saturn's history, may reside inside the moons. Because carbon dioxide is somewhat volatile at the temperatures of Saturn's moons, the original component of this ice probably is slowly leaking away and into space. It is also readily destroyed by exposure to the space environment of ultraviolet light and charged atomic particles. Perhaps some native carbon dioxide is slowly diffusing from the interiors of these satellites through a crustal layer and onto the surface. Further details of this picture remain to be clarified, but the discovery of two kinds of ice is opening a deep investigation of the chemistry of icy bodies in Saturn's extraordinary family.

Saturn's Rings

Saturn's most distinguishing characteristic is its rings, which can easily be seen with a small telescope and have fascinated scientists and the public alike since they were recognized in 1655 by the same Christian Huygens who discovered Titan.

Saturn's rings are aggregates of small pieces of ice ranging in size from that of a baseball to chunks a kilometer or so across, all moving in independent orbits around Saturn but in an ensemble that gives the rings the appearance of a solid sheet. The bits and pieces of ice collide with one another, sometimes stick together for a while, then break apart again, in a slow dance of changing partners that is orchestrated by Saturn's strong gravitational field. How can one count the number of moons of Saturn if all the lumps in the rings above some size limit would be considered moons if they were not always closely surrounded by billions of others? In the strictest sense, then, the number of moons of Saturn is indeterminate.

That the rings of Saturn are composed largely of frozen water was first hinted at by Kuiper's initial measurements briefly reported in 1957.[19] Their composition was confirmed in 1970, when better measurements became available.[20] Those improved measurement capabilities soon lead to the detection of the characteristic spectral indicators of frozen water on the largest moons. Those telescopic measurements have been superseded by close-range spectroscopic studies with the Cassini spacecraft, which entered an orbit around Saturn in 2004. With numerous close passes by each of the large moons over the thirteen-year lifetime of the spacecraft,[21] Cassini accumulated detailed images and spectra of the ices, not only confirming the ubiquitous presence of frozen water, but also revealing frozen carbon dioxide on several of them.

Cassini has revealed some extraordinary details about the geology of the icy moons of Saturn, pictures written in the craters, gouges, cracks, and ice flows on their surfaces. Nearly each one of the large moons is different from all the rest in amazing details, but a broad view emerges that demonstrates that these icy worlds are not static frozen lumps, but instead have changed over the more than 4 billion years of their existence. Enceladus, with its plumes of icy spray jetting into space from cracks near the south pole, is clearly changing before our very eyes. A robotic mission to Enceladus for a close-range study is under detailed study.

The Moons of Uranus

Twice as far from the Sun as Saturn, and in still-colder climes, Uranus shepherds a mixed family of five large icy satellites, and at least twenty-two smaller ones. As with the moons of Saturn, each of the Uranian satellites has its own story to tell from a geological viewpoint, but for our purposes we can concentrate on their icy characteristics. Because these moons are so distant, the challenge of finding clues to their compositions is even greater than for Saturn's moons. Also, while the Voyager 2 spacecraft flew past Uranus and its family in 1986 for close-range views of the craters, mountains, and varied geology of these remote little worlds, the instruments it carried were not suited for detecting ices or minerals on solid bodies. No mission comparable to Cassini's at Saturn has been to the Uranian system, so we must be content with the fleeting impressions gained as Voyager 2 flew by at nearly 50,000 mph (~80,400 kph) in January 1984. When Voyager 2 flew through the Uranian system, it gave us reliable measurements of the densities of the five largest moons, Miranda, Ariel, Umbriel, Titania, and Oberon.[22] They range from 1.2 g/cm^3 for Miranda to 1.7 g/cm^3 for Titania, in all cases showing that their bulk compositions are largely ice (about 40–60 percent), with varying amounts of rocky material mixed in.

What we know about the compositions of Uranus's moons comes from measurements made with large telescopes here on Earth, even though these bodies are about ten thousand times fainter than the large moons of Jupiter. Despite the limitations of the long-range view, we have some basic information. Water ice dominates their surfaces, and frozen carbon dioxide also has been found by near-infrared spectroscopy.

Thus at least two kinds of ice cover the surfaces of the Uranian moons—other frozen molecules may be there in small quantities, awaiting detection by a more-capable generation of astronomical instruments, either from observatories on Earth or space telescopes in Earth orbit—or perhaps a future space mission to Uranus. Such a mission is the dream of the current generation of planetary scientists, but that dream is currently nothing more than a few sketches and numbers on a sheet of paper.

The Moons of Neptune

Rather than moving deeper into the Solar System to describe Neptune's major satellite Triton, we save this extraordinary object for a section all its own. The spectroscopic studies of Triton that began in the late 1970s were a watershed in our understanding of the icy bodies in the outermost Solar System, and set the stage for later discoveries on Pluto and solid objects at even greater distances from the Sun.

Let us then sidestep Neptune, only to say here that like the three other giant planets it has a family of small satellites (thirteen as of mid-2015). The largest of these is Nereid, about which little is known except that water ice is found on its surface. Nereid's orbit around the planet is highly elliptical, and the satellite takes nearly a full Earth year, 360 days, to make one complete circuit. Kuiper found Nereid by photography in 1949 and gave it its name. Seen from Earth, it is nearly fifty times fainter still than the large moons of Uranus, and is thus an even greater challenge for a study of its surface composition. Only in 1997, using the Keck 10-m telescope, then the largest in the world, was the spectroscopic evidence for frozen water found.

Comets

Hundreds of comets have been observed over the centuries, both by astronomers and—in the special cases of bright comets—by the public. Comets are even now seen by some as omens, and despite our rather clear understanding of their nature, they still have the power to put fear in the hearts of some people. Comets have inspired artists and poets. In *The Childhood of Hiawatha*, Longfellow tells of the little boy with his nurse,

> *Many things Nokomis taught him*
> *Of the stars that shine in heaven;*
> *Showed him Ishkoodah, the comet,*
> *Ishkoodah, with fiery tresses.*

The association of fire with comets is a literary and cultural habit that has proven hard to break. In 1997, when Comet Hale-Bopp was a spectacle in the evening sky for several weeks, even a NASA news release referred to it "blazing across the sky."

Comets are not fire—they are ice and "dirt." The nature of their orbits around the Sun causes most comets to spend the great majority of their time in the cold, dark, outer regions of the Solar System, tail-less little lumps of matter left over from the formation of the planets some 4.5 billion years ago. As they approach the region of the inner planets on their way to perihelion, the icy components of comets warm and begin to evaporate. The vapors released by warming ices first envelope the nucleus in a gaseous coma, and then stream outward in the direction opposite that of the Sun as molecules and atoms of the vapor are gently pushed away by the pressure of the Sun's light. The stream of vapor from the comet's nucleus also carries along tiny dust-size particles of minerals, primarily silicate minerals that make up the second major component of a comet. Carbon-rich solid particles that include both simple and complex organic chemicals, the third principal cometary component, are also entrained in the escaping vapor plume and dispersed into space. That dust-and-organic-laden vapor plume is the comet's tail. It is weak to nonexistent when the comet is far from the Sun, gradually strengthens and lengthens as the Sun is approached, and in some cases at its maximum extent can stretch 100 million miles, more than the distance from Earth to the Sun.

Frozen water is not the only ice in comets. Carbon dioxide and carbon monoxide are also important components, and other frozen molecules are present as well, including hydrogen cyanide and methanol. While water ice dominates the mix, each of these other ices plays a role in the chemistry and volatile activity of a comet—carbon monoxide, for example, begins to evaporate before water ice as the comet warms on its approach to the inner Solar System, which starts the formation of the coma and then the tail. Other ices begin to evaporate and join the gaseous mix as the temperatures appropriate to their evaporation are reached in the nucleus.

Astronomers had little trouble grasping that a principal component of a comet is a combination of ices, acting as the glue that holds the whole thing rather loosely together—minerals, organics, and all. But only in 1950 did it become codified by the work of Fred L. Whipple, who coined the phrase "dirty snowball."[23] Since then, several space missions to comets, including Comet Halley, the most famous of them all, have bolstered this basic idea but suggest that at least for some comets, the descriptive nickname might instead be "icy dirtball." Why this different view? While the vapor emerging from a comet nucleus clearly comes from the ice component, close views of comets like 9P/Tempel (the Deep Impact mission), 67P/Churyumov-Gerasimenko (the Rosetta mission), and 103P/Hartley (the EPOXI mission) have shown only small patches of ice on the surface, which quickly evaporated when it came into sunlight.[24] Particularly for comets that have made several circuits around the Sun, most or all of the ice has been evaporated from the surface, leaving a layer of dust and dark, red-colored organic solids, and the vapor forming the tail comes from ice completely hidden from view underneath. Thus, when we fly close to a comet, we see mostly "dirt," a lumpy mix of various minerals with a small amount of tar-like complex organic molecules, all stirred up with the ices and held together until the comet is hit by another object or comes close enough to the Sun for the ices to begin to sublimate.

All three major components of comets are important. Dust dispersed into space by the gas escaping from evaporating ices creates dust streams that largely follow the comet's orbital path. As Earth passes through these streams, the tiny dust particles are swept up and enter our atmosphere at high velocities, causing them to flash as streaks of light across the sky, giving us meteor showers that are sometimes quite spectacular. More significant than the nighttime lightshow provided by the dust, the other two comet ingredients, ice and complex organic chemicals, bring to the planets the materials vital to the origin and evolution of life.

Comets traversing the inner regions of the Solar System cross the orbits of the planets, and occasionally collide with those planets (and their moons). The ancient cratered surfaces of our own Moon and numerous other solid planetary bodies are the permanent record of countless ancient collisions by comets and asteroids, and give us a quantitative picture of the complex and violent encounters between bodies, particularly in the earliest epoch of solar-system formation and evolution. But comets strike the planets in the modern era as well. Between July 16 and 22, 1994, some twenty fragments of Comet Shoemaker-Levy 9 (S-L 9) crashed into Jupiter's atmosphere. The pieces had previously all be part of the same comet nucleus, probably a few kilometers in size, that was torn apart by tidal forces on an earlier pass by Jupiter. Like all comets studied so far, S-L 9 was a fragile, low-density, and loosely compacted lump of icy dirt, and it broke apart easily in Jupiter's strong gravitational field. But each fragment stayed in the same original orbit of the parent piece, and together they flew along forming a faint streak in the sky that was detected on photographs by astronomers Eugene and Carolyn Shoemaker and David Levy on March 24, 1993. Once it was confirmed by later observations and an orbital path computed, two things became evident. First, it was not a single comet, but a string of them, and second, the whole ensemble was headed directly for a collision with Jupiter. Sixteen months later, after astronomers worldwide had made preparations to observe the events, the crashes began, extending over seven days and creating huge dark spots that persisted for many days in Jupiter's atmosphere.

The scientific bounty of this extraordinary event has given insights not only into comets, but also into Jupiter's atmosphere and interior structure. For our purposes, we recount the outline of this great story primarily to emphasize the fact that comets impact the planets, and to note further that by the nature of their compositions, they carry critical materials to their targets. Those comet materials bear on two profound questions: "What is the origin of Earth's water?" and "How did life start on Earth?" Comets are water-rich and they carry complex organic chemicals; on our planet these two characteristics appear to be closely related, since life consists of complex organic chemistry that needs liquid water. It is natural, therefore, to look to comets as we currently understand them as possible sources of both ingredients of the recipe that makes Earth unique in the Solar System, and apparently quite rare among all the planets currently being discovered elsewhere in the Galaxy. We cannot answer those two fundamental questions here; they are the subjects of numerous investigations looking into the detailed chemistry (the hydrogen isotope deuterium) of Earth's water, other plausible extraterrestrial sources of water, and the genesis of life from relatively simple, nonliving organic chemicals.

Because of the life-essential water and organic chemicals they bring to Earth, comets have been dubbed "the givers of life."[25] Yet gigantic collisions of comets and asteroids with our planet over time have in some instances harmed living organisms to the point of extinction. As various writers have suggested, ask any dinosaur, if you can find one. It is reasonable to consider comets as both the givers and the takers of life.

The most extraordinary space mission to a comet was the European Space Agency's Rosetta spacecraft. Launched in 2004 from the Guiana Space Centre in French Guiana, the spacecraft made a ten-year journey to the periodic comet designated 67P/Churyumov-Gerasimenko and began to orbit the Sun alongside it. Maneuvering around the comet at close range, the spacecraft viewed the nucleus from every angle and took spectra and other data for more than two years. A small landing craft called Philae detached from the main spacecraft and landed on the nucleus, where it sent images and other data from the surface. The entire Rosetta mission provided a wealth of information about the composition of this comet as well as its surface structure and the changes it underwent as it warmed up on its approach to the Sun. Ices and organic molecules were found, and the details of the organic composition and the behavior of the comet's large and small surface structures over the course of its orbit have yielded extraordinarily detailed information about how some comets are structured and evolve. Rosetta terminated its surveillance of the comet in September 2015 when it made a controlled (moderately) soft landing on the nucleus and ceased to operate. The two main ices that Rosetta observed, sometimes exposed for a brief time on the surface of the nucleus, were water and carbon dioxide. Mostly, the ices evaporated from below the craggy, dark, red-colored, crusty surface, sending jets of gas that formed the comet's coma. Water and carbon dioxide ices appear to be the principal icy components of comets, but other molecules in smaller amounts are also seen, including complex organics and even the simplest of the amino acids, glycine.

Asteroids

While the comets consist largely of ice, the asteroids, which have long been regarded as distinct from comets, have a mostly unexplored association with ices of any kind. Why are they thought of as different from comets? The first few thousand asteroids discovered, starting with the discovery of Ceres in 1801, unlike the comets, were all seen to be in roughly circular orbits and largely confined to the zone between Mars and Jupiter—the asteroid belt. They tumbled along their paths around the Sun without showing any signs of comet-like activity. This picture is now obsolete.

The discovery in recent years of wholesale quantities—tens of thousands—of asteroids by patrol telescopes and spacecraft shows that a vast number of the smaller ones (a few hundred meters to a few kilometers in size) travel paths that cross the orbits of Mars, Earth, and Venus, and thereby pose the risk of a planetary collision.[26] So comets are not the only hazard in our region of the Solar System. The population of Mars- and Earth-crossing asteroids is large now—more than fifty thousand estimated—but was much larger in the early days of the Solar System. Many of them have already been swept up by the inner planets, as the crater-scarred surface of the Moon faithfully records. Chips and fragments from violent collisions between the asteroids frequently fall to Earth as meteorites, providing free samples from which we can deduce their mineralogical composition. The break-up and fall of a piece of an asteroid the size of a bus in February 2013 near the Siberian city of Chelyabinsk is a recent example of a noteworthy meteorite strike on planet Earth. Window glass shattered from the blast wave as the meteor entered the atmosphere, sending a thousand people to hospitals, and at least one building collapsed.

Some meteorites also contain complex organic matter that has been sequestered in the parent asteroids since their formation during the origin of the Solar System. These organics and some minerals found in meteorites bespeak an ancient environment on the asteroid in which liquid water was present; and some contain small percentages of water trapped in their minerals even today. But if the present-day asteroids are mostly just dry rocks, they do not really bear on the question of ices or the delivery of water to the ancient Earth. Or do they? In 2006, Henry Hsieh and David Jewitt announced their finding of "activity" associated with an object formerly considered an asteroid because of its orbit. Since then, a handful of other objects that would normally be classified as asteroids have shown surrounding hazy dust clouds, and in some cases a tail-like projection. Not all these "active asteroids" are comets in the sense of evaporating ices because other phenomena could cause them to shed a puff of dust, but the presence of ices in at least some asteroids now seems to be generally accepted, although the definitive observations are still lacking. Still, the amount of water present in asteroids' water-bearing minerals make them a reasonable potential source of some of Earth's water, acquired in the earliest history of our planet and which for a substantial fraction of Earth's history covered the surface entirely.

Trans-Neptunian Objects

Whatever their current orbital characteristics, comets originally come from the region beyond the major planets—the trans-Neptunian region, where it has never been warm. Shielded by dust and gas from the light and heat of the newly forming Sun some 4.6 billion years ago, lumps of ice dust from the solar nebula coalesced to form planetesimals. Slow collisions caused them to stick together and grow to a size between a few tens of meters to several hundred kilometers, becoming trans-Neptunian objects (TNOs), which include one major population of comets. Comets comprise a mixture of minerals that must have condensed from hot gas very near the Sun and was somehow transported to the trans-Neptunian region where it was incorporated with the local ice and dust into comets and other TNOs. We know this from the high-temperature mineral grains found in comet particles collected near comet 81P/Wild by the Stardust spacecraft in 2004 and returned to Earth for examination and analysis.

How these mineral grains were mixed in the vicinity of the hot young Sun and transported to the outer icy regions is not understood, but is the subject of numerous theoretical investigations. The issue is of broader interest than just our own Solar System because planet (and comet) formation in the dusty, icy disks around newly forming stars is a common phenomenon in our Galaxy and in millions of others. Relatively nearby in our Galaxy is the star HL Tauri, estimated to be less than 100,000 years old and about 450 light-years distant. Images of this star made with the Atacama Large Millimeter/submillimeter Array (a kind of radio telescope) in Chile clearly show several luminous broad rings of gas and cold dust surrounding the central star in which planets are presumed to be forming. The structure is called a protoplanetary disk, and while HL Tauri is the first one to be imaged in such detail, the phenomenon is common. Water, carbon monoxide, hydrogen, and iron have been detected in this disk, and many more molecules and elements are surely yet to be identified. Ices, minerals, and metals feed growing lumps of rock and ice that are bumping, shattering, and colliding in the process of making planets and comets, even as we peer vicariously through our own planet's thick and relatively protective atmosphere at the maelstrom ensuing at HL Tauri.

Figure 11.1 This image of the star HL Tauri and the surrounding disk of dust and gas was made with the Atacama Large Millimeter/submillimeter Array in Chile. The dusty ring is a region of planet formation, and the gaps in the disk are probably formed by new planets that sweep up the material as they grow. The disk is comparable in size to the Solar System. European Southern Observatory image 1436a.

Back in our own neighborhood, we find ices in comets that are nudged out of their distant orbits by gravitational forces or collisions and make their way to the planetary region of the inner Solar System, but we can also find ices on those few TNOs that are large enough to come within range of our Earth-based telescopes. Straining to get the most out of observations of these objects with the largest telescopes in the world, astronomers have found ices of water, methane, methanol, ethane, and possibly nitrogen on the roughly ten TNOs large enough to be detected. Not all of these objects have the same ice inventories, and in some cases no evidence of ice exists at all. But on some TNOs, both those with and without obvious ice, the surfaces are colored brown, yellow, or even red. Since ices are almost always colorless, these hues must tell us that something else is present, a non-ice component.

We will return later to the colored non-ice components of otherwise icy bodies, but here we note that under certain circumstances, ultraviolet sunlight or charged particles (cosmic rays) from deep in space can induce chemical changes in simple organic molecules like methane, both in the gaseous and the frozen state. The molecule can be broken into fragments by absorbing the energy from ultraviolet light or cosmic rays, and these fragments can reassemble into more complex molecules. When fragments from nitrogen, water, carbon monoxide, and other relatively simple chemicals common in planetary environments also are present, some very large new organic molecules can result. This is how yellow-brown photochemical smog forms over heavily populated cities or industrial areas on Earth, resulting in pollution that is both unsightly and dangerous to health. Saturn's moon Titan may have the worst smog problem in the Solar System—its nitrogen-methane atmosphere is exposed to both ultraviolet and cosmic rays, resulting in the whole moon being enveloped in a haze of tiny yellow-brown particles that almost entirely obscure the view of the solid surface below. These colored particles gradually settle out of the atmosphere to accumulate on the surface, and the process appears to be continuous, even though we aren't sure how methane continues to be supplied to the mostly nitrogen atmosphere of Titan. These tiny colored particles making up the photochemical smog are called "tholin," a term coined by Carl Sagan and his associate, chemist Bishun N. Khare, as we will describe later.

Ice in Unexpected Places

In 1946, when U.S. Army Signal Corps scientists beamed a radio signal toward the Moon and then detected the return echo, it was both a noteworthy and a newsworthy event. In Project Diana, quarter-second pulses of radio energy surged through a specially made antenna aimed at the Moon, traveled for 1.25 seconds at the speed of light to the dusty lunar surface, bounced off, and made the return trip to Earth where they were detected with the same antenna. After a round-trip of nearly half a million miles and 2.5 seconds, the reflected signals demonstrated the first radar contact with a celestial body. The experiment was conducted to explore the use of the Moon as a passive reflector to beam radio signals around the curvature of Earth and also to eavesdrop on foreign radio communications half the world away. Project Diana also marked the birth of radar astronomy.

Some thirty years after the first radar contact with the Moon, scientists used the largest circular radar antenna on Earth, at Arecibo, Puerto Rico, to bounce radio signals from the moons of Jupiter—the same size as our Moon, but about seventeen hundred times farther away. Radar astronomy has proven to be a valuable tool for planetary scientists probing the nature of the surfaces of planets and their moons. Radar instruments have become an important component of several interplanetary spacecraft because they are configured to produce a map of the surface over which they are passing. These radar maps show topography and surface properties not detectable in ordinary wavelengths of optical cameras. Furthermore, radar passes through clouds and hazes that are otherwise impenetrable with ordinary cameras. We have mapped the entire surface of Venus from orbit around the planet using radar on the Magellan spacecraft from 1990 to 1994, although the planet's cloudy atmosphere is opaque to visual wavelengths. The Cassini mission made radar maps of Saturn's moon Titan from 2004 to 2017, again peering through a largely opaque atmosphere.

From Earth, the Arecibo dish and other antennas are used to make radar maps of the nearby planets Mercury and Venus, as well as the Moon, although these efforts are constrained by the limited aspects of these bodies as seen from our planet, a view less complete than that from an orbit around the body itself. Both the Moon and Mercury have distinctive radar "signatures," that is, the particular way their surfaces reflect radio frequency waves, and these signatures are different for different parts of the surface. Solid lava flows reflect radio waves differently than dusty plains, for example. When first exploring the radar signatures of various regions on the surface of Mercury, scientists found that the surface near the planet's north pole reflected the radar signal much better than all the other areas of the planet. Such high radar reflectivity had previously been seen only in similar observations of regions on Mars known to be icy, and also on the icy moons of the outer planets.[27] These observations led to the astonishing conclusion that in the polar regions of the planet nearest the Sun, ice is present in the permanently shaded walls of craters and other steeply inclined topography. Mercury's surface temperature near the equator at high noon is about 425° C (800° F), while in the depth of night on the side facing away from the Sun, the temperature is about –173° C (–280° F). This distinguishes Mercury as the planet with the most extreme temperature variations of all the planets in the Solar System, and this range occurs largely because there is no atmosphere to retain and distribute the daytime heat. Also, because Mercury's rotation period is very slow, the days and nights are very long, taking 176 Earth days to fully rotate.

While Mercury's icy polar regions were first detected from Earth-based radar observations, a close-range study was made with the Messenger spacecraft, which orbited Mercury between 2011 and 2015; it was intentionally crashed into the planet on April 30, 2015, after a highly successful mission of mapping the surface and studying the magnetic field. Messenger carried a mapping radar instrument that pinpointed the craters and surrounding areas at the planet's poles where the water ice is preserved in places of permanent shadow. Since Mercury's rotational axis is perpendicular to its orbital plane, both poles have zones where the shadows are perpetual and deep, and ice can be preserved.

Another surprising place where ice has been found is on our own Moon, again in the polar regions, and for the same reasons as Mercury. The discovery of ice on the Moon came a few years later and by another method of search, but was as shocking as it was for Mercury. Generations of scientists had perpetuated the view that the Moon is entirely dry, even though the ancient natural philosophers before the invention of the telescope saw dark regions and called them "maria," or seas.[28]

The story of water on the Moon is more complex and therefore more scientifically interesting than the simple hiding of ice in permanently shadowed regions near the poles. Attempts to find ice at the lunar poles by radar scans from Earth, much more powerful than the original Project Diana transmissions, have not succeeded. However, radio waves directed at the Moon from the Clementine spacecraft in its orbit were detected with a large antenna here on Earth, and gave encouraging but uncertain evidence of ice in the polar regions. Later measurements by other orbiting spacecraft have bolstered the identification of ice, and when the Lunar Crater Observation and Sensing Spacecraft intentionally crashed into the dark shadow of a polar crater, a measurable quantity of water was blasted out of the rock and detected by the companion spacecraft. The Chandrayaan-1 spacecraft of the Indian space agency sent a probe to the lunar surface and detected minute amounts of water vapor in a tenuous atmosphere above the surface. Other spacecraft, either orbiting the Moon or crashing into it, have given mixed results, but a new analysis of glassy bits in the volcanic lunar soil samples brought back to Earth by the Apollo astronauts shows that they contain tiny but measurable amounts of water that appears to come from the Moon's interior. Thus the Moon is no longer considered a totally dry planetary body, although no geological evidence has been found that liquid water was ever present on its surface.

Ice or other extractable forms of water on the Moon hold more than academic interest. Future human colonization of the Moon, or the use of the Moon as a waystation en route to other space destinations, could benefit greatly from a source of water that does not have to be carried all the way from Earth. With this intent in mind, future robotic missions to orbit the Moon and to land on its surface will survey for water resources, especially those that might be exploited by travelers in generations yet to venture into space.

The discoveries of ice in locations as varied as the chilly poles of the hot planet closest to the Sun and the icy lumps in the dimly lit outermost regions well beyond Pluto add a new and important dimension to our understanding of the history and evolution of the Solar System. They have demonstrated that water is a ubiquitous substance. The discoveries have also found that subtle differences in water's isotopic composition and the structures in which it freezes it offers clues to its ultimate origin and transport in both the primordial and modern Solar System. In its liquid form water provides the medium in which life originated and evolved, at least on Earth, and in its frozen form it profoundly shapes the landscapes of many planetary bodies, including our own.

Ices often dominate the compositions of small bodies in the outer, coldest regions, but it is also sequestered in places protected from direct sunlight, even on Mercury and the Moon. Water ice is by far the most abundant, but carbon dioxide, sulfur dioxide, nitrogen, methane, methanol, and many others play crucial roles in the chemistry and behaviors of many of the planetary moons, comets, trans-Neptunian objects like Pluto, and many more. The full story of the ices of Pluto is yet to come.

Recalling our brief earlier acquaintance with the glacial-ice-loving annelids in the genus *Mesenchytraeus*, posing this question seems reasonable: can icy environments on other planetary bodies, and even on comets, be habitats for creatures like *Mesenchytraeus* that somehow eke out a living in seemingly inhospitable places?

12

Why Ice on Pluto Matters

Implications for Mass, Size, Discovery—and Predictions in 1976

AT THE TIME OF the 1976 detection of frozen methane on its surface, estimates of Pluto's size and its mass ranged widely, yet, as we saw in chapter 8, Kuiper had placed a heavy burden on it as the planet that had scattered the comets throughout the Solar System. In the absence of a direct measurement of its size, the calculation of Pluto's diameter starts with its brightness as measured either on photographs or with an electronic light meter of some kind. In 1976, astronomers were measuring the brightness of stars, planets, and galaxies with photoelectric photometers, which were much more sensitive than photographic emulsions and gave numerical results more precisely and more directly.

The brightness of Pluto was thus well known. But exactly how does the planet's surface reflect the sunlight that falls on it from 3 billion miles away, thus making it visible from Earth? If the surface is black like coal, then Pluto must be enormous to reflect the measured amount of light. On the other hand, if it were covered with bright, fresh snow, then the measured amount of light could come from a much smaller Pluto. Still another possibility was already being discussed shortly after Pluto's discovery in 1930. Suppose that the surface of Pluto were shiny like a ball bearing. Then the light from it that we see from Earth would be a specular reflection coming from a small spot in the center of a much larger body. For a long time, no one had way of distinguishing from among these possibilities. Pluto might be as small as an asteroid or substantially larger than Earth. (Obviously, the latter fit better with the idea that Pluto was Percival Lowell's Planet X.)

As Lowell Observatory astronomers noted as soon as Clyde Tombaugh had found the moving speck that was Pluto on his plates, even in a large telescope like the 24-inch Clark refractor, Pluto appeared starlike—it was merely a point of light without discernible size. The only measurement of Pluto's disk that seemed to have succeeded was Kuiper's in 1950 using the Palomar 200-inch reflector, which gave a diameter that lay between the extremes. He found it to be 0.46 times the diameter of Earth, or about 3,650 miles (5,825 km). This was to remain the most credible measurement of Pluto's diameter until new information became available in 1976.

The mass of Pluto was similarly problematic, and largely depended on whether the residuals in Uranus's motion, on which Lowell had relied for his calculations, were valid. This was a highly technical question that exercised the talents of several leading experts on celestial mechanics without a unanimous conclusion. Most astronomers believed that Pluto's mass was less than the seven-Earth mass Lowell had calculated, and probably much less.[1] Kuiper's measurement of Pluto's diameter was consistent with this because— assuming a reasonable composition and density for Pluto—its small size meant a correspondingly small mass, one probably somewhat (perhaps considerably) less than that of Earth. Still, a lot of reasonable but unverifiable assumptions were involved. The 1976 methane discovery paper explicitly referred to the uncertainty of these mass determinations, and concluded that the mass was indeterminate from the existing data.[2]

Before 1976, Pluto research appeared to be in an inextricable cul-de-sac. But the discovery of methane ice allowed the forging of a new trail of logical inferences about Pluto's size and mass. This same trail could have been prompted by the detection of any of the three ices that the three filters might have shown because the essential point was that the reflectivity of Pluto's surface was high, characteristic of ice or snow, rather than low, characteristic of rock. Given the planet's brightness was already measured, a high surface reflectivity implied a small diameter rather than a large diameter. In fact, Cruikshank, Pilcher, and Morrison produced a shockingly small estimate of the planet's diameter: between 1,600 to 5,300 km (1,000 to 3,300 miles). This meant that it might even be smaller than Earth's Moon. Such a small diameter, together with a plausible guess at the density of the planet on the basis of the likelihood of an icy composition, also allowed the first calculation of Pluto's mass that didn't rely on the dubious data of perturbation theory. These authors wrote,

> Since the surface of Pluto appears to be that expected for a low-temperature condensate in the outer solar nebula, it seems likely that its composition is dominated (as is that of most outer-planet satellites) by frozen volatiles. If so, the mean density is likely to be in the range 1 to 2 g cm^{-3}. This density, together with a lunar-like diameter, yields a mass a few thousandths of that of the earth, much less than would be required to perturb the motions of Uranus or Neptune measurably. If this train of logic is basically correct, it appears that Tombaugh's discovery of Pluto in 1930 was the result of the comprehensiveness of the search rather than predictions from planetary dynamics.[3]

Though previously widely suspected, the conclusion now seemed inescapable. It might still be impolite or impolitic to say it in view of the stupendous mathematical effort that Lowell, Elizabeth Williams, and the other computers put forth, to say nothing of the painstaking search that Tombaugh carried out, but the discovery of Pluto was a happy accident. It was like Columbus's "discovery" of America, or Roentgen's discovery of X-rays. This, by the way, does nothing to detract from the honor due to Tombaugh. Though as a member of the Lowell Observatory staff who knew where his paycheck was coming from, he couldn't say it too openly, he quickly developed doubts about the precision of Lowell's prediction. Whether the planet he sought was Planet X or not, Tombaugh knew the predictions clearly would be of no real help in finding it, and its discovery would be forthcoming only from a thorough, labor-intensive search. When Pluto registered its pale image on his plates, it was much fainter than Lowell (whose calculations seemed to indicate the existence of a giant planet) had imagined, and the fact that it did not escape Tombaugh's meshes is a tribute to the latter's extraordinary alertness and care through a long, fatiguing, and discouraging search. After the discovery, Tombaugh came to be the object of increased attention and publicity as the world's only living discoverer of a planet (and only the third to do so in recorded history), as well as the only American to make such a discovery. But through it all, his genuine native humility and mild manners persisted. Though he accomplished a great deal of other important work in astronomy, and always regretted that he did not receive more recognition for his studies of Mars—the planet in which he was always the most interested—eventually he reconciled himself to his fate. As a legend in his own time, he accepted the fact that wherever he went, what people wanted to hear was the familiar story of his search and discovery. He must have told it thousands of times—always with the same disarming twinkle in his eye and a sprinkling of atrocious puns. ("When I was on the farm, that was my hay day." "I was the world's first Plutocrat.") Under the circumstances, no one wanted either to offend him or diminish the importance of "his" planet. Apart from a small (but steadily growing) coterie of planetary specialists interested in the outer Solar System, the tacit demotion of Pluto to a smaller-than-expected size and mass hardly rocked the scientific world.

In the mid- to late 1970s a few other astronomers were using infrared-sensing instruments capable of detecting Pluto through various color filters. Larry A. Lebofsky at the University of Arizona teamed up with astronomer and master instrument builder George H. Rieke to make some filter observations of Pluto similar to those of Cruikshank, Pilcher, and Morrison, but in their 1979 paper they noted some discrepancies between the two sets of results.[4] In the end, they too came down on the side of the methane identification, but it was hardly a firm corroboration. That, however, came soon afterward, when Gerry Neugebauer's group at the Caltech not only used a highly sensitive instrument but a larger telescope, the Palomar 200-inch, which collected 50 percent more light than the 4-m telescope on Kitt Peak. B. T. Soifer was the lead author on a 1980 paper from the Neugebauer group that reported the spectrum of Pluto measured at twenty-five wavelengths across the near infrared.[5] It confirmed the presence of methane in detail. That same year, Cruikshank and Peter Silvaggio published an improved spectrum of Pluto at twelve wavelengths in the region of one of the methane bands.[6] The initial identification made in 1976 was now secure.

The Methane Ice Surface Paradigm

Methane was the first frozen substance other than water ice found in the Solar System. Carbon dioxide frost was suspected in the polar caps of Mars, but not yet reliably identified, while the discovery of frozen sulfur dioxide at the foot of the volcanoes of Io was still a few years away.[7] Frozen methane was not much of a surprise to those who had bothered to think seriously about the chemistry of the outer Solar System, in part because of John Lewis's propositions and in part because methane gas had been a known constituent of the atmospheres of the giant planets since the early 1930s, when Rupert Wildt identified methane (and ammonia) as the culprits responsible for the absorption bands V. M. Slipher of the Lowell Observatory had detected in the 1910s in his spectrograms. However, until the identification of frozen methane was actually made, the possibility remained that—as often happens in science—something unexpected would turn up. As we shall see, this would indeed happen after a later discovery.

What is so special about methane that it was on the short list for chemicals in the outer Solar System? If we start with a list of the most abundant atoms in the Universe and then ask which simple combinations of them might be found in nature, methane, which combines hydrogen, the most abundant element in the Universe, with carbon, the most versatile, rises to the top. Hydrogen, consisting of one proton and one electron, is the simplest atom, and also the lightest. (It rises to the top literally: its property of lightness was exploited in the early days of ballooning to fill the gas bags of balloons, and later dirigibles; unfortunately, hydrogen's avidity to combine with oxygen, with the release of energy, makes it highly combustible. After the Hindenburg disaster in 1937, hydrogen fell out of favor for this purpose; nowadays helium, the second-lightest element, is preferred.)

Hydrogen, helium, and a trace of lithium were formed in the earliest epoch following the big bang. The first stars were massive and formed through the gravitational collapse of huge clouds of gas. Within these stars, hydrogen was burned in thermonuclear reactions to create helium and the heavier elements that are produced only in stars. When the stars used up their fuel, they exploded and threw their enriched entrails across intergalactic space, and new stars formed from the mixture of hydrogen, helium, and these recycled elements. Atomic physicists have worked out the details and calculated a ranked list of the most abundant elements in the current Universe, nearly 14 billion years after the big bang: hydrogen, helium, oxygen, carbon, neon, iron, nitrogen, and so on. Hydrogen still vastly outnumbers the rest of the elements combined by a ratio roughly nine to one. Some of the elements we've listed—the noble gases, including helium and neon—despite their abundance, don't readily combine with other atoms to make molecules, while iron doesn't readily combine with hydrogen. Thus, by a process of elimination, it follows that the simplest molecules we would expect to find would involve combinations of hydrogen with oxygen, carbon, and nitrogen, namely, water, methane, and ammonia. Especially under conditions of relatively low temperature, these are the prime candidates for the most basic components of the planets. Note that the exact composition of planets is dependent on the planet's mass and distance from the Sun. The two are not independent of each other, since larger planets tend to form at greater distances from the Sun, where the hydrogen and helium of the solar nebula contributed to the rapid growth of a few dominant planetesimals that became the giant planets. Closer to the Sun, the hydrogen and helium were largely lost because of the effects of the solar wind, and only the small rocky terrestrial planets could form. It follows that while hydrogen and helium are abundant in the atmospheres of the giant planets, where they are held by strong gravitational fields, they are disproportionally rare in the atmospheres of the smaller planets Earth, Venus, and Mars. The fact that Earth's atmosphere is 78 percent nitrogen, 21 percent oxygen (plus about 1 percent argon, which is also heavy but as a noble gas doesn't make molecules), and only a trace of water vapor demonstrates this effect.

Finding methane on Pluto bolstered the notion that a straightforward consideration of the most basic chemistry, as Lewis had done, would lead to correct predictions of the composition of outer–Solar System bodies. Furthermore, it helped to set the stage for future discoveries as planetary scientists continued to extend their reach.

Along with Pluto, the other body in the outer Solar System that seemed important to characterize was Triton, Neptune's largest satellite. Within weeks of the discovery of Neptune itself, Triton was discovered by William Lassell, a brewer and amateur astronomer in Liverpool who had just built an equatorially mounted reflector with a 24-inch mirror and was eager to demonstrate its power. (Lassell also thought he could see a ring, but that, at least, was an illusion produced by malalignment of his optics. Also, the name "Triton" was not chosen by Lassell; it was introduced by Camille Flammarion only in 1880, in his book *Astronomie Populaire*.)

Triton appears only as a feeble, starlike point of light in even a large telescope, and as with Pluto, its diameter could not be measured by the techniques available in the nineteenth and early twentieth centuries. Even skilled observers such as E. E. Barnard, who could measure with a micrometer such small objects as the asteroids Ceres and Vesta, did not attempt to measure Triton. With its diameter unknown, its mass, and hence its density, were also indeterminate. Thus Triton attracted little scientific interest except for its unusual property of orbiting around Neptune in a retrograde direction, that is, opposite to the direction of Neptune's axial rotation. This peculiar dynamical characteristic was especially difficult to understand because the orbit is otherwise normal: it is circular (as opposed to highly elliptical), and lies in Neptune's equatorial plane. This set of facts could be most parsimoniously explained by supposing that Triton had been captured gravitationally by Neptune sometime in the distant past. However, a more dramatic scenario was suggested by Cambridge mathematician Raymond A. Lyttleton, who toyed with the idea that Pluto had also been involved. At one time, both Triton and Pluto had been satellites of Neptune; however, they approached too near to one another, resulting in Triton's orbit being reversed while Pluto was ejected from the Neptunian system altogether and sent into a planetary orbit. Lyttleton's idea seemed attractive in some ways—among other things, it seemed to explain why at the closest point of its highly eccentric orbit to the Sun, Pluto veered just inside the orbit of Neptune. Kuiper himself championed the idea for a while. However, now that we know about the existence of the Kuiper Belt, the Lyttleton-Kuiper theory has fallen out of favor. According to the latest thinking, Triton was an ordinary Kuiper Belt object that somehow fell into Neptune's gravitational grip. This may have happened by what specialists in planetary dynamics call a three-body process; in the present instance, the capture of Triton by Neptune would have been facilitated if Triton originally had a companion of similar mass—a binary object, as we now know Pluto and its moon to be.[8] That putative companion has gone on to its own destiny, but Triton remains in its retrograde orbit around

Neptune. In the process, an enormous amount of gravitational and tidal energy must have been dissipated within Triton, first as it was captured, then as it was forced into its present circular orbit of low inclination. The details of how this happened remain perplexing even today.

Though faint, Triton was still about 40 percent brighter than Pluto during the mid-1970s,[9] and this small difference was enough to make it an easier object for the telescopes and infrared detectors of the day. Because of its greater brightness, observational work on Triton pulled somewhat ahead of Pluto, though obviously the two objects shared a lot in common. Thus at least some details of the Triton work are relevant here.

Coauthor Cruikshank's first observations of Triton were made in 1978 with an infrared technique that was a significant advance over that used for the 1976 three-filter observations of Pluto. Though the Kitt Peak 4-m telescope and the Blue Toad were used as before, a different kind of filter, with a higher spectral resolution than had been possible for fainter Pluto, could be used for Triton. The new data consisted of fourteen spectral points in the H- and K-band spectral region (1.44–2.52 µm), which better defined the characteristic signature of the methane molecule. Even with the 4-m telescope and a hopped-up Blue Toad, averaging the data from four nights (in April 1978) was necessary to get a rough spectrum showing methane. The data revealed one absorption band that was not well defined, but clearly present; a corresponding methane band lies at 1.7 µm, but is not accessible in that noisy region of the data. The question at the time was how to interpret the methane bands—were they from gas or ice? Partly because of a paucity of available methane ice spectra and knowledge of the methane ice properties, both possibilities had to be explored.

To make headway, some theoretical perspective was needed to interpret the meager lab data then available on methane. It turned out that Silvaggio at NASA Ames Research Center at Moffett Field, nestled in California's Silicon Valley, had already been working on band-model calculations for methane, which could in principle model the absorption bands seen in the new data. Silvaggio's calculations showed that the methane on Triton could possibly be largely gas, thus giving that moon an atmosphere. The pressure of a methane atmosphere at Triton's surface appeared to be broadly consistent with the vapor pressure of a gas above solid methane at a plausible surface temperature.[10] However, it later turned out that Triton's surface is more reflective (higher albedo) than originally supposed, so that the temperature is much lower and the quantity of methane gas in the atmosphere is far less than previously thought. Though the details still had to be worked out, the methane identification was clear. Moreover, the presence of one phase implies the presence of the other.[11]

Triton and the Discovery of Solid Methane

The next breakthrough took a few years and involved more observations, this time with two telescopes on Mauna Kea and a combination of data from both. Using spectra made with improved equipment on the 3.8-m United Kingdom Infrared Telescope (UKIRT) and the 3-m NASA Infrared Telescope Facility (IRTF), together with the earlier Kitt Peak observations, Cruikshank and Jay Apt could produce a complete spectrum from 0.8 to 2.6 μm much higher in quality than had been possible before. This broad wavelength coverage showed six individual methane bands. In addition, observations made on several nights demonstrated clearly that the band strengths vary as Triton's aspect changes with its axial rotation (period 5.8 days, the same as its orbital period around Neptune). Apt, a planetary astronomer at the Jet Propulsion Laboratory (and who later flew on four missions as a space shuttle astronaut), was working at the time on calculating random band models of methane gas and was invited to help interpret the Triton data. Combining knowledge of the bands, their variable strengths, and the 1979 Silvaggio paper, Apt computed a spectrum of methane gas over the full spectral range covered in the new data, and showed that the Triton bands did not match the theory. Two graduate students, Robert H. Brown and Roger Clark, at the University of Hawaii also helped interpret the data by making reliable spectra of methane frost in the laboratory, which they furnished to Apt for his study in progress. One was a spectrum of pure methane, another one of methane ice mixed with carbon black (which reduced the strength of the major bands as well as the albedo of the sample).

These six prominent methane bands and the variability of their strength with Triton's orbital (rotational) period indicated that they must arise from methane ice on the moon's surface and not from gas in its atmosphere. A methane atmosphere could not plausibly account for it. Triton spectra of much higher resolution might have distinguished solid from gaseous methane by fine details in the methane spectrum (as well as a shift in wavelength between solid and gas), but such high-resolution spectra were not then available. The widths and positions of the methane band alone showed the absorption could, in principle, be from either of the two phases, but an analysis of the full data set, in light of the points made above, led to the conclusion that the methane seen in the Triton spectrum existed as frost or ice on the surface, and further that its disposition varied with its location on that body. This result was announced at the conference Natural Satellites held at Cornell University, July 5–9, 1983, and then published in a 1984 paper.[12]

For the first time, Triton, hitherto an inscrutable telescopic speck of light, began to emerge as a world. Further, it could be related to another world, Pluto, by virtue of its inventory of methane ice. Since the perspective from Earth in the early 1980s included Triton's south polar region, the variability in the methane band strength clearly must have resulted from an irregular or patchy distribution in the satellite's equatorial latitudes.[13]

The spectrum of Triton compiled from data taken with three large telescopes finally clarified the situation with methane by showing six separate absorption bands of that molecule in the form of ice on the surface, while closer scrutiny unexpectedly showed something new. An additional absorption band centered near 2.16 μm was visible. It could not be attributed to methane, either as gas or solid. After searching through the chemical spectroscopic literature and discussing their finding with several colleagues, Cruikshank and Apt arrived at a probable identification, writing that the band "is not found in the methane spectrum and apparently represents a different chemical constituent. This important band is tentatively identified as a collision-induced (2–0) molecular nitrogen absorption."[14]

Nitrogen on Triton

The new spectral band at 2.16 μm was defined by only about three data points, and might have been overlooked or dismissed as a mere artifact of statistical noise. However, it had persisted in all the data sets acquired with different telescopes and spectrometers over two or three years. This meant that it had to be representative of some component on Triton that was not methane. As the ideas for the interpretation of the newly detected absorption band were explored, hydrocarbons other than methane were being considered as plausible candidates and eliminated. At some point, Tobias Owen, a one-time graduate student of Kuiper and then a professor of astronomy at the State University of New York, Stony Brook, mentioned that nitrogen has a spectral band at about the right wavelength, but that it had only been seen in a very large pathlength in nitrogen. As a homonuclear molecule (a molecule made of two or more of the same atoms), nitrogen has no traditional (permitted) energy transitions leading to an observable absorption spectrum, but when nitrogen is compressed to high density, perturbations of one molecule by its neighbor molecules can induce what is called a temporary dipole to the molecule. This is the basis for transitions that can result in an observable absorption spectrum. Laboratory work for gaseous nitrogen published years earlier left little doubt that the anomalous band in Triton's spectrum was indeed due to nitrogen, the same molecule that makes up 78 percent of Earth's atmosphere.

The detection of nitrogen on Triton through the absorption band at 2.16 μm was a breakthrough discovery that has had a broad effect on the study of icy bodies in the outer Solar System. Although the nitrogen story was still being developed, it was presented alongside the methane work at the Cornell satellite conference in July 1983.

The identification of the new Triton band with molecular nitrogen was soon accepted, but the questions of the abundance and phase (gas, liquid, solid) remained unanswered. In collaboration with Brown and Clark, Cruikshank and Apt considered what a dense gaseous nitrogen atmosphere would imply for Triton. To make the calculations, knowledge of Triton's size (to calculate its gravity) and its surface albedo (for the temperature calculation) was essential. Neither was yet known with certainty, but they did know that a nitrogen atmosphere could not exist on Triton thick enough to produce the absorption band seen in the spectrum. The gas possibility being eliminated, only solid or liquid nitrogen remained in contention.

When molecules form ices or frosts, they are normally highly reflective because the crystals are small. Thus the pathlength taken by the incident sunlight through the individual crystals before it emerges from the surface is also very small, on the order of a millimeter or less. If the nitrogen on Triton were in the form of frost or ice, a pathlength this short would be insufficient to give the absorption band observed in the spectrum. This line of reasoning seemed to rule out frozen nitrogen, leaving liquid nitrogen as the only remaining possibility.

Liquid nitrogen is a familiar substance in laboratories and is even used in some industries. It is familiar to elementary science students because it is widely used in demonstrations of extreme cold, producing almost instantaneous freezing of bananas or hotdogs (and by implication, fingers) that are then dramatically shattered to smithereens with a whack of a hammer. (Note: liquid nitrogen is usually harmless enough, though obviously it should not be allowed to contact exposed skin.) The possibility that the Triton spectral band was caused by liquid nitrogen could easily be tested in the lab by passing light through a container of it before admitting it into a spectrometer. A pathlength of several centimeters through the liquid was sufficient to produce a spectral band just like that seen on Triton. Since gas and frost had apparently been eliminated, the likelihood seemed to increase that part of Triton's surface might be covered with a liquid nitrogen sea. The nitrogen could not remain liquid at a temperature less than about 63° K (−210° C), since at that point it would freeze. However, the surface temperature of Triton wasn't known at the time. The nitrogen-sea possibility was published in 1984,[15] in a companion paper to the 1984 methane ice paper, though it had already been mooted at the Cornell satellite meeting in 1983.

The ink was hardly dry on this paper before the liquid-ocean concept was challenged by Caltech's David Stevenson and his graduate student Jonathan Lunine (now at Cornell University).[16] They based their argument on Triton's plausible surface temperature and the distribution of solar heating around the body of the satellite, arguing that the temperature could not be high enough to keep the nitrogen liquid. While the logic of their argument seemed unassailable, the problem of the necessary pathlength remained. In the end, the resolution came from an important property of nitrogen ice that was unappreciated at the time. Nitrogen ice was later shown to transform spontaneously from the small crystals in a frost to much larger, transparent crystals several centimeters in size.[17] The pathlength criterion could thus be met by sunlight passing through several centimeters of solid nitrogen condensed on Triton's surface as a polycrystalline mass. Thus we could have our cake and eat it too: the Triton band that seemed to indicate liquid nitrogen on the surface could also be produced by solid nitrogen on the surface. Given the low temperatures expected on Triton, solid nitrogen became the preferred option.

As these ground-based observations and discussions about Triton were taking place, a far-flung emissary—the Voyager 2 spacecraft, launched in August 1977—having passed by Jupiter, Saturn, and Uranus, was approaching Neptune and Triton for a closer look. It would make its closest approach to both the planet and Triton in August 1989 after an epic interplanetary voyage lasting twelve years. Passing within about 40,000 km of Triton's surface, Voyager sent back images of the icy expanses of the satellite's south polar region, determined the composition of the atmosphere, and discovered plumes of material jetting up 8 km from vents on the surface before being blown into horizontal streaks by Triton's winds. Voyager 2 clinched the case for solid nitrogen on the surface and further determined that the amount of nitrogen in the atmosphere was compatible with a surface temperature of about 38° K (−235° C),[18] well below 63° K (−210° C), where liquid nitrogen freezes into a solid.

Voyager 2 revealed Triton as an extraordinary planetary body with a unique geological record and a nitrogen atmosphere. But it made only a brief flyby of Neptune and Triton before heading onward out of the Solar System. No other spacecraft has studied Neptune and Triton at close range. After Voyager 2's triumph, astronomers had to return to what they could make out from Earth, and the absorption spectrum of Triton in even the largest telescopes remained both unique and intriguing. But the quest was still on. Astronomers continued to improve their instrumentation and within a few years would use it to obtain new data sets that would add critically important pieces to the Triton puzzle—and set the table for the next course in outer Solar System exploration, Pluto.

13

New Discoveries and a New Paradigm

Mauna Kea, the White Mountain of the Hawaiian Islands, stands as the tallest of all the peaks in the Hawaiian Archipelago, and indeed in all the world. Rising in a massive dome from the ocean floor 15,000 feet to sea level, it reaches nearly 14,000 feet more to join the other lesser volcanoes forming the Island of Hawaii. Rafted along the great Pacific Plate, Mauna Kea and its neighboring mountain masses glide slowly along the sea floor toward the northwest a few inches each year.

This tremendous mountain wears a white crown of snow through the winter months and offers visitors to its summit a barren, lunar-like landscape startlingly different from sea-level tropical Hawaii. The magnificent desolation of the cinder cones and lava flows of the upper elevations preserves the archaeological record of a millennium of Stone-Age Hawaiian activities and a quarter of a million years of glaciers with concurrently active volcanoes. Nowhere else in Polynesia are the extremes of volcanism, tropical glaciation, and ancient human habitation combined in a high-altitude alpine setting as they are on Mauna Kea. And nowhere else in the entire world do these vestiges of the ancient activities of Man and Nature set the stage for human exploration of the Universe through the gigantic eyes of the observatories built in the rarified air of Mauna Kea, where astronomers are seeking to gain a clearer view of what lies beyond the Earth.[1]

The Telescopes on Mauna Kea

Standing 13,796 ft above sea level on the Island of Hawaii, Mauna Kea, whose potential for Solar System astronomy was first discovered by Kuiper, has more than realized his dreams. Because of the pivotal role the telescopes eventually sited there have played in discoveries regarding Pluto and its moon Charon, as well as Triton, the mountain itself and its astronomical qualities deserve special note.

One can well understand why this extraordinary mountain should have been venerated by native Hawaiians as the home of two snow goddesses, Poliahu and Waiau (aka Waiaie), who are in perpetual conflict with Pele, the goddess of volcanoes—a fitting personification of the forces of fire and ice that have shaped it. This geologically young volcanic edifice pierces the moist, maritime air near sea level whose year-around warmth draws millions of tourists to Hawaii. As the tallest mountain in the Pacific Ocean, it reaches high into the cold, dry air that flows smoothly across the sea and over its summit. While the warm, humid air at sea level is soothing to beachgoers (and even to astronomers, when off duty), it is an impediment to the acquisition of data from the beyond. Moist air is particularly problematic for astronomers working in the infrared region of the spectrum, who require a very dry, cold environment for their telescopes and detectors to perform at optimal capacity, which is the first prerequisite for discoveries. After Kuiper first saw the potential of high and dry Mauna Kea for infrared observations in the early 1960s, his intuition was confirmed by a set of measurements of humidity, temperature, cloud cover, and air stability taken over two years. The results of this survey, together with experience using the University of Hawaii's 88-inch (2.2-m) telescope, which began operations in mid-1970, led Canada, France, the United Kingdom, and NASA to build even larger telescopes on the mountain. The largest of these, with a mirror 3.8 m in diameter, was built on behalf of the Science and Engineering Research Council of the United Kingdom, primarily for observations in the infrared. The telescope, UKIRT, was dedicated and put into regular operations in 1979.[2]

About ten years later, twin telescopes with mirrors 10 m in diameter were constructed for Caltech and the University of California, with funding from the W. M. Keck Foundation. Japan built its national telescope, the Subaru 8.2-m telescope, on Mauna Kea; it began operations in 1999. The Gemini 8.1-m telescope became operational in 2000.

A large telescope is an impressive thing. UKIRT, for example, weighs 6.5 tons and is an optical and mechanical wonder of precision balance, movement, and optical power. But a telescope does not stand alone. It is teamed with accessory instruments placed in its focal plane that contain the sensors that, after passage through special filters and other analyzing devices, turn the light and heat radiation captured by the giant mirrors into data with scientific significance. It is the job of astronomers to take these data and make sense of them. This is the science and the art of observational astronomy. If the whole enterprise is successful, the data, of no particular use in and of itself, is converted into something of value—new knowledge.

Triton and Pluto Come into Clear View

UKIRT was suitably equipped with sensing instruments to analyze the light and heat radiation collected by the big mirror. One of the most important instruments on any telescope is the spectrometer, and efforts are endless to improve and expand its performance. Over the years, each of the frequent improvements in the capabilities of the infrared spectrometers attached to the UKIRT meant that increasingly fainter galaxies, stars, and planets were brought into view. By 1991, both Triton and Pluto were within reach using an infrared spectrometer known as CGS4, the Cooled Grating Spectrometer 4, and a collaboration by planetary scientists in the United States and France, obtained the first high-quality spectra. Because Triton is a little brighter than Pluto, it was the first target. Then Pluto came into view, but the definitive observations of both objects were obtained in May 1992 and published in 1993. When the Triton spectra were obtained, the presence of frozen methane and nitrogen had already been established, both from ground-based observations and by the Voyager 2 spacecraft during its Neptune encounter in August 1989 (as described in the previous chapter). In addition to Voyager 2's detection of a thin atmosphere of nitrogen surrounding Triton, images of its surface clearly showed a frozen landscape, and a youthful one, to judge from the near absence of impact craters. Only a few million years—a mere blink of the eye in geologic time—had apparently elapsed since it was last resurfaced by flowing liquid nitrogen or some other process still not fully understood.

The new UKIRT spectra not only revealed the methane and nitrogen on Triton with great clarity, but showed the molecular signatures of carbon monoxide, carbon dioxide, and water as well. The molecular bands of these three newly revealed ices had been invisible in the earlier data, which were lower quality than UKIRT now provided. The three new molecules are all oxygen-bearing, and together previously observed nitrogen and methane, the four most abundant molecule-forming atoms in the Universe were now accounted for in Triton.

The first glance at the new Triton spectrum, even in its preliminary form and directly out of the computer at the telescope, was one of those rare "aha!" moments that provide the spice of a scientific career. The spectral fingerprints of carbon monoxide, carbon dioxide, and frozen water nearly jumped off the computer monitor That the result was new and significant was immediately apparent. It was as though a blurred image of an object on the distant horizon had come into focus with the mere turn of the knob on the telescope. What had been confusing and indistinct was suddenly seen with a sharpness that revealed all its dimensions and peculiarities. We were standing on a mountain top, both figuratively and literally, seeing a planetary body with a clarity not previously afforded to anyone.

The Ices of Triton and Pluto: A New Perspective

The groundwork laid in the earlier analysis of Triton's nitrogen band informed the interpretation of the new spectra of both Triton and Pluto, but especially Pluto. Since the Pluto nitrogen band is so weak compared to that of Triton, its certain identification would have been considerably delayed if the Triton band had not first been identified and studied for several years prior to the 1992 acquisition of the Pluto spectrum.[3]

The new high-quality spectra of both Triton and Pluto brought an additional important, and unexpected, discovery. While the telescopic observations were being made at UKIRT, laboratory studies of nitrogen and methane by planetary scientist Bernard Schmitt and his students in France had shown that in a frozen mixture of the two molecules, the spectral bands of methane are shifted in wavelength from those appearing in the spectrum of pure methane. The methane bands on both Pluto and Triton demonstrate this wavelength shift. This showed that the methane ice on these bodies is not pure. Instead, methane is dissolved in the nitrogen ice.

It is strange to talk of one ice dissolved in another. However, consider frozen carbonated water. This is an example of carbon dioxide dissolved in frozen water. Similarly, when nitrogen is a liquid, methane readily dissolves in it. When the mixture is then chilled below the freezing point, large crystals of the combined molecules begin to form. Solid nitrogen by nature forms large, transparent crystals several centimeters in size in the slow cooling process of annealing. Sunlight falling on the surfaces of Triton and Pluto passes through these large crystals before being scattered back to space where we see the reflected sunlight on Earth or from a spacecraft. The intrinsically weak spectral bands of nitrogen can only be seen if there is a long pathlength through these large crystals, and when the sunlight traverses several centimeters of transparent solid nitrogen, the small amount of methane incorporated into the crystal becomes evident. As an effect of being dissolved in nitrogen, the methane bands are slightly shifted to shorter wavelengths, just as Schmitt discovered.

With these unexpected properties of nitrogen and methane mixtures having been identified in the lab, and with the evidence for the mixture now appearing on both Triton and Pluto, a new picture of the surfaces of these bodies came into view. Rather than vast expanses of methane frost or ice covering the surfaces, as previously supposed, solid nitrogen is clearly the main component on the surfaces of both Triton and Pluto; methane is much less abundant, and all (or at least most) of the methane is dissolved in large nitrogen crystals. At once it became easy to conjure the image of a dimly lit, but beautiful and glistening, brittle polycrystalline surface crunching under the tread of a Pluto roving vehicle, or even under the boot of some future Plutonaut.

While Pluto was known to have a thin atmosphere from the stellar occultation observations of 1988, which will be discussed later, the detection of nitrogen ice on the surface immediately suggested nitrogen gas as the principal component of the atmosphere. This is because of nitrogen's high degree of volatility. Though it shares this characteristic with carbon monoxide, nitrogen is much more abundant—thus it must be the most abundant atmospheric gas.

The UKIRT spectra of 1992 yielded yet another important piece of information. In the case of Triton, the shape of the nitrogen spectral band was seen so clearly that even the temperature of the frozen nitrogen could be determined. Frozen nitrogen can exist with two different crystalline structures, the cubic phase called alpha nitrogen, and the hexagonal phase called beta nitrogen. Each phase has a preferred temperature range: alpha nitrogen exists below 35.6° K while beta nitrogen exists above that temperature. The two phases have different spectral-band shapes. For both Triton and Pluto, the shape corresponds to the higher-temperature beta nitrogen. The exact band shape in beta nitrogen is temperature-sensitive, so that a good measurement of the band shape leads to a refinement of the temperature of the frozen nitrogen surface. From the Triton measurements made in 1992, Caltech graduate student Kimberly Tryka determined that the satellite's surface temperature is 38° ± 1 K, in good agreement with an independent determination made with the Voyager 2 spacecraft in 1989. Later measurements of Pluto's temperature from the beta-nitrogen band shape by Tryka gave a slightly higher temperature of 40° ± 2 K.[4] As of 2016, all the nitrogen found on both Triton and Pluto is in the beta phase.

So far we have said very little about carbon dioxide, carbon monoxide, and water ice.[5] All three were found on Triton in the UKIRT data; on Pluto only carbon monoxide was detected. In principle, both carbon oxides should occur together, even though they freeze at different temperatures and evaporate at widely different rates. Carbon monoxide contributes to the atmospheres of both Triton and Pluto (see chapter 14), while carbon dioxide is a stiff and inert ice at Triton's temperature. It remains a curiously unsolved mystery why carbon monoxide occurs on both bodies but carbon dioxide has so far been found only on Triton.

Water ice is fundamentally important as a major constituent of most outer Solar System bodies, although it may be inconspicuous or even invisible when it is covered partly or totally by other ices. At the exceedingly low temperatures in the outer Solar System, water ice is the hard, unyielding bedrock that shapes the large-scale topographic features—essentially, as we discussed in chapter 11, it has all the characteristics of a rock. The more volatile and mobile ices of nitrogen, methane, and other minor constituents lie about as layered blankets that hide the underlying frozen water. In the case of Triton, although some water ice is exposed, the other ices apparently form a thick crust masking old impact craters that are thought to be preserved in the water ice mantle below. Pluto's surface is different, as revealed by the New Horizons spacecraft in 2015; we will hold off discussing that until chapter 19. While water ice was not reliably detected from Earth-based telescopic studies, New Horizons found it exposed in patches distributed around Pluto— bits of bedrock amid thick layers of frozen nitrogen and other components of the visible surface.

The discoveries of mixed ices on Pluto and Triton have provided a basis of interpretation for observations of other denizens of the outer Solar System, notably Kuiper Belt Objects (KBOs) and beyond. Because of their great distance and generally smaller sizes, KBOs are very difficult to observe from Earth, and to get any information about their compositions is a demanding task, even for the largest telescopes. The detection of the "stealth" molecule nitrogen, for example, is particularly difficult because of its weak signature in the spectrum. Methane, on the other hand, as a strongly absorbing molecule, is much more readily detected. Methane and nitrogen are mutually soluble in one another, and the presence of methane is a clue to the presence of nitrogen. We noted above that when methane is dissolved in nitrogen, its spectral bands are shifted to shorter wavelengths than for pure methane. Thus, in some cases, notably distant objects such as the curious (50000) Quaoar,[6] shifted methane bands are detected, and the presence of substantial amounts of nitrogen can be inferred, even if not seen directly.[7]

As we shall see later, the mixture of nitrogen, methane, and carbon monoxide ices on Pluto, in the approximate ratio one hundred parts nitrogen to one part each of methane and carbon monoxide, as found at the telescope has also been used to guide laboratory studies of these ices that simulate the conditions on Pluto's surface to explore the chemistry of the planet. In addition, these telescopic observations helped guide the design and development of instruments to be placed aboard the New Horizons spacecraft launched toward Pluto in 2006, the main subject of the latter part of this book. Thus, the confluence of capabilities and the persistence of astronomers working in the superb conditions on Mauna Kea, which led to the discovery of the ices on Pluto and Triton, truly opened vistas on these two large icy bodies at the outer margin of the planetary system. Thereby a new paradigm of chemistry in the coldest realms of the Solar System was first sketched, to which details continue to be added.

As we saw in chapter 11, under some conditions methane can be trapped in a cage of water molecules as clathrate. Methane clathrate occurs on Earth on the continental shelves of the ocean, where methane produced by microscopic organisms forms a frozen mass with sea water. When pieces of this methane-rich ice are dredged up from the sea floor onto the deck of a ship, they can be set afire as the gas comes out of the ice and burns in the air. Astonishingly, at oceanic depths of half a mile or more, worms an inch or two long called *Hesiocaeca methanicola* are found living in the methane clathrate, grazing on the bacteria in the ice.[8] The existence of this extremophile and other hydrocarbon-loving organisms piques the imagination about methane-rich environments elsewhere in the Solar System, which, as we have seen, include Titan, Triton, Pluto, and some KBOs, as the concept of habitability is broadened through new discoveries in our planetary system and in extrasolar planetary systems.

14

Ices Predict an Atmosphere

Pluto Must Have an Atmosphere

EVEN BEFORE THE PRESENCE of any ice beyond Earth and Mars had been detected from astronomical observations, in his brief 1950 paper about Pluto's diameter Kuiper foresaw the importance of not only frozen water, but ices composed of other molecules on both the surfaces and atmospheres of planetary bodies much more distant from the Sun. In the context of this chapter, we again quote part of the 1950 paper:

> Such a body must have some atmosphere, though most of its original atmosphere will have frozen out owing to the low equilibrium temperature for Pluto, 40°–50° K. Both the atmosphere and the condensation products will prevent the albedo from being extremely low: nearly all snows are white, the crystals being small (H_2O, CO_2, CH_4, etc.). On the other hand, the albedo need not be that of freshly fallen snow, 0.7–0.8, because of several effects, including grit deposited by comets and meteors, will darken snows over the ages.[1]

A Plutonian atmosphere was indeed predictable, although some years and some new observational approaches were required to find it. But Kuiper could not have foreseen that an utterly unique observatory bearing his name would be the platform from which the Plutonian atmosphere would reveal itself forty-two years later. Although Kuiper didn't know from direct observations that ices cover Pluto's surface, he correctly surmised that ices of various kinds would dominate the surfaces of small bodies in the outer Solar System. To see the connection between ices and atmosphere, some remarks on their mutual properties will help set the stage for the discoveries to come more than forty years after Kuiper's prediction.

Ices have the interesting property that under specific conditions they can evaporate directly from a solid to a gas. We are familiar with this property of frozen water because we see that ice cubes in the refrigerator freezer compartment slowly get smaller and smaller over a few weeks to a few months. The same phenomenon occurs on high, cold mountain tops, where snow and ice evaporate directly into the atmosphere without first melting. Dry ice (frozen carbon dioxide) also evaporates, or sublimates, from its solid condition to a gas when left in the open air at room temperature. Frozen water and carbon dioxide sublimate at ordinary temperatures in the open air where the air pressure is about 1 bar (14.7 psi, or 980 g/cm^2 at sea level). The rate of sublimation depends on the exact temperature and pressure of the environment—the lower the pressure and the higher the temperature, the faster it occurs.

In a closed container originally under a vacuum, a block of sublimating ice produces its own atmosphere. This can also happen in nature on a planet, creating an atmosphere above an icy surface. But in the vacuum of space, the gas released from sublimating ices can only create a long-lasting atmosphere if the planetary body is sufficiently cold and has a gravity field strong enough to prevent the gas molecules from escaping into space. The gravitational field of a planet depends both on its size and its mass—for example, smaller, asteroid-size bodies lose out. Even if ices in their interiors sublimate and release gas at the surface, the gas molecules quickly escape to space, unless there is some especially cold area on which they can freeze again. On the other hand, if there is a continuous supply of gas from sublimating ices on a planetary body and the rate of the loss to space is roughly balanced by the production rate, an atmosphere can exist for long periods. The temperature is also important—a warm gas escapes more quickly than a cold one, and

that is where the situation in a real planetary atmosphere becomes complicated, as we will see below.

The relevance of these ice and gas properties to Pluto (and Charon)
emerges from the discoveries of the various species of ice found from work
with Earth-based telescopes starting in 1976, when methane ice was first
suspected on Pluto. Though the details of our understanding of Pluto's icy
surface have changed with later discoveries, the identification of methane
ice enabled reasonable estimates of how much and what kind of atmosphere
would result from ice sublimation. Pluto's highly elliptical orbit causes its distance from the Sun to vary from 29.5 AU (perihelion) to 49.2 AU (aphelion)
over 125 years (half the orbital period), with an expected difference in the
surface temperature of about 13° C, depending on Pluto's albedo.[2]

Because methane sublimation depends strongly on the temperature of
the ice, predicting that the atmospheric abundance varies over a Pluto year
was possible. Early estimates suggested that the changing methane gas abundance would be different by a factor of about 250 from aphelion to perihelion, with an average value for the surface pressure of about 1.5 millibar.[3] This
is only 0.0015 of Earth's atmospheric surface pressure, and even this estimate
turned out to be about one hundred times too large, as we shall see in chapter 19. The pressure of a gaseous atmosphere depends on the vapor pressure
of the molecule, where vapor pressure is the pressure exerted by a gas that is
in equilibrium with its solid (or liquid) phases at a given temperature. The
vapor pressure is widely different for different molecular ices and liquids
and is also very strongly dependent on temperature. To calculate the pressure of a gas on a planet's surface, the vapor pressure as determined in the
laboratory must be known, as well as the temperature of the environment.
In the familiar examples noted above, if water ice and dry ice are put in an
ordinary refrigerator freezer, the dry ice will evaporate faster than the water
ice because its vapor pressure is significantly higher.[4]

The speculation that Pluto has an atmosphere of methane ran into difficulty because the methane ice should have evaporated and departed into
space long ago, according to considerations of the vapor pressure and Pluto's
presumed temperature and mass. That is, unless a heavier, but undetected
additional gas such as neon was in the atmosphere to help hold back the
escape of the methane. Neon doesn't combine with itself or with molecules,

and it is undetectable by spectroscopy. However, in sufficient quantity, the slightly heavier neon could help hold Pluto's methane atmosphere over an astronomical time scale. Even before the methane ice was discovered, researchers Michael Hart in the United States and G. S. Golitsyn in the Soviet Union had thought about a possible Pluto atmosphere, and both noted the need for heavy gases, such as neon. Low temperature, which is virtually guaranteed by Pluto's great distance from the Sun, and a relatively high mass, which was only estimated at that time, plus heavy gases were seen as necessary for the presence of an atmosphere.[5] The same basic conclusion was reached by University of Texas astronomer and planetary scientist Laurence Trafton a few years after the discovery and confirmation of methane ice, and he calculated more details on how a Pluto atmosphere would escape. The small mass of Pluto and the volatility of methane at Pluto's expected temperature were hard to reconcile with the presence of methane ice on the surface over the age of the Solar System. As Trafton calculated, "A heavier gas must exist, otherwise the CH_4 frost would have sublimated away long ago because of solar heat and rapid blowoff of gaseous CH_4."[6] More information was clearly needed.

Once the discovery of a volatile ice like methane was recognized by planetary scientists, the race was on to obtain observational evidence that Pluto's atmosphere actually existed. The logical way to pursue this quest is by spectroscopy, the tried-and-true method to detect gases both in the laboratory and in nature. But the spectroscopic approach was hampered by several factors, not least of which was the lack of adequate corresponding data for methane both as a gas and as an ice. The region of the spectrum available for observing the planets had largely been ignored by chemists, who were the main source for basic spectroscopic data about molecules of all kinds, and this paucity of laboratory spectral data for many materials of planetary relevance dogged planetary scientists for years (and still does, to a degree).[7] Even with good data obtained at the telescope, the interpretation of those data depends on knowing details of the behavior of molecules under a wide variety of conditions of temperature and pressure as measured in the laboratory. The faintness of Pluto, even in large telescopes, impeded obtaining high-quality spectral data at wavelengths most suitable for detecting atmospheric gases of any composition, certainly methane. This problem of Pluto's probable atmosphere clearly needed a new approach.

The Kuiper Airborne Observatory Chases Shadows

Observing Pluto with a telescope gazing out of a large hole in the fuselage of a four-engine jet airplane cruising at 45,000 ft seems defiant of basic aeronautics, not to mention a huge departure from normal astronomical observations in dome-topped buildings on cold mountaintops. In a conventional observatory, a chilly astronomer huddled at the business end of a telescope, coaxing the light from a faint, remote planet to yield its secrets, evokes a vaguely romantic image. Perhaps no less beguiling, and certainly exotic, is the image of a small team of astronomers huddled over their recording instruments while flying high in the stratosphere all night as their telescope, peering out of the side of the plane, bobs and weaves to keep a planet in view as they cruise along at some 400 mph. This is airborne astronomy.[8]

Airborne astronomy came of age in the 1960s, as it became clear that the cold, dry stratosphere would enable infrared observations that could not be done from the ground, even on the highest and driest mountain tops. The earliest effective airborne astronomy was done with a small telescope on a Learjet, and later with telescopes mounted in a modified four-engine Convair 990. With those successes, in 1974 NASA put into operation the Kuiper Airborne Observatory.[9] The KAO was a Lockheed C-141A Starlifter cargo airplane modified to carry a 36-inch telescope and instruments to find and track astronomical objects for several hours while flying at stratospheric altitude. This enormously productive facility, which NASA operated for twenty-one years until 1995, evoked romantic images within veterans of many observing flights. When the KAO was decommissioned, astronomer Edwin Erickson celebrated the plane's illustrious history in a poem:[10]

Aye, tear her gleaming ensign down!
  Long has it flown on high,

And many an eye has danced to see
That symbol in the sky.
Far down, bound by gravity,
The earthlings watched it soar;
The meteor of the stratosphere
Shall sweep the clouds no more.

The KAO made a critical set of observations related to Pluto on June 9, 1988, when it viewed in fine detail Pluto's shadow as it passed in front of a star. Every star in the nighttime sky sheds some light on Earth, and if an object gets in the way of that light, a faint shadow of that object is cast. As Pluto moves along in its orbital path, it occasionally covers, or occults, a star as seen by an observer. These shadows are not only faint, they are small, and their size depends on the size of Pluto.

The importance of observing occultations of stars by small Solar System bodies and planetary rings had been recognized several years earlier. Astronomer Robert L. Millis wrote in 1988 that "occultations of stars by planets, satellites, planetary ring systems, asteroids, and comets provide valuable opportunities to probe the Solar System in ways otherwise impossible from the surface of Earth. For example, one can precisely measure the size and shape of objects which are much too small to be resolved directly, accurately map the structure and transparency of ring systems, and *detect the faintest trace of an atmosphere.*"[11]

To observe a stellar occultation, the event must be predicted sufficiently early to plan the observations and to define precisely the path of the shadow across Earth. This can only be accomplished if the exact positions of the star and the planet are known, and determining these positions with the needed precision can be very difficult. Prediction occultations are often subject to last-minute revisions in the shadow's path by a hundred miles or more. These late revisions of the occultation path give an aircraft an advantage over telescopes on the ground; aircraft can accommodate changes in position and enable observations over the oceans that cover so much of Earth's surface. The 1988 occultation would never have been captured if it had been left to mere chance. Fortunately, by then Pluto's orbit was well-enough known and star catalogs had improved sufficiently to allow accurate predictions of

Pluto's path across the background stars. Astronomers calculated the occultation's zone of visibility, which was found to consist of a wide swath across the South Pacific, Australia, and New Zealand.

The 1988 occultation of a star designated P8 was observed both from the KAO over the ocean just south of Pago Pago, American Samoa, as well as with ground-based telescopes in New Zealand and Australia. It provided the first clear and direct evidence of Pluto's atmosphere. The ground-based team was led by Robert Millis and Larry Wasserman of the Lowell Observatory. They used seven telescopes, including several mobile telescopes placed in advance in remote locations along the occultation track in Toowoomba, Australia, as well as on Tasmania and New Zealand.

The observers on the ground and the team in the KAO watched as the edge of the planet approached the star; the starlight dimmed gradually over the course of a few seconds before Pluto completely covered it, the part of the occultation called ingress. If Pluto had no atmosphere, the star would have winked out almost instantaneously, but the gradual dimming showed that an atmosphere must be present. Furthermore, when the brightness of the star going behind Pluto's edge was plotted on a graph, an irregularity—a kink—was found in the curve. Then, a little over a minute later as the starlight began to emerge on the opposite side of the planet (egress), the same pattern was seen—a gradual return of the starlight with a kink at just the right time to correspond to that seen on ingress. The whole event lasted only eighty seconds, the time it took Pluto to move in its orbit by an amount equal to its own diameter.

Sitting at the observer's console on the KAO that night, the team leader James Elliot of MIT and his colleague Ted Dunham knew at once that they had seen the evidence for Pluto's atmosphere. The nature of the kink in the light curve was puzzling, however, and became the object of a much subsequent analysis by them and others. The kink could have been caused by a haze in the atmosphere surrounding Pluto that would induce an irregularity in the pattern of starlight absorption. Or it could have been caused by an abrupt change in temperature of Pluto's atmosphere at some moderately high altitude. Though a few other stellar occultations by Pluto were observed in the following years, this ambiguity in the interpretation of the nature of the planet's atmosphere remained until the New Horizons spacecraft flyby was accomplished on July 14, 2015. We return to this later.

The observations of the 1988 occultation from telescopes in New Zealand and Australia not only confirmed the KAO results, but they also provided information to help in the interpretation of the details of Pluto's atmosphere. Different locations on the ground watched Pluto take slightly different paths in front of P8, thereby sampling different swaths of the atmosphere.[12] Taken together, all the data from the occultation helped refine the diameter of Pluto, but uncertainties remained because of the effects of the atmosphere on the path of the starlight close to the planet's surface. Later occultations left ambiguities in the exact diameter of Pluto until the New Horizons flyby cleared up the effects of the atmosphere and from several concurrent techniques gave the planet's diameter as 1,187.6 km, with an uncertainty of about 1.8 km.[13]

In terms of the study of Pluto's atmosphere, May 1992 was a watershed month, and Mauna Kea Observatory in Hawaii was the locus of the discovery activity. On the nights of May 25 and 26, MIT graduate student Leslie Young and her advisor James Elliot were observing Pluto with a high-resolution spectrometer on NASA's Infrared Telescope Facility (IRTF). They were, of course, recording the spectrum of Pluto as seen through Earth's own atmosphere. While the air above Mauna Kea is thin and dry, the spectrum of our atmosphere seen in high resolution is full of both strong and weak absorption lines of the various gases that make up the air we breathe. The main gases are nitrogen and oxygen, but argon, carbon dioxide, and even methane also are present. Earth's small amount of methane gas comes from a variety of sources, primarily decaying organic matter, and while we usually don't care much about it, concern is growing about increasing levels in the atmosphere contributing to global warming.[14] In the data they took those nights, Young and Elliot could discern methane spectral lines on Pluto from those in our own atmosphere. They found for the first time the clear signature of methane gas in Pluto's atmosphere. Since frozen methane cannot have individual absorption lines displayed by the same molecule in a gaseous state, the detection of these lines clearly pointed to methane gas and not the ice found some years earlier.[15]

But, while the discovery of methane seemed entirely consistent with the earlier finding of methane ice, it did not solve the problem of the kink in the occultation light curve.

On the two nights immediately following Young and Elliot's detection of methane in the atmosphere of Pluto, May 27 and 28, Tobias Owen, Catherine de Bergh, and Thomas Geballe observed Pluto with a different instrument on a different telescope, the 3.8-m United Kingdom Infrared Telescope (UKIRT), also on Mauna Kea and just a few hundred meters away from the IRTF.

Those data showed near-infrared absorption bands not previously seen except on Neptune's satellite Triton. These bands new to Pluto were clearly identified as from two more surface ices, nitrogen and carbon monoxide. Both nitrogen and carbon monoxide are much more volatile than methane, and their occurrence as ices on the surface pointed immediately to their presence as significant components of the atmosphere. In fact, nitrogen was quickly understood to be the dominant gas (about 99 percent) in Pluto's atmosphere while methane and carbon dioxide are relatively minor constituents.[16] However, both methane and carbon dioxide are critical to the chemistry of both Pluto's atmosphere and its surface. For complicated reasons, the detection of nitrogen unseated neon as a contender for a starring role in Pluto's atmosphere.

The issues around the long-term escape of the atmosphere were reexamined considering the newly discovered molecules, but the same basic problem persisted. A discovery by the New Horizons spacecraft in 2015 helped explain the longevity of the atmosphere, as we shall see in chapter 19.

Structure and Chemistry in Pluto's Atmosphere

Just as the presence of volatile ices on Pluto's surface begets an atmosphere surrounding the planet, the mix of nitrogen, methane, and carbon monoxide gases leads to complex chemistry and a pallet of new molecules, fragments of molecules, and aerosol particles that form a haze of photochemical smog. And the tiny smog particles can aggregate into larger particles that fall to the ground and accumulate. Thus, both the atmosphere itself and the solid particles that precipitate to the surface tell of the complex chemistry that occurs in Pluto's surrounding gaseous envelope.

Not only do chemical reactions in the atmosphere produce new components, but they exert a controlling influence on the temperature from the surface to the top of the atmosphere, where it slowly (or quickly) bleeds into space. The variation of temperature and pressure with altitude above Pluto's solid surface defines the temperature-pressure structure of the atmosphere, which is controlled in part by molecules or fragments of molecules that form and combine at different altitudes. Other factors influencing the atmosphere include the light from the Sun, from deep ultraviolet to infrared, and the flux of charged particles in the solar wind and from deep space in the form of cosmic rays. Another important factor is rising and falling air currents, which actively mix the atmosphere. In the highest regions of an atmosphere molecules are ionized by sunlight and cosmic rays and form an extensive region called the ionosphere. Earth's ionosphere lies about 75 to 1,000 km above the surface.

The patterns of molecule fragmentation by the absorption of sunlight and the effect of charged atomic particles, and the tendency of molecules to recombine or attach to other fragments, are also major factors in establishing the temperature of the gas at different altitudes. Under circumstances where aerosols can form by the aggregation of accumulated molecules into tiny solid particles, a haze layer can result, as it does on Earth over some urban areas from auto or industrial exhaust. Volcanoes are natural sources of aerosols, as often occurs in Hawaii when Kilauea or Mauna Loa are active. These aerosols not only affect the temperature of the atmosphere, in some circumstances they can slowly rain on the surface. Hydrocarbon aerosols in the atmospheres of Titan and Triton precipitate onto the surfaces, affecting the surface chemistry.

Before the discovery of nitrogen and carbon monoxide on Pluto, little progress could be made in understanding the chemistry of the atmosphere. By the time those two molecules were found on Pluto, nitrogen was known on Triton and carbon monoxide was presumed present; the confirmation of carbon monoxide came at about the same time as its discovery on Pluto. The occurrence of these two molecules and methane on both Triton and Pluto not only enabled detailed calculations of atmospheric chemistry, but also stimulated comparisons between the two. Additional information about Triton and its space environment (specifically its relationship to Neptune's trapped radiation belts) had been obtained in 1989 by Voyager 2 as it flew through the Neptune system.

The best way to understand the details of a planetary atmosphere is to observe it up close, collect samples, and watch it change with seasons and other variable phenomena such as the activity on the Sun. This has worked well for Earth's atmosphere, where for centuries we have been collecting local temperature data, and in recent decades observing it on a global scale from orbiting satellites. This monitoring has helped us predict both short-term weather and long-term climate.

A great deal of attention has been given to the atmospheres of Mars and Venus, in part because they are relatively close to Earth and in part because they help illuminate several basic properties of atmospheres, some of which are applicable to our own. Satellites orbiting Mars collect data that we hope will give a clearer picture of conditions on ancient Mars, when life may have originated.[17] Beyond the local planets, scientists have sent probes into the atmospheres of Jupiter,[18] as well as Saturn's moon Titan,[19] to learn about processes that occur in atmospheres with a very different chemistry to that on Earth, Mars, and Venus.

Lacking any direct measurements within or even close to the atmosphere of Pluto, a few specialists used the known surface-ice components, the detection of methane in the atmosphere, and the stellar occultation light curve from 1988 to formulate models of the composition, temperature, and pressure within Pluto's atmosphere. The basic outcome of these models, which usually differed between scientists, was that new molecules are formed from the known methane, nitrogen, and carbon monoxide, some of which aggregate to form aerosol particles that precipitate to Pluto's surface. If undisturbed by geological forces, the aerosol particles can accumulate and form layers a few centimeters thick. These new molecules include other simple hydrocarbons such as ethane (C_2H_6), ethylene (C_2H_4), and acetylene (C_2H_2) that are made from fragments of dissociated methane molecules. Nitrogen atoms are also incorporated into new precipitating molecules such as cyanoacetylene (HC_3N) and hydrogen cyanide (HCN).[20] A real planetary surface, particularly on a planet with an atmosphere, will be subjected to significant changes over long stretches of time, with sublimation, precipitation, and perhaps even winds, all of which will affect the aerosols and the solid particles they deposit on the surface.

The calculations of the properties of Pluto's atmosphere, even before much was actually known about it, showed that the atmosphere extended very high, to an altitude where it can escape into space and will interact with the charged atomic particles in the solar wind.

Calculating a physical and chemical model of a planetary atmosphere requires knowing or estimating several quantities, as noted above, but the interactions of molecules and their ions is not well understood. These fundamental properties are usually measured, often with considerable difficulty, in the laboratory, and some measurements only give incomplete results that are later revised by better experiments. Because the number of possible reactions is in the hundreds, atmosphere models are often revised as new laboratory results become available. Some calculations depend on particular molecules in the atmosphere. Although carbon monoxide, for example, is not abundant, it has a large effect on the temperature structure of Pluto's atmosphere because of its cooling properties, and even small amounts lead to large effects on the composition of the ionosphere.[21]

Models and predictions about the properties of Pluto's atmosphere were occasionally updated as laboratory data on critical chemical reactions were improved and as additional stellar occultations were observed, but the real structural and compositional details had to wait for the New Horizons flyby of 2015.

Before New Horizons, we were left with a picture of Pluto's atmosphere that featured its very large size, extending out to the moon Charon, and rapid escape to space, although the details and rate of escape were subject to interpretation and modeling.[22] Here is the reasoning behind that expectation.

The Sun is not only a source of light and heat, but it also emits fragments of atoms in the form of electrons, protons, and alpha particles (the nuclei of helium atoms, consisting of two protons and two neutrons) that stream into space along paths shaped by the Sun's own magnetic field. This solar wind of electrically charged particles is ejected from especially active regions on the Sun's surface and from the gas above the surface, usually in bursts or flares that are often directed toward the planets at velocities averaging about 600 km/s. How the solar wind interacts with the planets depends on many factors, including the distance of the planet from the Sun and especially the presence of a strong or a weak planetary magnetic field. The magnetic field of Earth is sufficiently strong to shield the surface and lower atmosphere from most of the normal solar wind, but occasional huge bursts of particles can cause atmospheric effects that appear as brilliant aurorae. In some cases, the atmosphere at near ground level is so heavily electrically charged by a solar flare that power transmission lines are overloaded and fail. Some charged particles are trapped by Earth's magnetic field and form zones known as the Van Allen belts that shield the planet from all but the most energetic emissions from the Sun. Jupiter, Saturn, Uranus, and Neptune all have strong magnetic fields, and thus have Van Allen belts composed of trapped solar-wind atomic particles. For a planetary body with no magnetic field, or a very weak one, the solar wind interaction is also complex, but different.

The solar wind interaction also depends on the nature of an atmosphere. All the planetary atmospheres are slowly escaping into space because of the motions of the individual atoms and molecules at the highest elevations. Light atoms and molecules, such as hydrogen and helium, escape most readily, and the heaver molecules like nitrogen and methane escape more slowly.[23] The weak gravitational field of a small planet like Pluto also facilitates atmospheric escape, while Jupiter's great mass retards escaping molecules and atoms.

As nitrogen and methane "boil off" from the top of Pluto's atmosphere, they form a halo around the planet, and the molecules eventually acquire an electrical charge through ionization by ultraviolet sunlight. With their acquired electrical charge they begin to move in response to the solar wind, extending far from the planet's surface in a tenuous cloud of ions. Calculations of the expected escape rate for molecules of methane and nitrogen, their ionization by ultraviolet light, and their response to the solar wind initially suggested that the ion cloud around Pluto would be extremely large, extending even beyond Charon's orbit. This view changed with the flyby of New Horizons, as we shall see later.

15

Surprise!

A Moon Is Found

DISCOVERIES IN SCIENCE often come about in strange and unexpected ways. Though they may be the sought-after result of a thoughtful and focused search, they also can happen serendipitously as a byproduct of a study conducted for an entirely different purpose. Sometimes the discovery is made from acquiring new and better data, sometimes it is made when someone comes along with a fresh look at long-existing data and turns up what had simply been overlooked or missed.

So it is with the moons of the planets. When Pluto was discovered in 1930, 26 satellites of the planets were known, including Earth's Moon. Now there are 178.[1] Some of the moons lie close to their planets, such as Phobos and Deimos of Mars and Amalthea of Jupiter. Others are far away, such as Jupiter's eighth and ninth satellites, Pasiphae and Sinope, and Saturn's Phoebe. Those that lie close move in circular orbits that tend to hug the equatorial planes of their primaries; while the orbits of the far distant moons tend to be highly inclined and eccentric, and apt to move in retrograde. Each time a new planet has been discovered, its vicinity has been carefully scrutinized for possible moons. As noted earlier, the discoverer of Uranus, William Herschel, discovered two of Uranus's moons (and recorded a third, though his observation was forgotten before it was seen again), and William Lassell discovered Neptune's large satellite, Triton, soon after the planet itself had been sighted in Berlin by Johann Gottfried Galle. As soon as Pluto was discovered, astronomers were on the lookout for possible satellites, though the task must have seemed rather hopeless given that Pluto itself was faint and starlike even in the larger telescopes of the day. Though E. E. Barnard found the last visually discovered planetary satellite in 1892 (Amalthea, the fifth satellite of Jupiter), photographic searches with large telescopes have turned up many faint satellites in the years since. Notably, Charles Dillon Perrine added two distant moons of Jupiter, Himalia and Elara, with the Lick Observatory's 36-inch Crossley telescope in 1904–5. P. J. Melotte found yet another, Pasiphae, with the 30-inch reflector at the Royal Greenwich Observatory in 1906, and Seth B. Nicholson, at the time a graduate student at Lick who was following Pasiphae with the Crossley telescope to secure the observations he needed to complete his PhD thesis, came across Sinope in 1914. (Jupiter's satellites at the time were simply referred to as Jupiter V, VI, VII, VIII, and IX, etc. Their current names were adopted only in 1975.) In the 1930s, Nicholson was searching for additional faint satellites of Jupiter far from the planet with the 100-inch telescope at Mount Wilson. In fact, he was doing so at the end of 1930 as Jupiter shared the same field with Pluto. Had Clyde Tombaugh not found Pluto when he did, Nicholson would certainly have found it while searching for faint Jovian satellites. (He would find three more Jovian satellites, Lysithea and Carme in 1938 and Ananke in 1951.)

Though visual searches for Pluto satellites were undoubtedly undertaken, little was said about them because they were unsuccessful. Kuiper, who used the McDonald Observatory's 82-inch reflector to make the photographic discoveries of Uranus's then-innermost moon Miranda in 1948 and Neptune's far-flung satellite Nereid in 1949, surely scrutinized Pluto visually for moons with that telescope. He found nothing. He also searched with long-exposure photographs of the region farther from the planet in 1950, and a similar photographic search was undertaken by Milton Humason with the 200-inch telescope that same year; the results were again negative.[2] As successive estimates of Pluto's size gave smaller values, at some point continuing the search for a satellite that was unlikely to exist probably seemed useless. In 1950, the smallest Solar System body known to have a satellite was Mars (which has two, the smallest of which, Deimos, is of irregular shape and measures only 15 km in its maximum dimension).

Positional astronomy, the art and science of precise measurements of celestial body positions and motions, can be tedious because it depends on careful observations taken over many years before it answers any specific question. But despite being completely lacking in glamor, and its practitioners usually being unknown and unsung, it is among the most important areas of astronomy in a practical sense. The most immediate and tangible result for a long time was the determination of time and the correction of the clocks of the world by watching specific stars cross the local meridian. Time is not merely useful, it is a commodity. The Allegheny Observatory in Pittsburgh sold accurate time to the railroad system in the Eastern United States for many years such that "every railroad clock and watch, and the movement of every train [was] regulated from a single standard—that of the clock in the observatory."[3] Other nineteenth-century observatories did the same. In addition, the night watchmen of positional astronomy kept tabs on the clockwork-like motions of the planets relative to the background stars. During such routine work, positional astronomer Giuseppe Piazzi discovered Ceres and the large proper motion of the "flying star," 61 Cygni. Also, the precisely determined positions of Uranus made at the Paris, Cambridge, and Greenwich observatories underlaid the realization that Uranus's motion was not as it should be, even after the mutual gravitational effects of Jupiter and Saturn had been subtracted. It was this realization that led Le Verrier and Adams to Neptune. Still later, the same kind of data—pushed a bit farther than sustainable—inspired Lowell's quixotic quest for Planet X.

Many of the national observatories around the world, beginning with the Paris Observatory in 1667 and the Royal Greenwich Observatory in 1675, were established specifically for such "official" purposes as keeping time and assisting with longitude determinations in support of navigation. As Astronomer Royal, George Biddell Airy accepted that the Royal Greenwich Observatory's work should not stray from its mandated routines. Thus, instead of undertaking the planet search at Greenwich, he referred it to Cambridge, the results of which we discussed earlier. To the end of his life, he did not see how he could have done otherwise.

Similarly, the U.S. Naval Observatory (USNO) in Washington, DC—known also as the National Observatory for the first few years—was founded in 1825 with a similar mission when President John Quincy Adams, just before leaving office, signed the bill for its creation. It was established as the Depot of Charts and Instruments by order of the Secretary of the Navy in 1830, and from these humble beginnings gradually rose to a first-rate institution in possession of the instruments needed to accomplish its missions of timekeeping and positional astronomy. Its 26-inch Clark refracting telescope, built in 1873 and at the time the largest refracting telescope in the world, was originally sited in the malarial flats known as Foggy Bottom, a place where the observing conditions generally lived up to its rather inauspicious name; not until 1893 was it moved to its present location in Northwest Washington. While the observatory was still at Foggy Bottom, Asaph Hall used the great refractor to make its most famous discovery, the two tiny satellites of Mars, Phobos and Deimos. As long as poets (and astronomers) had spoken, as Tennyson did, of the "poles of moonless Mars," the planet's mass was only roughly known from an analysis of the mutual perturbations of Mars on the other planets, especially Earth and Venus. Once Hall had discovered the satellites, however, a reliable mass could be worked out from Kepler's third law. This meant an improvement in the detailed understanding of the other planets' orbital characteristics, including Earth's, and the computation of better astronomical data of the kind needed by astronomers, surveyors, and navigators that were produced by the human computers that sweated over their desks at the NAO and who produced the volumes of the *American Ephemeris and Nautical Almanac* (published for the years between 1855 to 1980). The NAO had, when founded in 1849, been a separate institution located in Cambridge, Massachusetts (where Simon Newcomb, who later became its superintendent, was first employed), but it moved to Washington, DC, in 1866. Eventually, after the move to Northwest Washington, they were co-located, with the NAO becoming a branch of the USNO.

Eventually, the Clark telescope was no longer regarded as a state-of-the-art instrument for positional astronomy work, and the USNO began to consider a suitable replacement. In 1934 a new telescope was installed, the pioneering 40-inch (1-m) Ritchey-Chrétien telescope, which was the first large telescope incorporating the now-classic design of American master optician George Willis Ritchey and French astronomer Henri Chrétien. By the 1940s, USNO astronomer John S. Hall began using the telescope to carry out breakthrough research on interstellar polarization, but by then the skies over Washington,

DC, were hopelessly affected by light pollution and a search for a darker site had begun. Hall himself selected a site just west of Flagstaff, and the 40-inch Ritchey-Chrétien was moved there in 1955. The next year Hall and another astronomer, Kai Strand, won funding for a 61-inch (1.55-m) "Astrometric reflector," and with this instrument, the Flagstaff Station, under the direction of Arthur A. Hoag, became a leading center for positional astronomy just as the demand for such data was becoming more urgent with the launch of the first artificial satellites and space probes.[4]

The fact that the elements of the orbits of the planets, which are used to calculate the ephemerides of their positions needed both by astronomers and by spacecraft-mission planners, are crucially based on precisely determined astrometric positions is not always sufficiently appreciated. This is no less so today than it was in the eighteenth and nineteenth centuries when such positions—painstakingly reduced by hand—made up the critical database whose acquisition and maintenance was the first priority of astronomers. In the case of Pluto, by the late 1970s several hundred precise astrometric positions were available, dating back to the prediscovery photographic plates of 1914, covering an arc of Pluto's orbit of seventy-five years—quite a third of its orbital period. (Note that in addition to plate-position measurement errors, further uncertainties had accrued because of systematic errors in star catalogs.) As with any orbit that has not been observed over an entire orbital period, the best predictions were available only along the observed arc. Predictions of Pluto were noticeably uncertain only a decade out, and became at least theoretically interesting to planners of future spacecraft missions (as we shall discuss later), higher-accuracy ephemerides were needed to reduce spacecraft pointing and arrival-time uncertainties.

James W. Christy and Some New Photographs of Pluto

To produce such higher-accuracy ephemerides, new positional observations and updated orbital parameters were needed. Among the astronomers involved in such work was James W. Christy, a native of Milwaukee, Wisconsin, who came to Tucson to study in the University of Arizona's astronomy program and who in 1962 was recruited to the astrometric program at the USNO Flagstaff Station. (Though he completed his bachelor of science degree, he never would finish his PhD, something he later admitted was a source of some inferiority feelings toward those that had.)[5] When Hoag went to Lowell Observatory in 1965, Gerald Kron of Lick Observatory, a pioneer of high-precision photometry using photomultiplier tubes, took over as director. Eventually, Christy had a mishap with one of Kron's prized photomultiplier tubes and was transferred from Flagstaff, which he loved, to the Astrophysics Division at the USNO in Washington, DC, where he assumed a new role under the direction of Robert S. Harrington measuring the positions of objects on photographic plates with an automatic measuring machine.

Visual—and later photographic—measurements of the positions of Uranus and Neptune and their satellites had been included in the USNO's astrometry program since Simon Newcomb's time, but for some reason Pluto was only added in the mid-1970s, at Christy's urging. For some reason, the Flagstaff Station astronomers in charge of obtaining the plates did not immediately take up the request. Only after a year and a half of delay, in April and May 1978—the months when, from Christy's own experience, the observing conditions at Flagstaff were the best for the purpose—did they finally obtain his plates of Pluto, taken with the 61-inch astrometric telescope. However, to the astronomers who had taken them, the images of Pluto did not appear very good, so the plates were marked "defective." They were then sent off to Washington, where they ended up in Harrington's desk, awaiting further attention.

In June, Christy was about to leave on a much-needed vacation. He hadn't taken one for a year, and just then he and his wife, Charlene, were also in the process of moving from an apartment into a small house of their own. Christy had cleared everything on his desk, and was looking for something relatively simple to do before he left. He approached Harrington, who pulled the Pluto plates from his desk. Christy's first reaction was to feel a little miffed: what was Harrington doing with "his" plates? In fact, however, so much time had passed since he had made his request that he had forgotten about them. He didn't get around to measuring them until the next morning.

The plates showed eighteen images of Pluto in all, and when he began to examine them under a special microscope designed to determine Pluto's precise position on each plate, Christy could see why the plates had been marked "defective." The images were distorted, as if the telescope's guiding had been faulty during the exposures or the seeing had been particularly bad. Under the circumstances—and eager to get out of the office to start his vacation—Christy would hardly have been blamed had he decided not to dither over such unpromising materials. However, as he continued to examine the plates, Christy realized that it was only Pluto's images which were distorted: the stars were just as they were supposed to be, perfectly round. The astronomers and technicians in Flagstaff had done their work flawlessly. Whatever was affecting Pluto's image had to do with Pluto itself.

Christy's key observation was that on the April 13 plate, Pluto's image appeared elongated to the south; however, on May 12 it was elongated to the north. Christy began to run through every possibility he could think of to explain this strange state of affairs. First he thought perhaps Pluto's image appeared elongated because it had fallen, by chance, on background field stars. That could be ruled out by checking the *Palomar Sky Atlas*. No stars bright enough to produce a blended image with Pluto were near it on those dates. Next he began to wonder whether Pluto had a gigantic mountain sticking up from its surface, or whether there was a volcano erupting on the surface (but a volcano whose eruption had lasted a whole month?). Finally he realized, with appropriate astonishment, that he must be seeing a satellite.

To make sure he had not overlooked anything, Christy invited a colleague, Jerry Josties, into the darkened room. Josties agreed that Christy had found something interesting, and they tried to brainstorm all possible alternatives: a flaw on the plate, a speck of dust. At last they came back to the satellite notion, and Christy tracked down Harrington. According to Christy, when he told Harrington that he thought his plates had revealed a satellite of Pluto, Harrington responded, "You're crazy, Jim."

It's Not a Flaw, It's Pluto's Moon

After Harrington examined the evidence and went through the same mental checklist of what might be affecting the images, he also agreed that the only possible conclusion was that Pluto has a moon, barely but repeatedly visible as an elongation of the planet's image on several photographic plates taken over a considerable span of time.[6] Instead of taking the next few days off, Christy returned the next morning and looked up plates of Pluto going back decades in the USNO's archive. A set from June 1970 not only showed the elongation, it showed that it appeared to shuttle from one side of Pluto to the other in about three days. Since this agreed with the 6.387-day rotation period that had been discovered in the 1950s from the variability of Pluto's brightness, Christy suspected that Pluto's rotation period might match the satellite's precise orbital period—in other words, that the satellite was in tidal lock with Pluto. According to MIT astronomer Richard P. Binzel, then working as a summer student at the USNO, Christy asked Harrington to independently calculate, using the 6.387-day period, on which side of Pluto the bump would be on the plates taken eight years earlier, in June 1970, while Christy did the same. Their results agreed with each other and with the images on the old plates. A couple hours later, Christy found two more plates, taken by Otto Franz at Lowell Observatory in 1965. They also showed the elongation in the right place. Ironically, Franz had noticed the elongated images at the time, but had assumed they were a result of poor observing conditions. Similarly, Henri Camichel, the Pic du Midi Observatory astronomer who invented the disk-meter and was a highly skilled planet photographer, told coauthor Sheehan that distorted images of Pluto often were noted in the plates taken at Pic du Midi. Again, no one had paid any attention. As with the first exoplanet discoveries ("hot Jupiters"), no one had expected a large satellite so close to Pluto.

The 6.387-day orbital period was not only useful in tracking down the satellite's presence on earlier plates, but it also led to the first reasonably definitive calculation of Pluto's mass. Combining the assumption that this was the orbital period of the satellite around Pluto with the assumption of a circular orbit and an estimated distance between Pluto and the satellite of 17,000 km, Harrington and Christy used Kepler's third law to find that Pluto's mass is 0.0017 times that of Earth.[7] The idea that Pluto was Lowell's Planet X, already weakened and brought to its knees by the discovery of methane ice on Pluto two years earlier, was finally finished once and for all by this result. Its mass was tiny. Combining Harrington and Christy's mass with the 1976 estimate of the diameter of Pluto at 3,000 km,[8] they derived a bulk density of

~0.7 g/cm³. (The value of the diameter was later found to be somewhat too large; a smaller Pluto raises the bulk density, now known to be 1.86 g/cm³.)

A Name for the New Moon

The presumptive satellite, given preliminary designation 1978 P1, needed a proper name. Giving it one was Christy's prerogative—subject, of course, to ratification by the official body charged with such matters, the International Astronomical Union (IAU), which had (and has) strict rules about such matters and stipulated, among other things, that it be chosen from Greek or Roman mythology. His first choice had been "Oz," but he realized this was not likely to be accepted since it wasn't Greek. Next, while at dinner, he told his wife, Charlene, he wanted to name it for her, "Char-on" (by analogy to the neutron, proton, and electron). This was awkward at best, and not likely to survive IAU scrutiny. Meanwhile, "Persephone," the name of the wife of Pluto, was being floated by other astronomers. Two days after he discovered the object, Christy had moved into the new house with his wife. The power hadn't yet been turned on, and they turned into bed, exhausted. Nevertheless, Christy's mind was still working on the problem of the satellite's name. "About 1 a.m.," he told coauthor Sheehan, "I woke up, and thought I have to tell Charlene I can't name it after you after all. I don't know whether I would have done this in the daytime, but I had a flashlight, and went downstairs and started looking through boxes. I found a dictionary, and went through the mythology names, and at once came across Charon." Charon, of course, was none other than the grim boatman of Greek mythology who ferried dead souls across the river Styx into the underworld. Christy insists he had never encountered the name before. Certainly he did not consciously remember it, but it could hardly have been more perfect. Not only did it boast a perfect mythological pedigree—and thus would be acceptable to the IAU—but it would also honor Christy's wife, Charlene, provided only that one ignored the Greek pronunciation, "Khar-on,"[9] but instead pronounced it as Christy does, "Shar-on," as in "Shar-lene." Many astronomers have apparently followed Christy's preference, though there are a few of us who insist on the Greek.

Figure 15.1 Christy speaking at a Pluto conference in 1980, reporting on his discovery of Charon two years earlier. The conference in Las Cruces, New Mexico, commemorated the fiftieth anniversary Pluto's discovery and honored its discoverer, Clyde Tombaugh. Photo by Dale Cruikshank.

An Extraordinary Opportunity Is Seen

Subsequent observations refined the separation distance and the orientation of Charon's orbit. Serendipitously, it turned out that shortly after Charon's discovery, Pluto and Charon would soon begin a rare series of mutual transits and occultations as seen from Earth. This stunning possibility was first recognized by a young Swedish planetary scientist, Leif Andersson.[10] Before becoming an astronomer, Andersson had earned notice as a child prodigy who won a Swedish television quiz show twice, the first time at sixteen.

He began his studies of astronomy in Sweden, but came to the Indiana University to complete his PhD in 1973 and was hired by Kuiper as a postdoctoral research associate at the University of Arizona. (Unfortunately, Kuiper died later that year, so they never actually worked together. Andersson did, however, work under legendary selenographer Ewen A. Whitaker, who had come from the Royal Greenwich Observatory in England to join Kuiper at Yerkes Observatory in 1955, when Kuiper first became seriously interested in the Moon, and later followed him to the Lunar and Planetary Laboratory in Tucson.)

Eventually Andersson's interest shifted from the Moon to Pluto, and in 1979, he came to realize that sometime in the next few years Pluto and Charon would experience a series of mutual transits and occultations. Because of uncertainties in the orbital parameters, when these events would start was not clear. Andersson and other astronomers thought they might already be taking place, though the expectation soon shifted to the early 1980s. These mutual events are very rare, occurring for only a few years once every 124 years, around the time that the Sun and Earth pass through Charon's orbital plane.

The fact that the series of mutual events began to happen so soon after Charon was discovered was sheer good luck. Mother Nature was extremely generous on this occasion, since the last Pluto-Charon mutual events had happened during the American Civil War, when Percival Lowell and W. H. Pickering were still children, while the next ones would not occur until the next passage of the Sun and Earth through Charon's orbital plane, in 2107–8.[11]

As Andersson and others realized, these mutual events had the potential to define the properties of the Pluto-Charon system in an utterly unique way. Accurate timing and precise brightness measurements of these mutual events from Earth-based telescopes could be used to determine the diameters and masses of both Pluto and Charon much more accurately than previously possible, would serve to refine Charon's orbital parameters, and could (in principle) even produce a rough map of the distribution of bright and dark regions on both the planet and its satellite. Sadly, Andersson did not live to see any of this. He died from lymphatic cancer in 1979, at the age of only thirty-five. No one doubts that had he lived, he would have become one of the leading planetary scientists of his generation.

As soon as the imminence of the mutual events was realized, an international effort was quickly organized to observe them from around the world. A transit would occur when Charon's orbit carried it in front of Pluto as seen from Earth, an occultation would occur half an orbital period later (~3.2 days) when Charon would disappear behind Pluto. In the early 1980s, Pluto and Charon presented a merged image that could not yet be resolved in ground-based telescopes,[12] so the transits and occultations could only be detected by changes in the total measured brightness of the pair. This meant that large telescopes were needed. Observers included Richard P. Binzel and J. Darrel Mulholland at McDonald Observatory in Texas, David J. Tholen at Mauna Kea Observatory in Hawaii, Carlo Blanco at the European Southern Observatory in Chile, V. M. Lyutyi and Vera P. Tarashchuk at the Sternberg Astronomical Institute's telescope in Crimea, Robert L. Marcialis at Steward Observatory in Arizona, and Edward F. Tedesco and Bonnie J. Buratti at Palomar Observatory in California.

Even in exceptionally good seeing conditions (steady air) with excellent image quality and the knowledge of exactly where to look, Charon can barely be seen with a 2.2-m telescope, and then only when it is in positions farthest from Pluto as viewed from Earth.[13] Later developments in telescope optics have mostly erased the atmospheric smearing to enable the light from Pluto and from Charon to be studied separately, giving rise to a deeper understanding of both objects.

The Transits and Occultations Begin

Though at first only null results were reported, monitoring of Pluto's light curve did improve the orbital parameters and a first suspected mutual event was observed by Tedesco and Buratti using a CCD detector at Palomar on January 16, 1985; this was followed by the definitive detection by photometry of a partial transit of Charon across the disk of Pluto on February 17, 1985, by Binzel at McDonald, with the corresponding occultation observed 3.2 days later by Tholen at Mauna Kea. These detections provided the first confirmation of the existence of the satellite (previously known only as a bump on Pluto's image and still officially known, pending this confirmation, by its preliminary designation 1978 P1; henceforth it would be known, always and forever, as Charon).

As Pluto continued along its distant orbit and the Pluto-Earth geometry slowly changed, the transit and occultation events became both more accurately predictable and more easily observed. The first events were "shallow" in a photometric sense because only small parts of either Pluto or Charon were either in front of or behind the other, resulting in a change in their combined brightness by only a few percent. As the geometry changed over the next few years, larger fractions, and eventually all parts of the surfaces, were involved, with changes in total brightness of the pair reaching about 50 percent. The events were also of longer duration, lasting up to four hours in some cases.

The game of celestial hide-and-seek, which began in 1985, centered on the 1987–88 passage of the Sun and Earth through Charon's orbital plane, continued until 1990, when the orbit of Charon no longer carried it in front of or behind Pluto as seen from Earth. By then, an enormous amount of accurate information had been gleaned about the Pluto-Charon binary (as it had come to be called).[14] The separation between them was put at 19,640 ± 320 km, with the period refined to 6.387245 ± 0.00012 days. Pluto's diameter, so long uncertain, was now well defined at 2,302 ± 14 km, with Charon's at 1,186 ± 20 km.

Thus—unexpectedly—Charon turned out to be the largest satellite in the Solar System relative to its primary, with a diameter fully half that of Pluto. This was a distinction formerly given to Earth's Moon. (No wonder the implications of the distorted image on Pluto plates had so long failed to register with astronomers. No one had ordered anything like that!) The pair was seen to be tidally locked such that Charon's orbital period was identical to the rotation period of Pluto and both the planet and the satellite kept the same hemispheres always directed toward one another. The center of the pair's rotation-revolution, called the barycenter, is located outside of Pluto; by comparison, the barycenter of the Earth-Moon system is inside Earth, but offset from the geometric center by 4,671 km (Earth's radius is 6,378 km).

The mass of Pluto was found to be only a fifth that of Earth's Moon; Charon's mass came out at a seventh of Pluto's. With its mass known, Pluto's density could be recalculated: it was 2.0 g/cm³, which meant that its composition was likely about two-thirds rock and one-third ice. This distinguished it from most of the outer-planet satellites, which are mostly ice—except for Triton, which has a density similar to Pluto's. Pluto also proved to have a highly tilted axis (118°), wherein it resembles Uranus. The south pole is shrouded in darkness, as it will continue to be so for another century or so.

The transits and occultations allowed astronomers to draw rough maps of albedo features on Pluto and Charon. One set of maps was produced by Eliot F. Young of the Southwest Research Institute (SWRI) in Boulder, Colorado, and Richard Binzel of MIT; another was the work of Marc Buie, then at Lowell Observatory and now at SWRI, and his collaborators. They were in general agreement, as both showed Pluto to have bright polar caps and a slightly reddish equatorial zone. Charon proved to be much more uniform in color and albedo, and less reflective. In contrast to ruddy Pluto, Charon was gray.

Ice on Charon

During the cycle of mutual events that had been so critical in allowing astronomers to calculate all the basic information about the sizes, masses, orbits, and other characteristics of the Pluto-Charon binary, Charon was occasionally completely hidden behind Pluto for about an hour on each of its orbits around the planet. Then its orbital motion brought it out from behind Pluto and both bodies were in full view of the telescope. We should recall the technology of telescope optics at that time made it nearly impossible to see or photograph the images of Pluto and Charon separately—they were too close to one another and their images blended into one bright object. This was the problem in finding Charon in the first place. The brightness and the spectrum of the blended objects could, however, be measured accurately. By measuring the spectrum of the Pluto-Charon blended image, and then that of Pluto alone (with Charon hidden behind it), deriving the spectrum of Charon required the straightforward subtraction of one measurement from the other: (Pluto + Charon) − Pluto = Charon.

The success of this approach depended on knowing the timing of the eclipse events, but this was predictable from the precise measurements of events from the first years of the brief epoch. Success also depended on the availability of suitable telescopes and good observing conditions at the right times in the right locations.

The first successful observations that gave spectral information on Charon were made on March 3, 1987, by Robert L. Marcialis, George H. Rieke, and Larry A. Lebofsky of the University of Arizona. The measurements made at four critical wavelengths in the near-infrared gave the tentative result that unlike Pluto, Charon showed no evidence of methane ice, but rather had a surface of primarily water ice.[15] Efforts to view the same event by Marc Buie and his colleagues at Mauna Kea Observatory were thwarted by high wind and otherwise bad observing conditions. Buie's next opportunity at Mauna Kea came on April 23, 1987, and it was entirely successful. He recorded the spectrum of Charon at thirteen wavelengths across the span from 1.5 to 2.5 μm, and the spectrum of water ice was entirely clear, confirming the results of Marcialis, Rieke, and Lebofsky.[16]

As the last of the mutual events passed into history in 1990, a feverish period of Pluto discovery, which had begun in 1976 with the detection of methane and continued through 1978 with the discovery of Charon, came to an end. In 1976, Pluto was little more than a tiny speck in all but the largest telescopes. By 1990, it had become one component of a fascinating binary system, and both Pluto's and Charon's surfaces had at least roughly been mapped. But the rate of technological advance has been so fast that these results were surpassed within only a few years. New maps, resulting from computer enhancement of Hubble Space Telescope (HST) images of Pluto's tiny disk, superseded the maps painstakingly built from the mutual events data. The HST maps have roughly the resolution of the Earth-based maps of the Galilean satellites of Jupiter drawn by visual observers such as Bernard Lyot, Henri Camichel, and Audouin Dollfus, and they show a ragged north polar cap and various light and dark spots, including a large dark patch that at the time was thought to be a likely impact feature. (New Horizons would show exactly what it was in just a few short years.)

In addition, new ground-based techniques, combining advances in fast computers with the use of flexible optical components (adaptive optics), greatly reduced telescopic-image distortions due to atmospheric turbulence. These techniques generated much clearer views of celestial objects, allowing observers to clearly separate the image of Charon from that of Pluto and also permitting studies of Charon's spectrum without interference from Pluto's. The spectral signature of frozen water was confirmed, but, surprisingly, the shape of the water band showed the ice to be in the crystalline state.

At the low temperature of Charon, about 40° K, water was expected to be in a non-crystalline or amorphous phase,[17] but instead on Charon (and many other cold Solar System bodies) it exists as hexagonal ice. Though some amorphous ice may be on Charon, the predominance of the hexagonal phase was—and remains—a mystery. If Charon had been heated by some past event such that the surface reached the amorphous-crystalline transition temperature, the conversion of original amorphous ice could have occurred. However, the bombardment of the surface by energetic atomic particles from space should have broken the crystalline structures and returned the ice to an amorphous state. Why this has not happened remains unclear.

In addition to the puzzle of the crystalline water ice, Charon's spectrum revealed another mystery molecule. At a wavelength between strong absorption bands of frozen water, a weaker absorption band was a detected characteristic of frozen ammonia, or more likely a chemical combination of ammonia with water (e.g., NH_4OH or $NH_3 \cdot nH_2O$).[18] Ammonia and its chemical compounds are not expected to be on very old surfaces in the Solar System, since they readily are destroyed by ultraviolet light from the Sun and other stellar sources. The fact that they do exist on Charon suggests that some kind of surface or subsurface chemical activity is refreshing the supply.

One of the discoverers of Charon's putative ammonia, Jason Cook, then at Arizona State University, offered a novel explanation for how this might work. The ammonia could be refreshed by geological activity carrying material from Charon's interior to the surface through a process similar to volcanic activity on Earth.[19] Volcanic activity on cold, icy bodies is cryovolcanism, a set of processes by which cold liquids or slushy ices are disgorged from the interior a planetary body to the surface. As applied to Charon—a relatively small, extremely cold body at the edge of the Solar System—it presents a conundrum: has Charon somehow retained its heat from its formation 4.5 billion years ago, or is another mechanism heating it today?

As we shall see, both Pluto and Charon, on closer examination, will afford many another conundrum.

16
More Than Ice

Some Extraordinary Chemistry

What Colors the Ices of Pluto?

BOTH PLUTO AND CHARON are covered with ices—not just frozen water, but ices of varied chemical composition. Pluto has a thick mantle of water ice, possibly overlying a subsurface ocean (as is the case on Saturn's moon Enceladus). However, the water ice is largely covered with layers of frozen nitrogen, methane, carbon monoxide, and ethane, and it only shows through in visible expanses in limited regions of the planet's surface. In contrast, Charon's surface displays its water ice so prominently that it was discovered from Earth with telescopes and spectrometers as early as 1987.[1] Its interior was assumed also to be largely water ice, consistent with the satellite's low density.

Ices are not colored, but as we will see in chapter 19, some pigment clearly tints the landscapes of both Pluto and Charon with shades of yellow, orange, and a dingy red, and in some areas of Pluto the colored material seems thick enough to blanket the underlying topography. To be sure, the translucent ice of Earth's glaciers often displays an appealing crystalline blue tinge, sometimes quite intense, but this only arises when bubbles are forced out by the pressure in a thick ice sheet allowing the ice to form large, transparent crystals. In these transparent blocks, sunlight penetrates deep into the ice, and the short-wavelength light in the blue to violet region of the spectrum is efficiently reflected, or scattered, by very tiny particles in the ice. This is called Rayleigh scattering, and is the same basic process that makes the clear sky appear blue—the shortest wavelengths are scattered by tiny particles and even the molecules of oxygen and nitrogen that make up the air. The coloring of Pluto is clearly different.

The reddish tint of Pluto was known in advance of the New Horizons mission. However, no one expected the rich diversity of Pluto's surface and the range of colors—yellow, orange, red, and very dark red—seen in the flyby pictures. As we saw in the previous chapter, Charon's surface was known from ground-based observations to be largely gray. New Horizons showed a few scattered hints of red as well as an anomalous large brownish-red region centered on the north pole. We will return later to the possible cause of the polar spot on Charon.

What can be the source of these unexpectedly warm colors on these cryogenic worlds?

Clearly, the colors are produced by some non-ice component added to the colorless ices on the surface. Whatever that component is, it is most likely consistent with what we know about the compositions of other bodies in the outer Solar System that show similar colors: the icy moons of the giant planets, other Kuiper Belt objects, asteroids, and even comets. The surfaces of some of these bodies consist of a mixture of rock and ice. A few unusual cases also display similar colors—notably, Jupiter's satellite Io, whose riot of colors was described by Voyager mission scientists as resembling those in a pizza. Its surface is coated with sulfur and silicate lavas from active volcanoes.

Organic Chemistry in the Solar System

For the colors of Pluto and several of the other icy bodies, we must turn for insight to organic chemistry and look for organic molecules that are produced in the environment of the outer Solar System. Several candidates are available, including the organic molecules found in comets, some of which formed primarily in the outer Solar System as the terrestrial and giant planets were coalescing closer to the Sun. Some organic materials may have been inherited from the solar nebula; some solid mineral grains found in meteorites certainly come to us relatively unaltered.

We will return to the question of Pluto's colored material below, but first we digress a bit to explore the organic world because it is so closely related to life. Organic molecules are indeed basic to living organisms (hence the name *organic*). Until 1828, all organic compounds had been produced from organisms, living or dead. However, that year the German chemist Friedrich Wöhler synthesized an important molecule previously known only in living things, urea, which has the chemical formula $CO(NH_2)_2$, while treating the inorganic materials silver cyanate with ammonium chloride. Wöhler's synthesis has been taken to mark the beginning of the field of organic chemistry. As with many scientific milestones, it was somewhat accidental; at the time, it was widely believed that urea and ammonium cyanate were the same compound, and Wöhler was trying to synthesize the latter and ended up with urea instead.

For our purposes, organic chemistry is meant in this wider sense—the same as will be familiar to any student who has suffered through an organic chemistry course. For the chemistry of living things, we reserve the term *biochemistry*. Thus life is not necessarily implied by complex organic chemicals. They can be any compounds that arise from the combination of carbon (C), hydrogen (H), oxygen (O), and nitrogen (N)—CHON, for short. The processes that involve such combinations in living organisms, though for obvious reasons of greatest interest to biochemists (and more recently to planetary scientists), are a special case; those processes also require the elements sulfur (S) and phosphorus (P).

With the inert gas helium (He), CHON are the most abundant elements in the Universe. But unlike helium, CHON atoms are highly social and gregarious, combining in virtually limitless ways to make molecules ranging from the very simple to the highly complex. Under certain circumstances, some of the highly complex molecules show strong colors similar to those on Pluto, Charon (the north polar region), and several other Solar System bodies. Thus they are of great interest to us here.

To understand what these molecules may be, it is useful to first recall the efforts of scientists to understand the origin of life on Earth.

The Origin of Life: A Chemical Perspective

The origin of life—along with, perhaps, the origin of the Universe itself and the nature of consciousness—is among the ultimate questions people have asked themselves since earliest times. For a long time, such questions seemed simply unanswerable, and to require, perhaps, some kind of divine intervention (which involved hiding one mystery behind another). Now, however, the answers are beginning to come into view. Though many details have yet to be filled in, we now have a promisingly good idea how life first formed on Earth—and can speculate that it would have followed a somewhat similar course if it developed in suitable niches on other planetary bodies. (Such niches might include the ancient lakebeds of Mars, which are being explored even now by the Curiosity rover, or the frozen-over water oceans of Jupiter's Europa or Saturn's Enceladus, or even the hydrocarbon polar lakes of Saturn's large smoggy satellite, Titan.) Each (or none) of these places, which we are eager to explore—and reexplore—with our spacecraft, may harbor life. We say nothing of the exoplanets around other stars, the numbers of which have increased exponentially since the first (51 Pegasi b) was discovered in 1995, and which seem to be virtually limitless. Of these, a nonnegligible number exist in "habitable zones," and primitive life forms may exist on some (or many).

That the origin of life might not require divine intervention but be a matter of straightforward chemistry was first clearly set forth by Charles Darwin in 1871 (the same year he published *The Descent of Man*). Darwin had just read about an experiment showing that some molds were able to survive boiling in water; he speculated in a letter to his friend, the botanist Joseph Hooker,

> It is often said that all the conditions for the first production of a living organism are now present, which could ever have been present.—But if (& oh what a big if) we could conceive in some warm little pond with all sorts of ammonia &

phosphoric salts,—light, heat, electricity &c present, that a protein compound was chemically formed, ready to undergo still more complex changes,
at the present day such matter w^d be instantly devoured, or absorbed, which
w^d not have been the case before living creatures were formed.[2]

Darwin's "warm little pond" would later inspire the "primordial soup" of
the pioneering investigator of the origin of life, Soviet biochemist Alexander
Ivanovich Oparin, who coined the term in 1924. At that time, or perhaps a
little later, the compositions of the atmospheres of Jupiter, Saturn, Uranus,
and Neptune were regarded as primordial (still unevolved). They appeared
to be heavily permeated with hydrogen and hydrogen-rich compounds such
as methane and ammonia, and Oparin posited that this had been the state
of Earth's early atmosphere as well.

Hydrogen and hydrogen-rich compounds are called reducing compounds.
Earth's current atmosphere is made up mainly of nitrogen and oxidizing
compounds—oxygen, and carbon dioxide. Oparin, from basic chemical
principles regarding oxidizing and reducing molecules, saw that living matter
could never have formed under oxidizing conditions. Following what was
then believed about the evolution of the planets, he assumed that Earth's
atmosphere had once resembled those of the giant planets, that is, it had once
been much more reducing than oxidizing.

Oparin's concept, which was independently developed about the same
time by the eminently quotable British scientist J. B. S. Haldane (who said,
for instance, "the universe is not only stranger than we imagine, but stranger
than we can imagine" and "God must have loved beetles, he made so many
of them") involved several basic assumptions. Chemist Robert Shapiro summarized these assumptions as follows:[3]

1. The early Earth had a chemically reducing atmosphere, like those presently found on the giant planets.
2. This atmosphere, exposed to energy in various forms, produced simple
   organic compounds ("monomers").
3. These compounds accumulated in the "soup," which may have been
   concentrated at various locations on the early Earth (shorelines, oceanic
   vents, etc.).

4. By further transformations, more complex organic polymers—and ultimately life—developed in the soup.

We might add a fifth assumption: as Earth's atmosphere evolved toward its
present oxidizing state, life adapted through a series of reactions (metabo-
lism) to sustain itself.

Solar System Chemistry in the Laboratory

These ideas all hold together nicely—so long as one assumes that Earth's primordial atmosphere was, like those of the giant planets, rich in hydrogenous
compounds such as methane, ammonia, and hydrogen gas. That was still the
consensus in February 1951, when Stanley Miller, who had been known as a
"chem whiz" at Oakland High School and completed his bachelor of science
degree at the University of California, Berkeley, began graduate studies at
the University of Chicago. He was first interested in theoretical problems in
chemistry and spent a year working with Edward Teller. This was an unproductive year for Miller, and since Teller was preparing to leave to work on
the hydrogen bomb, Miller had to find another professor to supervise his
PhD research. He had the good luck of attending a seminar by Harold Urey,
who as we noted in chapter 9 had won the 1934 Nobel Prize in Chemistry for
his discovery of deuterium (a form of hydrogen that has a neutron added to
the atom's nucleus). Urey had become deeply interested in the origin of the
Moon and planets, and thus had entered the same field that Kuiper, then his
colleague as the University of Chicago, had made his own.[4] In his seminar,
Urey spoke about the possibilities of the synthesis of organic molecules in
Earth's primeval atmosphere. Miller left inspired, and in September 1952 he
approached Urey with the request that he might test the organic synthesis
hypothesis in the laboratory.

Urey is said to have at first regarded the experiment as too iffy for a
doctoral thesis, but eventually came around and helped Miller with the
experimental design. First a flask was filled with hydrogen, methane, and
ammonia gases to simulate the putative primitive atmosphere of Earth. Then
the mixture was exposed to an electric spark, simulating lightning. Earth's

primitive ocean was represented by water vapor circulated through the glass apparatus and recondensed. Any molecules produced by the reaction fell into a trap at the bottom of the apparatus—a refuge from their destruction by the next spark.

Miller's experiment was successful far beyond his (and Urey's) expectations. Within a few days, the liquid in the trap at the bottom of the apparatus became pink; after another week it turned deep red. Miller noted that these colors, as well as a yellow-brown material, were due to the presence of solid organic compounds generated in the experiment. When the liquid was analyzed, it was found to contain at least two amino acids, glycine and alanine (both alpha and beta), which are fundamental to living organisms. (The analysis of the colored solids was not to come until much later. See below.)

Miller published a short note on his results bearing the modest title, "A Production of Amino Acids Under Possible Primitive Earth Conditions," in the May 15, 1953, issue of *Science*.[5] He was only twenty-three at the time. Though the experiment is often referred to as the Miller-Urey experiment, it is gratifying to note that Urey—who is not always depicted as a pleasant personality—crossed his own name off the paper. "I already have a Nobel Prize," he is supposed to have said. Miller's paper appeared less than a month after James Watson and Francis Crick had dropped their own bombshell, a brief paper in *Nature* that laid the foundation for understanding the self-replicating molecule deoxyribonucleic acid (DNA).[6] (Though Miller went on to have a distinguished career, and was often nominated, he never would receive the Nobel Prize.) Life, hitherto the epitome of inscrutability, was suddenly revealing secrets that it had long concealed.

Miller's experiment has been repeated many times, including by Miller himself. Many variations on the starting conditions were used, including using ultraviolet light instead of electric sparks as the source of energy to stimulate the chemical reactions. Details of the results vary, but in general they found that whenever simple CHON molecules were in the mix, amino acids were almost always the result.

Since Miller did his first experiments, we have learned a great deal about the evolution of Earth. Many long-held ideas have been overthrown; in particular, assumptions about the early atmosphere have changed. Though Earth's original atmosphere may have been captured from the solar nebula and consisted, like those of the giant planets, of molecular hydrogen, ammonia, methane, and water, it wouldn't have survived long. Asteroid impacts blasted away the original atmosphere. Its successor was primarily oxidizing, produced not by captured gas from space but by volcanic outgassing from within Earth, and consisting of oxidized carbon in the form of carbon monoxide and carbon dioxide, water vapor, the acrid-smelling sulfur dioxide, and the putrid-smelling hydrogen sulfide. Nitrogen, which is neither oxidizing nor reducing, was also a major component of the replacement atmosphere; today nitrogen is 78 percent of the atmosphere we breath. Despite the presence of oxidized carbon and sulfur, as well as water vapor, the early atmosphere had little or none of the free oxygen that enriches the modern atmosphere, making up 21 percent of the total. That came later, when the first photosynthesizing organisms, the cyanobacteria, appeared on Earth about 2 billion years ago.

Our knowledge of the first billion years or so of Earth's history, including the composition of the oceans and atmosphere when life first appeared, is still sketchy in the details. Very little geological or geochemical evidence survives. We do know that Earth's surface was being bombarded by primordial asteroids and comets, and that it was the scene of massive volcanic eruptions on a global scale. Because the young Sun was 20 or 30 percent fainter (in visible light) than it is now, Earth may have gone through periods of deep freeze—though never for long, as heat generated within would have produced huge volcanic eruptions, with the release of massive amounts of greenhouse gases. In addition, the primordial asteroids pummeling Earth's surface would have provided regular inputs of extraterrestrial energy.

During the violent earliest millennia of its youth, Earth and the other terrestrial planets were subjected to a withering assault by asteroids. As described in the previous chapter in connection with how Pluto came to be trapped in its unusual orbit, the icy debris of the Kuiper Belt perturbed the giant planets, which migrated out of stable resonances into new positions.[7] (We are learning, to our astonishment, that the distant Kuiper Belt may well have had a significant effect on the inner Solar System, including Earth itself.) Over a few million years, Jupiter was pushed slightly inward to its present location, and the other giant planets were pushed outward. Eventually—perhaps after a giant planet was ejected from the Solar System on a Voyager-like trajectory into interstellar space—the giant planets resta-bilized through additional interactions with the Kuiper Belt, but not before sending, as a byproduct of this process, barrages of asteroids careening through the inner Solar System, including those that produced the Moon's large basins, such as Mare Imbrium and Mare Orientale. Doubtless Earth bore similar scars, though none from that era have survived. Only near the end of the period of heavy bombardment did microcontinents of crust begin to form the seeds of the future continents—and provide a palimpsest on which later impacts could leave their imprint.

A few crystals of the mineral zircon found in 4-billion-year-old sedimen-tary rocks from Australia show that Earth then had, at least intermittently, liquid oceans. They would have been shallow, since primordial water would have been rapidly destroyed by reaction with iron. On the emerging micro-continents of that time, scattered examples of Darwin's "warm little pond" may have formed. However, if life happened to appear in one of them, it might have been quickly destroyed by impacts or by the intense ultraviolet light of the young Sun. (The ozone layer did not yet exist, and Earth's surface was thus unprotected from lethal ultraviolet radiation.)

In fact, for the reasons just given, the "warm little pond" has fallen out of vogue. Instead, the prevailing opinion since at least 1977 is that when life appeared, it did so in a more covert and protected location. The unexpected discovery that year of abundant life associated with "black smokers" off the Galapagos Islands—deep volcanic vents lining mid-ocean ridges, where magma wells up to form new ocean crust—led to the vents becoming the favored location for life's origin about 4 billion years ago. As noted in a recent review article, the deep volcanic vents are "geochemically reactive habitats that harbor rich microbial communities. . . . There are striking parallels between the chemistry of mutual reactions between H_2 and CO_2 that occurs in hydrothermal systems and the core energy metabolic reactions of some modern organisms that get their carbon from CO_2 and use light in the pho-tosynthesis process as their energy source. Some organisms can get energy from inorganic chemicals. Together, these are the prokaryotic autotrophs."[8] To put it perhaps more colorfully and accessibly: if the origin of life were a murder mystery, we have, in these deep volcanic vents, something approach-ing means, reason, and opportunity.

How life might have gotten from the prokaryotic autotroph stage to complex organisms like reptiles, amphibians, birds, and mammals is a question that we cannot pursue further here. Obviously, much of that has to do with conditions—and steps—that may have been taken uniquely under specific conditions that existed on the early Earth. By contrast, Stanley Miller's experiment, involving the production of amino acids under the relatively simple conditions, seems to pertain broadly to what has taken place in many environments in the Solar System.

For instance, consistent with his experiment, many amino acids have been found in carbon-rich meteorites that regularly fall onto Earth's surface. Chemical analysis has revealed in the carbonaceous meteorites nearly eighty different amino acids, including most of the twenty varieties used by life.[9] These amino acids are thought to have formed in the presence of liquid water in the interiors of the meteorites' parent asteroids. Finding amino acids in meteorites that are the age of the Solar System makes plausible the ideas that life could originate by entirely natural processes, that it might have arisen elsewhere in the Solar System, and that (almost certainly) it exists elsewhere—perhaps abundantly—in the Milky Way.

Of course, amino acids alone do not make life, but we do know that under certain conditions, amino acids link together to form peptides, and vast strings of peptides make proteins. It isn't easy to envision natural conditions under which these complex transformations occur (though scientists studying the processes occurring around the deep ocean vents are making progress), and we do not yet know whether these complex transformations occurred only once—here on Earth—or whether they occur commonly, as long as certain preconditions (such as those exemplified by the hydrothermal vents) are fulfilled. As the late astronomer and astrobiologist Tobias C. Owen put it at a recent conference, one can stir a pot of amino acids for many, many years, and nothing will crawl out. But perhaps as our understanding continues to grow, in the near future something *will* crawl out.

Miller's experiments ultimately inspired the origin of an active and growing field of study called astrobiology. The experiments resonated with his near contemporary, the astronomer and planetary scientist Carl Sagan, who had a lifelong compelling interest in not only the origin of life but the possibility of life, including intelligent life, existing on other planets.

Carl Sagan and Chemistry in the Solar System

Sagan was born in Brooklyn, New York, in 1934, the son of an immigrant garment worker from the Ukrainian city of Kamianets-Podilskyi. Like many bright students before and since young Carl found his high school classes unchallenging and uninspiring, but he did what he could on his own: he set up his own chemical laboratory and took up astronomy as a hobby. As an undergraduate at the University of Chicago—he entered the same year as Miller, 1951—he boldly sought out as advisers the geneticist Hermann Muller and Urey. Sagan's senior thesis, written under Urey, was on the origin of life. Muller introduced him to Miller, and eventually Sagan visited Miller's lab and attended the seminar in which Miller presented his results. Sagan's impression was that the other professors didn't appreciate the importance of what Miller had done, though Miller, long afterward, disagreed. "What Carl didn't realize," he explained, "was, it wasn't that they didn't *understand* the importance, it's just that they couldn't believe the *results*! All those amino acids!"[10] Sagan himself justly regarded the experiment as "the single most important step in convincing scientists that life is likely to be abundant in the cosmos."[11]

As a graduate student, Sagan worked under Kuiper, who supervised his dissertation ("Physical Studies of the Planets") during Sagan's occasional visits to Yerkes Observatory. Kuiper also appreciated that Sagan's was a unique and eclectic talent (not unlike himself). "Some persons work best in specializing in a major program in the laboratory," Kuiper said. "Others are best in liaison between sciences. Dr. Sagan belongs in the latter group."[12]

Eventually Sagan went on to establish his own laboratory at Cornell University, where for nearly three decades he and chemist Bishun N. Khare conducted a wide range of experiments of astronomical significance, always focused on the question that fascinated Sagan, the possibility of life in the cosmos.

Tholin

Sagan and Khare were particularly interested in the solid material formed from energetic processing (such as electrical discharge and ultraviolet light) of gas mixtures found in planetary atmospheres. This solid material, a mixture of complex organic molecules that are stable at room temperature, appeared to be akin to the pink, yellow, and red solids that had formed in Miller's experiment. When, for example, Sagan and Khare subjected to an electric spark a mixture of methane and nitrogen gases like that found in the atmosphere of Saturn's satellite Titan, they found a brownish-red solid material deposited on the walls of the glass reaction flask. These and similar materials Sagan dubbed "tholins" after the Greek word for "muddy."[13] The colors of this tholin and others made with variations on the gas and energy mixture intrigued Sagan and Khare because of the similarity to the yellow-brown color of Titan's atmospheric haze. However, the tholin materials were intractable; at that time, making definitive chemical analyses of such complex molecular materials difficult. When Miller had first produced his red and brown material in his flask in Urey's lab, analytical techniques could only reveal its composition in terms of vague generalities, such as "intractable polymers" and "macromolecular carbonaceous complexes." Miller himself could identify only five amino acids and other molecules of unknown structure that he had created, though after his death in 2007, more amino acids were found in sealed containers that he had made but not fully analyzed. Analysis of the preserved samples revealed twenty-two amino acids.[14] Another challenge Sagan and Khare faced in demonstrating the relevance of tholins to planets and satellites was the difficulty of simulating the extreme conditions of the space environment in any realistic way in the lab.

There are many kinds (and colors) of tholin, depending on the starting materials and what phase they are in, and how they are processed by ultraviolet light or charged atomic particles. The one unifying component in all tholins is carbon, which can come from methane, carbon dioxide, or other carbon-bearing molecules.

Despite ambiguities in the analysis of the colored material, the Titan work marked a significant turning point for tholin studies. Sagan and Khare's tholin not only exhibited the colors of Titan's smog but they also showed agreement with the more detailed optical properties of its atmosphere (i.e., their complex refractive indices, which describe mathematically the interaction of light with solid materials, even if the exact chemical structure is unknown). The Titan work had involved a tholin made from gases because Sagan and Khare were trying to simulate chemistry in the atmosphere. They went on to extend their work to the mixtures of ices presumed to exist on the very cold surfaces of satellites and comets in the outer Solar System, irradiating these ices with ultraviolet light and electrical discharges. Though these experiments on ices were more difficult to perform, the results were similar: small molecules containing carbon and other abundant elements were transformed into more complex organic materials that were stable at room temperature and colored. In other words, into a tholin.

This brings us back to the origins of the colored non-ice material that tints the surface of Pluto. As the inventory of ices on other Solar System bodies, including Triton and Pluto, became known through astronomical observations with Earth-bound telescopes, realistic laboratory experiments to synthesize tholins of direct relevance to these bodies could be carried out. (Obviously, the laboratory experiments involve a suitable reduction in scale!) Experiments in the Astrophysics and Astrochemistry Laboratory at NASA's Ames Research Center in California, and in other laboratories, have now simulated Pluto's surface using a mixture of the three principal ices found there, nitrogen, methane, and carbon monoxide, in the proportions of about 100:1:1. The ice-mixture creation process begins with mixing the three gases in a container that is then connected to a high-vacuum chamber with a cold surface inside. As the gas is admitted, it freezes on the cold surface in a layer that can be observed through a glass window. Once a suitable ice layer is formed, it can be irradiated with high-intensity ultraviolet light to simulate the ultraviolet light from the Sun. Alternatively, it can be exposed to a beam of charged atomic particles, either electrons or protons, with a suitable apparatus connected to the chamber. Whatever the source of irradiating energy, chemical changes in the ice begin immediately. Broadly, the result after many hours of ice irradiation is a residue of complex chemicals, stable at room temperature, that can be recovered and subjected to detailed analysis. This is a tholin. When made from a combination of ices known to occur on Pluto, it can be called a Pluto ice tholin.

On the scale of these laboratory studies, the quantities of tholins produced are very small—only about a milligram in an experiment running for about a month. Still, that is enough to analyze, and we can now do better than refer to tholins as made of "intractable polymers" or "macromolecular carbonaceous complexes." The Pluto ice tholin is found to consist of numerous combinations of the CHON starting ingredients. Some are relatively simple, and incorporate only five or ten atoms; others are built of dozens of atoms, and have atomic weights up to a few hundred. Among the recognizable molecules are alcohols, carboxylic acids, amines, aldehydes, ketones, and even urea. Even though all these organic molecules have been made abiotically, some of them, such as the carboxylic acids and urea, are molecules "of interest for prebiotic and biological chemistries."[15] Although amino acids are not found in tholins, its precursor molecules are. Thus the most rudimentary chemical building blocks of life appear to form readily even on such a seemingly inhospitable body as Pluto. Of course, they also exist on many other Solar System bodies, as they must on the countless trillions of other planetary and interstellar environments across our Galaxy and beyond.

The last chapter has yet to be written about this colored sludge-like material that has tantalized and vexed researchers since first appearing in the trap of Stanley Miller's apparatus in 1953. Clearly, this is a simultaneously attractive and repulsive material—attractive because it provides a clear pathway to understanding the chemistry of atmospheric smogs and surface colors on planets and satellites; repulsive because the intrinsic complexity of these "intractable polymers" eludes clear and concise molecular description and stands as a barrier to the awareness of tholins by scientists who are not particularly involved in this specialized area of research (to say nothing of the general public). Nevertheless, these yellow, pinkish, or reddish-brown tholins undeniably somehow represent a fundamental stage in the processes that lead to the origin of life in the Universe. And there is really no reason to think that those processes should be simple and universal. The processes that have led to life have undoubtedly been as messy—and at critical points as "intractable"—to our current understanding as tholin itself.

Tholins are a recurring motif throughout the outer Solar System. Nature seems never to tire of Miller's experiments: the flask of the outer Solar System forever and lovingly creates these intractable polymers and macromolecular carbonaceous complexes of yellow, brown, and red material that derives its name from "muddy." Human as we are, and with blinkered sight, it is not easy for us to work such a motif into a grand scheme. Yet we sometimes suspect that if we truly understood things to their depths, this humble stuff might appear to us as sublime as the giant redwoods.

17
Genesis of a Flight to Pluto

Space Beckons

THE AVIATION-INSPIRED POEM "High Flight," written soon after the pilot John Gillespie Magee Jr. had completed a flight reaching 33,000 ft in a Spitfire Mark I, includes these inspiring lines:

> *Oh! I have slipped the surly bonds of Earth*
> *And danced the skies on laughter-silvered wings;*
> *Sunward I've climbed, and joined the tumbling mirth*
> *Of sun-split clouds . . .*
> *Up, up the long, delirious, burning blue*
> *I've topped the wind-swept heights with easy grace,*
> *Where never lark, or even eagle flew—*
> *And, . . . with silent, lifting mind I've trod*
> *The high untrespassed sanctity of space.*[1]

The unrelenting human urge to slip Earth's surly bonds to fly in the sky and even to leave the home planet has pervaded literature and science for centuries. The invention of sustained powered flight in 1903 and the development of rockets in the first decades of the twentieth century have enabled those ancient aspirations—grandly.

The story of the invention of airplanes is well known, but the development of rockets for space travel is perhaps less familiar. Rockets were used for warfare centuries ago, but the idea of rockets destined to cross the space frontier can be said to have spawned in the minds of three principal players—Konstantin Tsiolkovsky (1857–1935) in Russia, Robert H. Goddard (1882–1945) in the United States, and Hermann Oberth (1894–1989) in Germany. The rich and detailed story of rocketry has been admirably told by Willy Ley, an early arrival on the scene as it unfolded in Germany and a member of the Verein für Raumschiffahrt (VfR, or the Society for Space Travel, usually referenced in English as the German Rocket Society).[2] Without recounting the whole history of this fascinating subject, we jump to the time when liquid-fuel rockets had become a practical and relatively reliable means for carrying heavy payloads over long distances.

Practical rockets with the potential for spacefaring came onto the scene in the 1950s, largely as refinements of the V-2 rockets developed by German engineers and used as weapons during the late (and, by then, desperate) stages of World War II. At the end of the war, Werner von Braun and other leading German rocket engineers surrendered to the Americans rather than allow themselves to be captured by the Soviets, and an American rocket program was quickly organized under their leadership. Several captured V-2s were fired from the White Sands Missile Range near Las Cruces, New Mexico, and improved rocket designs were developed, including a multistage rocket consisting of a V-2 lower stage and a smaller WAC-Corporal upper stage. It was successfully fired in 1949 and reached the then unheard-of altitude of 250 miles (400 km). (As noted in chapter 9, Clyde Tombaugh played an important role in this work.) Thenceforth no one could doubt that space was within human reach.

During the 1950s, the United States and the Soviet Union, allies in World War II, found themselves opposed across the ideological divide of the Cold War. The use of rockets for peaceful purposes was subordinated to their use in intercontinental ballistic missiles (ICBMs), designed to carry not instrument packages for the peaceful investigation of other worlds but the lethal cargo of nuclear warheads. At the time, the West believed that the United States had an insurmountable lead; in July 1955, the United States announced plans to launch the first artificial satellite, or manmade "moon," into orbit around Earth. However, the Soviets were also taking a keen—but secretive—interest in rockets. Though by the time they arrived at Germany's V-2 research center in Peenemünde at the end of World War II everything had already been captured by the Americans and the facility itself had been destroyed, the Soviets succeeded in designing their own ICBMs based on the V-2 model. The leading role was taken by a brilliant engineer, Sergei Pavlovich Korolev, whose own experiments with rockets—inspired by writings of Russian rocket visionary Konstantin Tsiolkovsky—had begun in the 1930s. On October 4, 1957, Korolev and his team of engineers launched Sputnik 1, the first artificial satellite, atop an R-7 ICBM. Both the launch site and even Korolev's identity were carefully concealed; he was referred to only as the "chief designer."

The success of Sputnik came as a shock in the West, and the Space Race was on. While the Soviets built on their success a month after launching Sputnik 1 by orbiting a much larger craft, which in addition to scientific instruments carried the live dog, Laika, into space, the Americans suffered the embarrassment of having their first Vanguard rocket blow up on the launch pad on December 6, 1957. This humiliation was partially redeemed when, on January 31, 1958, von Braun and his team put up the small satellite, Explorer 1. Though smaller than the Sputniks, it sent back results of great scientific value, furnishing the data from which University of Iowa scientist James Van Allen first identified the radiation belts surrounding Earth that now bear his name.

Though the uses of rockets for destructive purposes always remained in the foreground, Van Allen's discovery showed the existence of purer motivations. The early rocket pioneers were motivated primarily by curiosity, and had dreamed of reaching and exploring directly the extraterrestrial destinations that had hitherto only beckoned, blurry and indeterminate, in vistas through telescope eyepieces. The first extraterrestrial destination to lie within reach was, of course, the planetary body closest to home, the Moon.

Plate 1 Pluto was imaged at high resolution with the New Horizons spacecraft on July 14, 2015. The image combines blue, red, and infrared images taken by the Ralph/ Multispectral Visual Imaging Camera (MVIC). Pluto's surface displays a remarkable range of subtle colors, enhanced in this view to a rainbow of pale blues, yellows, oranges, and deep reds. The image resolves details as small as 1.3 km (0.8 miles). NASA photo PIA19952.

Plate 2 New Horizons captured this high-resolution enhanced color view of Charon just before closest approach on July 14, 2015. The image combines blue, red, and infrared images taken by the spacecraft's MVIC; the colors are processed to best highlight the variation of surface properties across Charon. Charon's color palette is not as diverse as Pluto's; most striking is the reddish north (top) polar region. Charon is 1,214 km (754 miles) across; this image resolves details as small as 2.9 km (1.8 miles). NASA photo PIA19968.

Plate 3 This high-resolution image captured by New Horizons shows the western lobe of the "heart," called Sputnik Planitia, a sea of frozen nitrogen that also includes frozen methane and carbon monoxide. At the lower left is a series of linear fractures that cut across the highlands and some craters. The bottom of the frame shows the heavily cratered ancient highlands with scalloped cliff faces and other highly mixed terrain types. Red tholin deposits cover large expanses of the highlands at the lower right of the frame, while jumbled angular blocks of (presumed) water ice are piled up against the shoreline of Sputnik Planitia. NASA photo PIA20007.

Plate 4 Water ice is abundant on Pluto but is mostly covered by tholins and other ices. In some regions, the frozen water is exposed, as shown in blue in this image. NASA photo PIA19963.

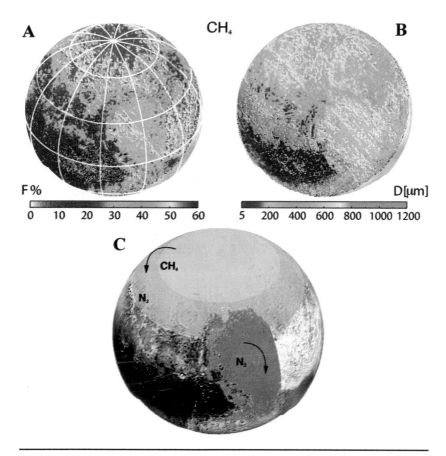

Plate 5 A. The distribution of Pluto's methane ice; red is the highest concentration (at the north pole) and magenta is the lowest concentration (in Cthulhu Regio). B. The size of the methane ice grains varies over Pluto's surface, with the largest ones (about 1,000 μm, or 1 mm) colored orange and the smallest grains in Cthulhu Regio. C. Methane is currently evaporating from the north polar region and is condensing at lower latitude. On Sputnik Planitia, nitrogen is migrating from north to south. Panels A and B from S. Protopapa et al., "Pluto's Global Surface Composition Through Pixel-by-Pixel Hapke Modeling of New Horizons Ralph/LEISA data," in "The Pluto System," edited by Richard P. Binzel et al., special issue, *Icarus* 287 (2017): 218–28, reproduced with permission by Elsevier. Panel C courtesy of S. Protopapa.

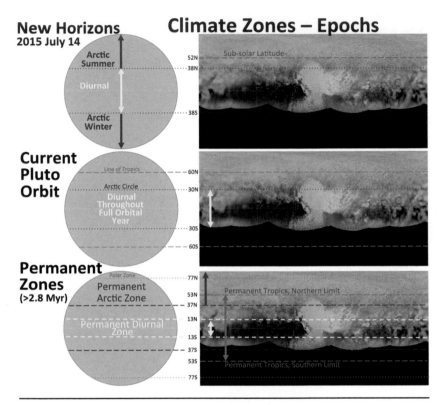

Plate 6 On timescales of millions of years, precession of Pluto's elliptical orbit results in large regions of the surface experiencing more sunlight on average than other regions. These inequalities result in the long-term migration of volatile surface ices to the regions that are colder on average. Image from Binzel et al., "Climate Zones on Pluto and Charon," *Icarus* 287 (2017): 30–36, used with permission of Elsevier.

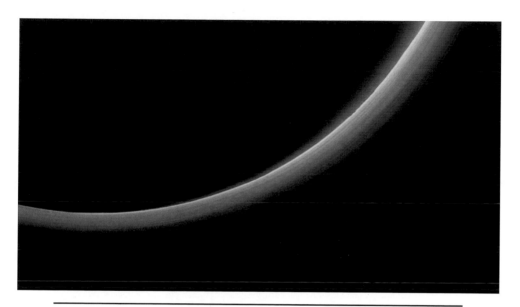

Plate 7 Several discrete layers of photochemical haze particles are evident in this view of Pluto's atmosphere on July 14, 2015, just after New Horizons passed beyond the sunlit side of the planet. The haze particles are composed of hydrocarbons such as acetylene and ethylene, and their small size causes them to preferentially scatter sunlight in the blue end of the spectrum, giving this hazy halo a distinct blue color. As they settle down through the atmosphere, the haze particles form as many as twenty or more horizontal layers, some covering hundreds of miles around Pluto. The haze layers extend to altitudes of more than 200 km (120 miles). Irregularities in the silhouetted limb of Pluto are mountainous terrain. NASA photo PIA20362.

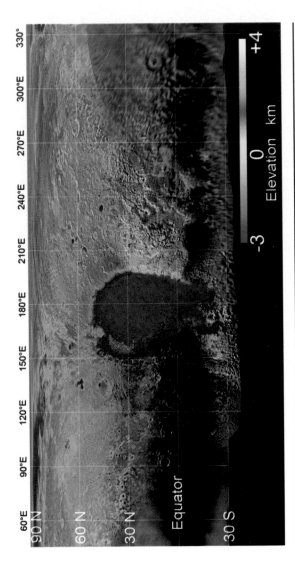

Plate 8 Stereo imaging of Pluto from the Long Range Reconnaissance Imager (LORRI) camera system on New Horizons was used to create this topographic map showing the elevations across the landscape, with a spatial resolution of 300 meters (985 ft). Terrain south of about 30° S was in darkness during the flyby, and is shown in black. Sputnik Planitia is the large area centered at 180° E and about 20° N, and sits 2–4 km lower than the surrounding region. Adapted from NASA image PIA21862.

To the Moon and Beyond

The exploration of the Moon by robotic spacecraft began with the Luna 2 probe launched by the Soviets, which crashed into the lunar surface just west of the crater Autolycus on September 14, 1959. Thus it accomplished what American rocket pioneer Robert Goddard had been ridiculed for proposing only four decades earlier. Less than a month later, on October 7, Luna 3 sent back the first images of the far side of the Moon. These images were the first ever received from space. Prompted by Soviet advances in space science, the United States began a surge of lunar exploration by unmanned exploratory probes, culminating in the first human landing on the Moon on July 20, 1969. In retrospect, the scientific aspects of lunar exploration appear to have been among the most significant achievements of that turbulent era, though at the time, the political and military considerations with which they were interleaved often seemed more important.

Within a few years of the first successful lunar probes, the first steps were also being taken, both by the Soviets and the Americans, in sending unmanned robotic spacecraft to the planets. The first interplanetary probe was the Soviet Venera 1, which was launched toward Venus in February 1961. Unfortunately—as was common in those days—radio contact was lost long before it reached the planet. The first successful interplanetary probe was the American Mariner 2, which made a successful flyby of Venus on December 14, 1962. Though it did not send back pictures, it provided data showing that the planet had no magnetic field and that beneath the clouds the planet sweltered in unimaginable heat (a first salutary warning of what can happen on a world subject to a runaway greenhouse effect).

On July 14, 1965, another American spacecraft, Mariner 4, made a flyby of Mars and sent back images of the planet's surface. (Memorize that date, for fifty years later to the day, it would again prove significant.) The images, with their coarse TV scanning lines, were not great even by 1965 standards; a single photo with a good 35mm camera contains about twenty-five times more information bits than the entire Mariner 4 photo set. Furthermore, the images had surprisingly low contrast; they appeared murky and hazy because designers had not counted on the dust, haze, and mists that often veil Mars—though it is also possible that the problem was something more mundane, such as a light leak in the camera system. Nevertheless, poor as they were, they were the first spacecraft images of another planet—and they were of the planet that at the time was by far the most interesting because, as Percival Lowell had dreamed, it might harbor life. These pictures provided the first view of real topographic and geological features on Mars instead of the broad shadings and shifting clouds that had been the domain of generations of telescopic observers. The features included nary a canal-like line but scores of impact craters, something like three hundred in all, including a feature 120-km wide in the dark area known as Mare Sirenum. It now bears the name "Mariner."

At the time of the Mariner 4 flyby, coauthor Cruikshank was one of Kuiper's graduate students at the Lunar and Planetary Laboratory in Tucson, Arizona. Coauthor Sheehan was a grade-school student in Minneapolis, Minnesota; fifty years later he wrote down how he felt on seeing those stark images from the Red Planet:

> A little bit of me died that day. Mars—and Percival Lowell's theories of the canal-builders—had meant so much to me. It was a shock to see Mars as a battered, lifeless, lunar world, almost as if someone had just proved that there was no Santa Claus. For a long time afterwards I was rather depressed. Eventually, of course, I recovered, but I took away from the whole experience a sense of just how tricky "seeing" is, and how far even eyewitness testimony, at least when it came to making out the features of a far away world in glimpses through a moving and blurry atmosphere, is not to be trusted. Also the way our human wishes enter in.[3]

Additional feats followed in rapid succession. The Soviet Venera space probes achieved the first successful soft landings on another world, Venus, in the early 1970s. In 1971, Mariner 9 entered orbit around Mars. Once the global dust storm that greeted its arrival cleared, Mariner 9 revealed that Mars was more interesting than it had appeared in the wake of Mariner 4 (and two further flybys, Mariners 6 and 7, in 1969). Mariner 9 unveiled huge shield volcanoes, gigantic canyons, and riverbeds. Clearly it had been shaped by the forces of volcanism, wind, ice, and—at some point in the past—running water. In 1974, Mariner 10 became the first spacecraft to fly by Mercury, returning stunning images of its cratered surface. Jupiter and Saturn were reached by the two flyby Pioneer spacecraft in the early 1970s as they followed a trajectory that would carry them out of the Solar System. This set the stage for the twin Voyager spacecraft, perhaps the most productive planetary missions ever conceived and implemented, that were launched in 1977.

Voyage to the Outer Planets

The Voyager story began in 1965, when the brilliant American aerospace engineer Gary A. Flandro, then at the Jet Propulsion Laboratory, was assigned the task of studying techniques for exploring the outer planets by spacecraft. Like everyone else who has ever wielded the mighty tools of celestial mechanics, he stood on the shoulders of giants. The classical analytical approach, first developed by Lagrange and Laplace, and still used by Adams, Le Verrier, and Lowell in their calculations, was to develop perturbations to each planet's orbit into a periodic series representing a function of the time. By the 1960s, this approach had been largely replaced by numerical integration methods, which were well suited to digital electronic computers.[4] While working through the calculations, Flandro discovered that a rare alignment of Jupiter, Saturn, Uranus, and Neptune would occur in the late 1970s (not to be repeated until the 2150s). Not only would this produce an interesting series of conjunctions of these planets in the night sky, but—more to the point—it would allow a spacecraft to be slung by gravity assists between one planet and the next, thereby shortening by many years the time of the trip and greatly reducing the cost by allowing a single spacecraft to survey several planets in series. It would be, in fact, rather like the post-Oxbridge tours of multiple European countries in search of art, culture, and the roots of Western civilization that were once undertaken by wealthy British youths as a necessary adjunct to their education, a Grand Tour. But this would be a tour on a much larger scale—a Grand Tour of the outer Solar System.

Pluto was not included in Flandro's original Grand Tour opportunity outline, but it was added in another outline by James E. Long the following year. As wildly successful as the Grand Tour concept would prove to be, NASA resisted it at the time, instead favoring a Jupiter-intensive mission. The Jet Propulsion Laboratory became the main advocate for the Grand Tour option, and promoted it largely because it would save money. Even though this was the period when NASA experienced what Tom Wolfe in *The Right Stuff* called "budgetless financing"—a situation that would never be repeated—the funding priority was the Apollo program, and planetary exploration did not necessarily receive the same largesse. An insight into the psychology is provided by NASA Planetary Programs Office Director Donald P. Hearth, who recalled what he thought when he first heard about the Grand Tour concept:

> You've got to remember selling a new start is a bitch. Even then [1960s]. It's almost as hard to sell a hundred million dollar project as it is a billion dollar project. And a hell of a lot more work to sell two $100-million projects than one $200-million project.[5]

Even the enthusiasm for the Apollo program did not long outlive its own success. By the time the last Apollo landed on the Moon in December 1972, public interest was severely on the wane, and NASA began casting around for a new, exciting, and nationally important mission to justify its continued existence. Among the possibilities were a manned mission to Mars, a space station in Earth orbit, and a space shuttle to transport people and materials to the space station and to move objects in space. Of course, the realization of any of these depended on the national purse. Unfortunately, by the early 1970s, Vietnam and economic issues had overtaken the space program as a national priority. Then, between 1973 and 1975, the country suffered an economic recession.

Fortunately, before the budget axe associated with these emerging economic strains fell with full severity, the new, exciting, and nationally important mission had already been approved. The Grand Tour, involving two Voyager spacecraft to the outer planets, narrowly escaped Nixonian cancellation, having been approved in May 1972, before the recession began to bite. In the end it would prove to be a bargain on the same scale as Queen Isabella of Spain subsidizing Columbus's three ships. The total cost to the

U.S. taxpayer of the two Voyagers through the Neptune encounter would be only $865 million, which—as someone has calculated—comes to only 8 cents per U.S. resident per year, much less than the cost of a single candy bar. These two missions—Voyager 1 and especially Voyager 2—would complete an epic voyage of discovery that will never be forgotten. The revelations they sent back from the planets would stir the imagination as no other mission ever has.

Soon after the launch window for the Grand Tour opened, the two spacecraft set out. Voyager 2 went first, on August 20, 1977, aboard a Titan-Centaur rocket; Voyager 1 followed on September 5, 1977, but followed a trajectory on which it would overtake its sister craft. Voyager 1 went only to Jupiter and Saturn, while Voyager 2 went not only to those two but on to Uranus and finally Neptune, where it arrived in August 1989 after a twelve-year journey. Few will need to be reminded of the breathtaking images sent back from the two Voyager spacecraft, showing the riotous clouds of Jupiter; Io's dramatic, tidally powered, sulfur-spewing volcanoes; the complex rings of Saturn; the spidery, dark rings of Uranus and the icy canyons and cliffs of its moon Miranda; and Neptune's deep-blue oceans of methane and the peculiar cantaloupe-rind skin of its satellite Triton.

No Pluto

The only outer planet not included in the Voyager itinerary was Pluto. It was initially on the program for Voyager 1; however, Saturn's ring system and its fascinating large satellite Titan were the main mission priorities, and to make a close flyby of Titan, the Pluto flyby had to be abandoned. Pluto was never on the Voyager 2 itinerary. The spacecraft might have been slung there from Neptune, but to do so would have involved a harrowing—and perhaps fatal—maneuver, requiring it to pass so close to Neptune that it would have passed inside the orbit of Triton. This would also have meant missing out on Triton itself—a fascinating world in its own right. Under the circumstances, no one could argue with the choices made. Pluto would have to wait.

After their flybys of the giant planets, the two Voyagers continued—still in radio communication with Earth thanks to the power delivered to them by their plutonium generators—to head ever farther from the Sun. Voyager 2 is predicted to remain in radio contact with Earth well into the 2020s. After its encounter with Saturn in 1980, Voyager 1 was thrown out of the plane of the ecliptic at an angle of 35° north, in the general direction of the solar apex (the direction of the Sun's motion relative to the nearby stars), and in August 2012, thirty-five years after launch, passed beyond the heliosheath, the outermost layer of the heliosphere. Here the solar wind is resisted by the pressure of the interstellar medium. It had thus officially crossed the boundary into interstellar space. In forty thousand years it will glide within 1.6 light years of AC +79 3888 (also known as Gliese 445), a tenth-magnitude red dwarf star in Camelopardalis, not far from Polaris.

Voyager 2, after its encounter with Neptune in 1989, veered south, in the direction of the constellations Sagittarius and Pavo. In about forty thousand years it will pass within 1.7 light years of the red dwarf star Ross 248 (also known as HH Andromedae or Gliese 905), a twelfth-magnitude star in Andromeda.

Voyager 1 is the most distant manmade object from Earth (and the Sun) at more than 20 billion km (134 AU). Voyager 2 is more than 16 billion km (116 AU) out.

Whither Pluto Bound, Then?

The scientific treasure trove brought back from the Voyager missions— which notably included Voyager 1's last image, the Solar System's "family portrait" taken from beyond the orbit of Pluto in 1990, which shows Earth as what Carl Sagan called a "pale blue dot"[6]—was so immensely rich and mind-bending that only the greediest or most fanatical Plutophile could possibly have complained at the time that Pluto had been overlooked.

In fact, there weren't many Plutophiles at the time. Any allure that the ninth planet possessed was muted and utterly swamped in the glare of what the Voyagers were finding out about the giant planets. For ten years either side of its perihelion passage in 1989 Pluto was temporarily inside the orbit of Neptune, but it was still a long way off, and because of its small size, it didn't seem very important. It received little attention or was assigned a low priority in the grand strategies being worked by scientists and managers at NASA as they began to prepare (and try to fund) the post-Voyager phases of the scientific study of the Solar System.

Nevertheless, by the late 1980s Pluto was gaining more respect. By then, it was known to be a binary planetary body with frozen methane on its surface. Moreover, it was shown to have yet another planetary feature: an atmosphere.

In the days when popular astronomy writers could only speculate about what spacecraft in the far-distant future would find when they visited other planets, Pluto was almost unimaginably aloof and nondescript. In his 1955 book *Guide the Planets*, Patrick Moore (the most gifted astronomy writer of his generation, and a perennial favorite of both the present authors), after describing the vivid impressions the giant planets Uranus and Neptune would present as viewed from some future spacecraft, the best he could do for Pluto was as follows:

> The scene is utterly unlike anything we have pictured, even in our dreams. This time there are no choking clouds of ammonia or methane, no brilliant planet looming large in the sky and shedding its radiance across the rocks; no breath of wind stirs the thin, icy hydrogen atmosphere, and the whole world is bathed in a deathly gloom. Pluto is a planet of eternal half-light. The Sun can be seen as an intensely brilliant point, but it is small and remote, and the shadows gather blackly around us.
>
> . . . Beyond all doubt, Pluto is the loneliest and most isolated world in the Solar System—cut off from its fellows, plunged in everlasting dusk, silent, barren, and touched with the chill of death. Nature seems to have passed it by, and it can never have known the breath of life. It marks the frontier of the Sun's kingdom.[7]

Another generation on, Pluto would prove to be far more interesting than Moore—and other astronomy writers of that era—could possibly imagine.[8] Moore could not have known that Pluto's loneliness and isolation were assuaged by the companionship of its large satellite (as well as four smaller ones), but at least he guessed that Pluto might have an atmosphere, though he was wrong about its composition. We have described the actual discovery of Pluto's atmosphere from observations on June 9, 1988, from the Kuiper Airborne Observatory and from telescopes on the ground along the occultation path. The identification of nitrogen as the main atmospheric constituent came four years later, in 1992, as we have also described.

All this intriguing information about Pluto, its atmosphere, and its outsized satellite, was available by the time Voyager 2 swung by Neptune before continuing its solitary trek across the Galaxy. If the Voyager missions to the giant planets had been the main course, Pluto was still waiting as the dessert. Already, in 1989, a NASA planning document noted that a mission to Pluto would complete the first spacecraft reconnaissance of the Solar System. The presidentially appointed administrator and the managers of NASA have always liked to achieve "firsts" in the exploration of space, a sure way of keeping up the public's interest and support, but while a mission to Pluto was programmatically and intellectually appealing, no source of funding within the space agency was identified.

Nevertheless, the ink was hardly dry on the planning document before more landmark discoveries about the outer Solar System were being announced. Pluto continued to get "curiouser and curiouser."

Kuiper Belt Objects

One of these discoveries pertained to the postulated, but still unconfirmed, Kuiper Belt. As noted in chapter 8, to account for short-period comets, Kuiper and others were led to surmise the existence of a ring of icy material outside the orbit of Neptune from which a few bodies were released into the inner Solar System each year. As of the 1990s, however, no Kuiper Belt Objects (KBOs) had ever been observed directly. Among known objects, Pluto, and its large satellite Charon, still marked the "frontier of the Sun's kingdom." Otherwise, the outer Solar System appeared to be strangely—disquietingly—empty.

Searches were made for objects beyond Pluto, but each time the searchers came up empty-handed. Clyde Tombaugh's exhaustive survey after discovering Pluto, ending only in 1943, showed that any such objects that might exist in the outer Solar System were bound to be much fainter than Pluto. In the 1970s, Charles Kowal, who suspected the existence of a tenth planet, carried out an even deeper search than Tombaugh using the 48-inch Schmidt telescope at Palomar Observatory. This time the limiting magnitude was nineteen. Though Kowal found Chiron, the first of the so-called centaurs orbiting between Saturn and Uranus, the space beyond Pluto continued to appear strangely empty.

Both searchers had used photographic plates, which were still state-of-the-art at the time. By the late 1980s, however, a new technological breakthrough presented new possibilities. The charge coupled device (CCD), first introduced at the end of the 1960s, was evolving as technological innovations always do, and the CCD detector arrays were becoming both sensitive enough and large enough to replace film for most astronomical applications. Among the first to seize on CCDs for deep solar-system survey work was David Jewitt, a native of England who had graduated from University College London before moving to Caltech for graduate studies. While at Caltech in 1982, he and G. Edward Danielson used a CCD on the 200-inch telescope at Palomar to make the first detection of the inward-bound nucleus of Halley's Comet, headed to its perihelion passage in 1986. At the time, Halley was a twenty-fourth-magnitude object, about a billion miles out and 1.5 AU beyond the orbit of Saturn. In the next few years, Jewitt became increasingly preoccupied with the problem of the "emptiness of the outer Solar System" and determined to find an object beyond Pluto, if any existed. In 1987, Jewitt, then at MIT, encouraged Jane Luu, a graduate student at MIT and the University of California, Berkeley, to join him in the search.

Luu had entered astronomy through an unusual path. She was a native of Vietnam, where her father had served as a translator for the U.S. Army and had escaped to the United States in the last wave of refugees before the fall of Saigon. Her interest in astronomy had been stimulated by a family visit to the Jet Propulsion Laboratory.

Jewitt and Luu carried out the first searches using photographic plates with telescopes at Kitt Peak in Arizona and Cerro Tololo Inter-American Observatory in Chile. They were no more successful than Tombaugh and Kowal had been. However, by 1988, Jewitt had moved to the Institute of Astronomy in Hawaii, and he and Luu were using a CCD camera with the 88-inch (2.2-m) telescope on Mauna Kea. Over five years of searching, the available CCD detector arrays increased in size. By 1992, they had reached 1,024 by 1,024 pixels, expanding the field of view and covering more of the sky in a single exposure. This greatly expedited the search, and led to success. In August 1992 they struck pay dirt with an object, magnitude 22.8—more than three thousand times fainter than Pluto—moving at a rate that showed it to be trans-Neptunian. Until an accurate orbit could be worked out, it received the provisional designation 1992 QB_1.[9] As the discoverers, it was Jewitt and Luu's prerogative, as soon as it had been observed over a long enough arc for an accurate orbit to be worked out, to give it a name. Untutored in the sometimes-inscrutable rules of the IAU, they had already decided on Smiley, for George Smiley, the fictional spymaster of British writer John le Carré's novels. Unfortunately, the name had already been assigned to an asteroid (in honor of Charles Hugh Smiley, a Brown University astronomer who had died in 1977). The IAU frowned on Smiley, since it involved a duplication of names, and subsequently, inertia set in. QB_1 remains QB_1.

That which we call QB_1 by any other name would be—trans-Neptunian. Not only trans-Neptunian, but vastly trans-Neptunian. At 43 AU from the Sun, or 1.2 billion miles (2 billion km) beyond Neptune, and with a period of 289 years, it resembled some of the putative planets that Todd, Gaillot, Pickering, and Lowell had posited from the tables of comet aphelia. Clearly, it belonged to the Kuiper Belt, and in a short time it lost its singularity. The following year, Jewitt and Luu found five more, and a further dozen turned up in 1994. As with asteroid discovery in the middle of the nineteenth century, the floodgates were opened. Now, more than fifteen hundred are known, divided into several categories. There are "cold classical" KBOs (where "cold" here applies not to their temperatures—they are all cold!—but to the orderly and unperturbed orbits in which they move), "hot classical" KBOs (whose orbits are more eccentric and highly inclined), "resonant" KBOs (orbiting in resonance with Neptune), and "scattered" KBOs, whose extremely elongated and inclined orbits, sometimes ranging from inside the orbit of Neptune out to distances hundreds of astronomical units from the Sun, are perhaps a result of approaching too close to Neptune in the past.

Some objects don't seem to fit with the KBOs at all—for instance, Sedna, with a diameter roughly half that of Pluto and in an orbit farther from the

Sun than any other known KBO. At perihelion, it is 76 AU from the Sun; at aphelion, 1,000 AU. It completes each orbit in twelve thousand years.

The term *resonance* deserves an explanation. A resonance involving two planets occurs when their orbital periods are in the ratio of two small integers, such as 2:1, 4:3, or, as in the case of Neptune and Pluto, 3:2. In the latter case, this means that in the time it takes Neptune to make three orbits of the Sun (i.e., $3 \times 165 = 495$ years), Pluto makes two orbital revolutions ($2 \times 248 = 496$ years); the numbers aren't precise.

The result of this correspondence of orbital periods is that the two planets exert a significant gravitational pull on one another every 495 years, and this locks them into this orbital resonance. The orbits of Jupiter's large moons Ganymede, Europa, and Io are locked in resonances with one another; Europa's period is twice that of Io's and Ganymede's period is four times that of Io. Many other resonances exist throughout the Solar System.

Orbit Patterns Beyond Neptune

At the same time Jewitt and Luu were finding the first KBOs, a mathematical researcher, Renu Malhotra, now at the Lunar and Planetary Laboratory at the University of Arizona, was trying to understand the implications of the 3:2 average resonance between Pluto and Neptune, which had first been recognized by C. J. Cohen and E. C. Hubbard in the mid-1960s via pioneering numerical integrations methods using computers.[10] Malhotra, the daughter of an aircraft engineer at Indian Airlines, had been born in New Delhi in 1961 and completed a master's degree in physics at the Indian Institute of Technology Delhi in 1983, before coming to the United States. She entered the PhD program at Cornell University, where she was introduced to nonlinear dynamics by the celebrated mathematician Mitchell Feigenbaum. In 1988, she completed her PhD with a dissertation on the moons of Uranus, and then obtained a postdoctoral fellowship under Peter Goldreich at Caltech, who was well known for work in which he had shown that Saturn's narrow F ring was maintained by shepherd moons and that Uranus's narrow rings were maintained by similar shepherds. The Uranian shepherd moons were later found by Voyager 2 during its Uranus flyby in 1986. Malhotra was intrigued by the oddness of Pluto's orbit. Unlike the orbits of the other major planets, which are well separated, nearly circular, and share roughly the same plane, Pluto's distance from the Sun at perihelion differs by almost 20 AU from its distance at aphelion, so that for twenty years around perihelion (last passed in 1989) it lies inside the orbit of Neptune. It also makes excursions of 8 AU above and 13 AU below the plane of the ecliptic—one of the circumstances that made it hard for the early searchers to capture it. Additionally, Pluto is effectively a binary planet with its large satellite, Charon. How did this binary planet in its very peculiar orbit in the outer reaches of the planetary system form, and how did it manage to remain stable?

Earlier astronomers, such as Raymond Lyttleton of Cambridge and later Kuiper, had speculated that because Pluto's orbit crosses that of Neptune, and Neptune has a Pluto-sized satellite, Triton, that moves in a retrograde direction around its primary, Pluto might have been an escaped moon of Neptune. However, as knowledge of the characteristics of the Pluto-Charon system improved, the escaped-satellite scenario lost its plausibility. Instead Pluto seemed to have formed, like other planets, in the outer Solar System (presumably as one of many small icy planets to form there) outside its current 3:2 resonance with Neptune and in a nearly circular, low-inclination orbit. Somehow it had managed to capture Charon and move into the very narrow region of phase space where it is protected by the 3:2 resonance.

Malhotra's model of how this came to be was regarded as implausible when she published it in 1993, but its basic principles are now generally accepted, bolstered by the discovery of some exoplanets with unexpected properties.[11] Malhotra proposed a process whereby, instead of remaining in the positions where they had formed, the giant planets had been subject to drastic migrations during the early Solar System. Beginning with the formation of the giant planets from the flattened disk of dust and gas that circulated around the Sun 4.6 billion years ago, she pictured the following sequence:

> The giant planets' gravitational perturbations cleared out their interplanetary regions by scattering the unaccreted mass of planetesimals. Some fraction of this mass now resides in the Oort cloud of comets . . . but most has been lost from the planetary system. A planetesimal scattered outward gains angular momentum, while one scattered inward loses angular momentum at the

expense of the planets. The back reaction of planetesimal scattering on the planets caused the planetary orbits to evolve.... Most of the inwardly scattered objects enter the zones of influence of the inner Jovian planets (Uranus, Saturn, and Jupiter). Of those scattered outward, some are lifted into wide, Oort cloud orbits, while others return to be re-accreted or re-scattered; a fraction of the latter is again re-scattered inward where the inner Jovian planets control the dynamics. In particular, Jupiter, due to its large mass, is very effective in causing a systematic loss of planetesimal mass by ejection into solar system escape orbits. Therefore, as Jupiter preferentially removes the inward scattered Neptune-zone planetesimals, the planetesimal population encountering Neptune at later times is increasingly biased towards objects with specific angular momentum and energy larger than Neptune's. Encounters with this planetesimal population produce a negative drag on Neptune which causes Neptune to gain orbital energy and angular momentum; as a result, its orbit expands. Jupiter is in effect the source of this angular momentum and energy; however, owing to its much larger mass, its orbit shrinks by only a small amount.[12]

Malhotra's scenario is correct in broad outline, but has been refined with additional study and calculations folded into what is now referred to as the Nice model. Its name comes from the fact that it was initially formulated by astronomers specializing in planetary dynamics at the Nice Observatory in France.[13] This new and more complete understanding of the evolution of the orbits of the four giant planets through mutual interactions has profound implications for the dynamical history of the primordial small bodies in the outer Solar System, especially Pluto. As Neptune migrated outward, a series of resonances swept across the outer Solar System. If Pluto had formed initially with an orbital radius such that the 3:2 resonance was the first major Neptune resonance to sweep by, Pluto would have been captured into this resonance lock with Neptune. Its binary companion, Charon, was likely captured at an earlier stage, when Pluto was in its original low-eccentricity, low-inclination orbit, and the binary would have been pulled into the 3:2 resonance with Neptune.[14] Numerical computations using computers have shown that evolution within the resonance would eventually lead to its present high-eccentricity orbit. (The high inclination of Pluto's orbit is trickier, and apparently due to multi-planet perturbations rather than being a result of the 3:2 resonance alone.) Within its 3:2 resonance zone, Pluto and Charon are in a protected orbit. They are not the only small icy planets that exist in the outer Solar System—they are simply among those that have survived where they formed; the rest, including the other KBOs, were removed from their original positions by collisions involving the outer planets.

New Horizons

The existence of the KBOs, as well as oddball objects such as Sedna, shows conclusively that the outer Solar System is neither as straightforward nor as bland as had been supposed. In fact, it is a realm of fascination, and has become an area of active research in its own right.

Of all these icy worlds, Pluto is the chief. In some ways, it is so much so that it merits the ninth-planet status it used to have when it rounded out the recitation of the planet names as the authors learned them: Mercury, Venus, Earth, Mars, Jupiter, Saturn, Uranus, Neptune, *and* Pluto. Mercury was the bookend on the inside, Pluto the bookend on the outside of this planetary library.

Pluto was still a bookend in 1989, when the NASA planning document mentioned above identified the need for a mission to Pluto to complete the first reconnaissance of the Solar System. Though no source of funding was identified, S. Alan Stern, a newly minted PhD in planetary astronomy with a preternatural—and already decade-long—interest in Pluto as a potential target of exploration, was already beginning to push the envelope at NASA. In May 1989, as Voyager 2 was still months away from Neptune and Triton, Stern approached the head of the Solar System Exploration Division at NASA Headquarters to make the case for a mission to the ninth planet. That official, Geoffrey A. Briggs, listened carefully, and instead of rejecting the idea encouraged Stern and other Pluto enthusiasts within the planetary-science community to flesh out the possible goals and technological requirements for such a mission. Briggs, himself a planetary scientist who had worked on numerous Mars missions, first as principal investigator on the Mariner Mars 1971 imaging team and later Voyager, in his official capacity at NASA Headquarters was always interested in new ideas for planetary exploration.

What followed from Stern's approach to Briggs was a long series of studies establishing science goals, drafting mission profiles, and writing proposals for funding. A mission to Pluto would have to fit with NASA's grand plan, frequently revised and updated, for the exploration of the planets in which Mars was most often heavily favored because it's only an eight-month trip and the edge the Red Planet has held (and continues to hold) for the first discovery of extraterrestrial life. From the start, Stern was the central figure in all planning for possible missions to Pluto, from large and capable Voyager-type spacecraft to tiny devices with only a camera as the scientific payload. The technical difficulties of a mission to Pluto included, but were hardly limited to, finding a power source able to operate the spacecraft so far from the Sun, assuring sufficiently hardy hardware for such a long voyage, and procuring a rocket large enough to get the spacecraft on a fast trajectory.

Over several years, with programmatic and funding issues, conflicts over the size and scope of a Pluto mission, and internal politics within NASA and other government entities to which it is responsible (e.g., the Office of Management and Budget), a clear choice for the desired trip to Pluto was not forthcoming.

An important factor on the way to the eventual approval of the mission to Pluto was the recommendation to NASA from the Decadal Survey panel, convened by the National Research Council, to define a new mission class to be called New Frontiers. This new class was to consist of principal investigator-led missions with a cost cap of $650 million. This distinguished it both from lower-cost missions in the Discovery class and higher-cost Flagship missions.[15]

For the purposes of the Decadal Survey study, Pluto was lumped together with comets and asteroids, small planetary satellites, KBOs, and interplanetary dust, all of which were considered by the Primitive Bodies committee, chaired by coauthor Cruikshank. Meanwhile, NASA had already selected New Horizons for flight, but funding and management obstacles were in its path. A recommendation from the Decadal Survey was clearly needed to move the mission into full development. After a full review of the importance of a mission to Pluto and the Kuiper Belt, consideration of the science objectives and the feasibility of achieving them, and consideration of available radioisotope power systems and the state of space-qualified technology, the panel recommend to the Steering Group as its first priority a Kuiper Belt-Pluto Explorer mission. This was ranked higher than a comet-nucleus sample return mission, a Trojan asteroid reconnaissance, an asteroid lander with rover and sample return, and another Triton-Neptune flyby. The panel's recommendations were carried to the Steering Group, and after lengthy debate and comparison with proposals from other panels, the Kuiper Belt-Pluto mission was recommended as the top priority for NASA's newly defined New Frontiers mission class. This recommendation, together with many other issues considered by the panels and Steering Group, was presented to NASA, announced in 2002, and published the next year in an official document of the National Research Council.[16]

The sequence of events that eventually led to success had begun with NASA's announcement of the opportunity to propose for a mission to the Kuiper Belt and Pluto in March 2001. Stern assembled his science team, cemented the relationship with the Applied Physics Laboratory at Johns

Figure 17.1 Members of the original New Horizons science team, San Antonio, Texas, November 2003. Back row: Ralph L. McNutt Jr., Marc Buie, William B. McKinnon, Darrell F. Strobel, Dennis C. Reuter, Donald E. Jennings, Andrew F. Cheng, Richard Terrile, Dale P. Cruikshank. Middle row: Jeffrey M. Moore, Richard P. Binzel, G. Leonard Tyler, William M. Grundy, Harold A. Weaver. Front row: Fran Bagenal, Michael E. Summers, Leslie A. Young, S. Alan Stern, Bonnie J. Buratti, G. Randall Gladstone. Photo by Dale Cruikshank.

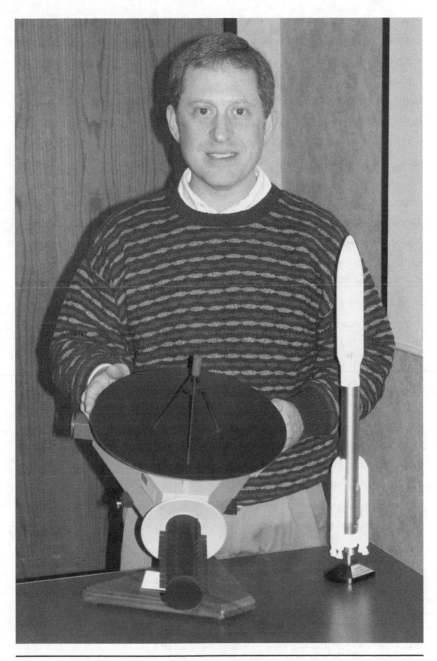

Figure 17.2 S. Alan Stern, principal investigator of NASA's New Horizons mission, with a model of the spacecraft and the Atlas 5 rocket on which it was launched. Photo by Dale Cruikshank.

Hopkins University—which would build and operate the spacecraft—and crafted the proposal document. It was submitted the next month, but the mission was blocked once more. A resubmission of the proposal in September 2001 led to the selection of New Horizons on November 29, 2001, for a Phase B study with the goal of a launch in 2006.[17] Yet another cancellation was ordered by the George W. Bush Administration in favor of a redesign of the mission, incorporating a huge, nuclear-powered spacecraft being considered by NASA at that time. As impressive as it sounded, no one expected such a spacecraft would ever be built. Fortuitously, in July 2002, the National Research Council announced the recommendation of the Decadal Survey to put a New Frontiers–class mission with the generic title Kuiper Belt-Pluto Explorer at the top of the queue for NASA funding and implementation.[18] While this recommendation appeared to clear the way for the development of New Horizons, which had already been extensively reviewed and approved for development, other hurdles had to be cleared. Additional delays were imposed as alternate technological approaches were first demanded and then withdrawn, and the selection of the launch vehicle also faced various obstacles. After all this, mission development finally got underway with the goal of a launch in early 2006. That goal would be reached, with the project on target, on time, and on budget.

In his study of the origin and eventual execution of the New Horizons mission, space historian Michael J. Neufeld notes that this mission "reveals a shift in the balance of power around (year) 2000 among the important players: NASA senior management, the planetary-science community, the space technology community and industry, Congress, the sitting presidential administration, and public advocates and lobbies."[19]

Is Pluto a Planet or Not?

Meanwhile, steady news was coming from the outer Solar System. KBOs continued to tumble out of an inexhaustible cornucopia. In 2002 the team of Michael E. Brown, Chadwick Trujillo, and David L. Rabinowitz, using the 48-inch Schmidt telescope at Palomar—the same telescope that Charles Kowal had used in his 1970s-era search for planets—discovered a somewhat large KBO, (50000) Quaoar, which had a satellite of its own, Weywot. In 2003 they discovered Sedna, described above, and then Eris in January 2005. Eris had a satellite, Dysnomia, and was thought to be larger—or at least more massive—than Pluto itself. This made a huge impression at the time, and demonstrated conclusively that the planetary system did not end at Pluto. Moreover, it added to the argument being made by at least some members of the astronomical community that Pluto ought not to be classed as a planet at all and should be demoted to a dwarf planet along with other KBOs and the two largest "classical" asteroids, Ceres and Vesta.[20] In this context, dwarf planets were seen to be essentially the same as minor planets, the accepted designation for the asteroids.

The timing of Eris's discovery clearly influenced what happened next, coming as it did shortly before the IAU held its 2006 congress in Prague, at which the status of Pluto was a topic of high interest on the agenda. Under circumstances that, in the view at least of those who disagreed with the result, resembled those associated with the election of a Renaissance pope, Pluto lost its planetary status and was officially declared a dwarf planet. (Harvard historian of astronomy Owen Gingerich, who participated in the deliberations, points out the semantic inconsistency of Pluto being a "dwarf planet" yet not a "planet.") The IAU's decision was immediately controversial, has been much discussed since, and remains unpopular with many (but not all) planetary scientists, and particularly those specializing in Pluto studies. It is safe to say that some have not, and never will, completely accept it. Since the subject has already been argued to death elsewhere, we will say no more about it here. Suffice it to say, Pluto is planet-like enough to have a large satellite, seasons, and an atmosphere. However, the astronomers, mostly dynamicists, who remained in Prague after the majority had left decided that all these factors were overridden by the dynamical (gravitational) interaction of an object with its space environment—for an object to be a planet, it had to clear out material in the zone in which it travels. This, then, being the sine qua non for planetary definition, Pluto was found wanting. It was a modern-day case of handwriting on the wall: *mene, mene, tekel, upharsin*.

So Pluto was demoted and ousted from the planetary club (or, to use a word that caught on for a brief period, it had been "plutoed"). One result—entirely unintended—of the IAU decision was how it affected public opinion.

People who knew very little about astronomy and would have glazed over with discussions of methane ice and tholins did not hesitate to express an opinion about Pluto's status. In general, and especially among school children, public opinion ran heavily in favor of its retaining ninth-planet status. Pluto not only remained in the public eye, but it achieved pop-cultural celebrity status.[21]

Ironically, since the IAU vote did not take place until August 2006, Pluto had still been the ninth planet when New Horizons launched in January. By the time it arrived at Pluto, nine years later, it was a dwarf planet.

Launch to Pluto

Coauthor Sheehan was lucky enough to be at Cape Canaveral on January 19, 2006, watching, in the company of David Tholen, from a site on the west side of the Banana River, six miles from Launch Complex 41 on which the powerful Atlas V 551 stood, a tiny finger of silvery light catching the fleeting sunlight of a piebald day, poised for liftoff. The launch had been twice postponed, the first day because of high winds, the next because of a power outage caused by severe winter weather at Maryland's Goddard Flight Center. On the third day, January 19, the winds had dropped but broken clouds continued to sail overhead. The launch was delayed for two suspenseful hours until a patch of blue arrived big enough to thread a rocket through.

At last, at about 2 p.m., that patch of blue arrived. The hold was lifted, and the countdown proceeded: "T minus 10, 9, 8 . . . 3, 2, 1." On cue, a puff of smoke appeared on the distant horizon, followed by a burst of orange flame. The rocket was an inverted candle—with fire spewing out the rostral end—and it rose slowly, majestically, above the pad, along a smoke-wreathed easterly arc across the sky and out over the Atlantic. The loud rumble of the rocket engines lagged behind its visual position, and as the rocket ascended, fell farther and farther behind.

The great journey to the edge of the planetary system was on. Everyone present knew that history was being made, and felt perhaps as those Spaniards did who watched as the Niña, the Pinta, and the Santa Maria set out on their voyage of discovery from Palos in 1492. At least the New Horizons, for all its deafening roar, went in peace, and without malintent; it sought not gold and earthly riches, but scientific findings from another world.

In Columbus's time, it was still believed not that Earth was flat but that the Sun traveled around Earth rather than the other way around. The Renaissance variant of the "what is a planet?" controversy would begin with the publication of Nicholas Copernicus's *De revolutionibus orbium coelestium* (The Revolution of the Celestial Spheres) half a century later, when Earth would be demoted from its rank as center of the Universe to "mere" planetary status.

Columbus's three ships took two months to cross the Atlantic, including a layover of a full month in the Canary Islands. The New Horizons probe, accelerated to a speed of 61,000 kph (38,000 mph) and leaving Earth's gravity at the highest velocity of any civilian launch, made the same passage in minutes. Just eight hours and thirty-five minutes after launch, it passed the orbit of the Moon, and arrived at Jupiter thirteen months later, on February 28, 2007. Thanks to Jupiter, it received an additional boost in speed, to 83,600 kph (52,000 mph) relative to the Sun at closest approach to the planet. As it flew by, Jupiter pulled back, and the speed was ultimately reduced to 50,400 kph (31,300 mph) for the remainder of the trip. Even so—such is the vast scale of the Solar System   it would take nine and a half years to reach Pluto. It was scheduled to make its closest flyby of the dwarf planet at 11:40 UTC (Coordinated Universal Time, the time standard used around the world), July 14, 2015, and then fourteen minutes later make its closest approach to Charon.

All of this would happen fifty years to the day after the Mariner 4 flew by Mars and sent back hazy pictures of its cratered surface, beginning the revolution in our thinking about the Red Planet.

18

The Flight of New Horizons

NEW HORIZONS MADE the trip to Jupiter faster than any other spacecraft before or since. From liftoff at Cape Canaveral to flyby of the Solar System's largest planet, the spacecraft gradually accelerated as it came under the influence of Jupiter's gravitational pull. Just thirteen months after launch, on February 28, 2007, with the instruments on board recording images, spectra, and other data, New Horizons sped past Jupiter about 2.3 million km (1.4 million miles), and about 2.2 million km (1.4 million miles) from its satellite Io. The path of New Horizons was nearly at a right angle to the orbit of Jupiter, so the encounter was brief, giving the spacecraft a boost in speed from the slingshot effect of its close approach to the planet and its quick exit from Jupiter's gravitational field. Jupiter's gravity briefly pulled back as New Horizons sped away, but the net effect of the encounter was to increase the speed of the spacecraft by nearly 14,000 kph (9,000 mph), roughly equivalent to another stage on the launch rocket. With the velocity boosted to 50,400 kph (31,300 mph), the remainder of the trip to Pluto was shortened by more than three and a half years, as planned.

A close flyby of Jupiter was an opportunity that could not be missed, not just because it was a convenient way to speed up the trip to Pluto, but also because it was a chance to do some good science at Jupiter. This meant the spacecraft and its instruments had to be in good working order, with all the operations thoroughly rehearsed and debugged. In fact, the preparations for the Jupiter encounter began almost immediately after launch. By the time Jupiter came into view, everything was ready.

During its brief flyby, New Horizons made close-range observations of Jupiter's dynamic and beautifully textured cloudy atmosphere, recording wave motions, convective movements, and the formation and rapid dissipation of ammonia ice clouds in regions of upwelling. On the night side, it captured views of brilliant polar auroras but found the airglow, which had been so prominent when the Voyagers flew by in 1979–80, largely absent, indicating a high degree of variability in the interaction of the atmosphere with its space environment.

As the largest planet in the Solar System, and the planet with the strongest magnetic field, Jupiter sports the most extensive magnetosphere, the region around the planet filled with high-density plasmas (atoms stripped of some of their electrons). Extending far beyond the planet itself, in a direction away from the Sun, is the magnetotail, shaped by Jupiter's magnetic field and the flux of charged particles from the Sun. The particle-detecting instruments on New Horizons measured the energy of particles in Jupiter's magnetosphere. After passing Jupiter, the spacecraft's unique trajectory took it right down the magnetotail for about 175 million km (109 million miles), finding clumps of plasma and other structures and phenomena never seen before.

If one were to choose one highlight, however, it would have to be the extraordinary eruption of one of the volcanoes on Jupiter's moon Io. The volcano, called Tvashtar after the Hindu god of blacksmiths, is one of many intermittently active Io volcanoes, and it serendipitously erupted during the New Horizons flyby, shooting a plume of gas and particles 350 km (220 miles) above the surface. Time-lapse images furnished evidence for the condensation of particles of sulfur and sulfur dioxide from the gas plume. Meanwhile, images of other glowing volcanoes on Io's night side indicated lava temperatures consistent with basalts.

As Jupiter shrunk in New Horizons' rearview mirror, the spacecraft settled in for the long haul to Pluto ahead, which even at its cruising speed of 50,400 kph (31,300 mph) would take another eight-and-a-half years. Mostly the spacecraft remained in a state of hibernation, although at various times it was awakened by radio contact with Earth to check on the status of the power source, the instruments, and other onboard systems. A few small mid-course corrections were made to the trajectory toward Pluto as the space-craft navigation team refined knowledge of the position and velocity of New Horizons through radio contact. The celestial mechanicians had done their work supremely well. Incidentally, calculations of the masses and motions of the outer planets based on measurements by the Voyager 2 spacecraft had revealed that the orbits of Uranus and Neptune were just fine on their own, with no Planet X required to explain their motions.

During New Horizons' long hibernation, studies of Pluto with telescopes on Earth and with the HST, launched from a Space Shuttle in 1990 and now working well after being initially plagued by a case of myopic optics, continued apace. Among the goals of this work was to monitor the surface composition for changes in the amount or distribution of ices as Pluto, having passed perihelion in 1989, began the long sojourn in its orbit away from the Sun. One major goal was to take advantage of additional stellar occultations to monitor the state of Pluto's atmosphere, and—the province of the HST alone—to search for additional satellites or possible rings around Pluto. Yet another priority for ground-based astronomers, who by now were rapidly increasing the tally of known KBOs after the initial breakthrough discovery by Jewitt and Luu in 1992, was to find one or more KBOs that could be targeted for an encounter with New Horizons after the 2015 Pluto flyby.

Even before New Horizons left Cape Canaveral, these investigations had begun to bear fruit. Two small satellites were discovered with the HST seven months before launch.[1] They were named, in accordance with the Plutonian underworld theme, Nix, for the Greek goddess of night, and Hydra, for the multiheaded dragon that guarded the entrance to the underworld and whose poisonous breath and blood were so virulent that even its scent was deadly. Still more could be found, as well as rings, if the depth of the search were increased further. Unfortunately, this search—as well as that for suitable KBOs to target beyond Pluto—was hampered by Pluto's position in the sky. Pluto's orbit crosses over the plane of the Galaxy twice during its 248-year orbit, and in the 2000s, the planet moved in front of the clutter of stars near the galactic center, the densest part of the Milky Way. With Pluto surrounded by an almost continuous carpet of faint stars, the search for faint satellites or rings was made supremely difficult.

The interest in inventorying satellites and rings was, of course, partly a matter of scientific curiosity, but a more practical reason existed: they could be a hazard to New Horizons as it flew on its intended trajectory close to Pluto and Charon. Though the chances of a collision with a specific body were so small they could be ignored, satellites could be sources of dust. In grinding collisions with meteoroids or comets over eons, satellites could eject fragments ranging in size from dust grains to boulders. Moreover, under some circumstances, the dust grains could congregate around the planet, at least temporarily, as a ring. At the high speed at which New Horizons would fly through the Pluto-Charon system—ten times the speed of a rifle bullet—any collision, even with a tiny grain of dust, had the potential to cripple one or more critical spacecraft components and jeopardize the entire mission. That the whole enterprise depended on the ability of the spacecraft to survive its passage through the Pluto-Charon system undamaged cannot be emphasized enough, since all the data collected during the brief encounter phase—more than fifty gigabits—would be stored on board for later transmission to Earth. A knockout blow from an encounter with a grain of dust—or larger particle—had the potential to cut off New Horizons' song after the first few notes. That would be almost worse than getting no information back at all. A breakdown of the spacecraft under such heartrending circumstances would be almost guaranteed to lead about four and a half hours later (the time required by a radio beam to traverse from Pluto to Earth) to the pernicious breakdowns of some of the mission scientists.

As the spacecraft continued to approach the flyby date of July 14, 2015, the urgency of evaluating any potential dust hazard rose. Meanwhile, Pluto's motion against the background clutter of stars was slowly carrying it away from the densest parts of the Milky Way. A few opportunities did arise, which do not seem to have been taken advantage of, when Pluto moved across some of the dark clouds of interstellar dust that interpose before more distant star clouds.

The gradual drop in the numbers of faint stars in the field of view improved the chances of finding faint satellites or a ring, and so it turned out that in June 2011 a fourth satellite of Pluto was found. It was named Kerberos, for the three-headed dog of the underworld in Greek mythology that guarded against unauthorized entry into Pluto's gloomy realm. Yet another satellite was picked up in June 2012 and named Styx. This raised the total number of Pluto satellites to five. For a small planet—or a dwarf planet, as it had to be officially called according to the IAU's August 2006 decision—this was an impressive retinue; Pluto alone surpassed the satellite tally of all the terrestrial planets combined. One might have expected that this would have entitled it to some respect.

These small satellites were a potentially interesting subject for scientific investigation in their own right, but at the moment their existence only served to heighten concerns among the worry-prone members of the New Horizons team about the potential dust hazard. To allay their concerns, the science and spacecraft teams decided to take a series of images at intervals as the spacecraft closed in on the planet. About fifty days before the encounter, the quality of the images with the spacecraft camera (the Long Range Reconnaissance Imager, or LORRI) began to exceed the best that could be obtained from Earth (those taken with the HST). Simultaneously, contingency plans were made for a possible change in the spacecraft's trajectory as it passed by Pluto should a dust hazard be found or suspected. A retargeting of the aim point somewhat farther from Pluto meant that the science results would be compromised—it would mean fewer high-resolution images and spectral data, and the potential to lose critical data, such as that involving occultation of the radio signal, as the spacecraft briefly passed behind Pluto and Charon. This was not the desired option; no one wanted it activated unless it became absolutely necessary. In that case, it would have to be done with a rocket firing just ten days before the flyby. An alternative, last-minute, and perhaps less effective mitigation strategy was to turn the spacecraft so that the large circular antenna would face in the direction of motion and thereby shield most of the spacecraft's critical components. Since that scenario also would reduce the scientific data obtained, it was also highly undesirable.

With the encounter date looming, the rate at which deep images of Pluto and its surroundings were captured with LORRI increased. Frantic calculations were underway to better understand the dynamical properties of dust particles in the Pluto system. Team members were relieved when they realized that Charon's gravity would clear dust particles away from the zone through which New Horizons would pass on its original planned trajectory at the desired distance of 12,500 km (7,767 miles) above the surface—provided that the dust had been created by an impact at least one hundred years ago. Gravitational clearing takes some time. Though the likelihood of a collision in the last century was not zero, the odds had greatly improved.

Science Team member John Spencer, of the Southwest Research Institute in Boulder, Colorado, led the hazard search group over the many months of deep imaging by both the HST and LORRI, and on June 22, 2015, just three weeks from the flyby date, he reported that no evidence of a ring or other dust signature could be seen in images taken by the spacecraft. At that point, the final decision was made to follow the original optimum trajectory through the Pluto-Charon system, which also meant proceeding with the pre-set observing sequences that had been thoroughly tested and approved.

Another urgent matter that had occupied Spencer and a team of at least twenty other astronomers was the search for suitable target KBOs for a potential New Horizons rendezvous after the Pluto-Charon encounter. The volume of space to be surveyed was small because retargeting the spacecraft after the Pluto flyby would be severely limited by the remaining onboard fuel for the trajectory corrections. This limited the set of possible suitable objects, all of which would inevitably be very faint and therefore camouflaged by the high density of background stars. Nevertheless, the search was begun in 2011. Large ground-based telescopes with wide-field cameras and the HST were used to search for a suitable target, but none was found. This was frustrating, since the later any potential discovery was made, the further limited was the volume of space to which the spacecraft could be redirected with the available fuel—and the fewer objects that were likely to be suitable. That no Kuiper Belt objects fitting the prerequisite criteria would be found was a distinct possibility.

Finally, in 2014, after the enormous efforts of many people, the search with the HST succeeded in discovering three potential targets, given the informal provisional names 2014 PT1, 2014 PT2, and 2014 PT3, with PT denoting "potential target."[2]

All three of the 2014 objects were recovered a year later, and their orbits were sufficiently determined to ensure that they would not be lost. When their orbits and potential sizes, and the fuel required for New Horizons to divert toward them, were calculated, 2014 PT1, later given the official designation 2014 MU_{69}, was chosen as the best target.

By the time New Horizons crossed the orbit of Neptune, radio signals traveling at the speed of light between Earth and the spacecraft took about four hours each way; when New Horizons was close to Pluto, the one-way travel time had become four hours and twenty-five minutes. This long delay meant that the spacecraft commands that would ensure that the required measurements were made during the few hours of the closest approach to the planet and its satellites had to be preset and stored in the onboard computers. The complexity of the commands to the spacecraft and its seven scientific instruments, together with the requirement that everything had to happen at exactly the right time, meant that the sequence for closest approach had to be written, tested several times, and sent to the spacecraft at just the right moment. Redundancy was a hedge against failure. In fact, the close-approach sequence was only one of several developed and tested in the years leading up to the encounter. As a hedge against possible onboard computer glitches, the sequences were stored in several locations in the spacecraft so that New Horizons could operate reliably in an autonomous mode for several hours at a time. Both the science team and the spacecraft-engineering team contributed to the effort. The science team worked out the exposure times for all the cameras, scan times for the spectrometers, and other details for the collection of the scientific data. The spacecraft-engineering team dealt with data storage, transmission back to Earth, spacecraft attitude, and dozens of other issues critical to the success of the once-in-a-lifetime encounter with the Pluto system. Everything humanly possible that could be done to assure success was done.

The spacecraft-engineering and navigation teams held drills to ensure that command loads were sent from Earth and acknowledged and tested on board. Spacecraft navigation was reviewed and evaluated many times to ensure New Horizons would hit the aim point at the right distance from Pluto and then Charon at exactly the right time. The science team held three rehearsals, called operational readiness training (ORT) in the mission center at the Applied Physics Laboratory in Laurel, Maryland. The first was in January 2015. The purpose was in part to test the team's ability to extract data from the format coming from the spacecraft and to process that data to derive a first impression of whatever scientific information they might reveal. Blocks of synthetic data in the spacecraft's format were prepared by an outside party to appear as reasonable facsimiles to the data that would come back from Pluto. The subgroups in the science team received the fake data and tested their processing and modeling techniques. New computer routines were then written as needed. An additional goal was to work with public-relations people so that the scientific results could be conveyed quickly, accurately, and—the members of the team hoped—engagingly to the media, and thence to the public.

The three ORTs served their purpose well, and by the time the real data were streaming back to Earth from New Horizons, most of the needed tools were in place and a protocol for communicating with the public had been developed and tested.

The spacecraft-navigation team that charted and maintained the long voyage of New Horizons to Pluto and Charon accomplished this task with extraordinary skill and precision. When New Horizons launched, Pluto's orbit was well established. However, the exact position of Pluto, because of the imprecision of the techniques by which it could be observed from Earth, was still uncertain by a few thousand kilometers. Obviously this was not good enough to plan a spaceflight trajectory, so Pluto's position had to be further refined during the flight. As the spacecraft approached Pluto, the onboard camera was used to discern Pluto's exact position. The exact velocity of the spacecraft also was carefully monitored by radio signals received from it. Thus the relative positions of the spacecraft and the planet it was approaching were eventually known to a high degree of precision.

The launch of New Horizons in January 2006 had itself been so nearly perfect that only eight very small corrections to its path to Pluto were needed in-flight. These corrections were applied by brief bursts of hydrazine gas from thrusters on the spacecraft. In preparation for the encounter, a very small ninth correction was needed. On June 29, 2015, a 23-second burst of hydrazine increased the spacecraft's velocity by just 27 cm per second (just over 0.5 mph). This tiny correction was needed to put New Horizons on track to pass over Pluto's surface at the exact distance of 12,500 km (7,750 miles) at exactly the right time. Without the correction, New Horizons would be more than 100 km off target and about 20 seconds late for the essential measurements of Pluto's atmosphere as the spacecraft passed behind it. In addition, this would have thrown off all the spacecraft command sequences, which had been carefully preplanned, tested, and set for execution at precise moments at various points along the trajectory.

With the flyby of Pluto and Charon set to occur on July 14, 2015—fifty years to the day after Mariner 4's flyby of Mars and twenty-five years after Voyager 2's flyby of Neptune—humankind was about to experience another first encounter with a planetary body. All the other planets had been visited, most of them more than once. Despite unproductive debates about the planetary status of Pluto, the scientific and cultural importance of this remote world was, and has remained, deeply etched in the human psyche. Whatever the astronomical community chose to call Pluto, interest could hardly have been greater. All of humanity—or at least that part of humanity with an interest in the wider Universe—stood tiptoe and awaited the next revelation of what lay over the hill.

Recognizing the special place that Pluto occupies, NASA, the Applied Physics Laboratory (APL), and Principal Investigator Alan Stern did everything they could to promote the mission and its impending success to the public. The public-relations campaign and educational-outreach activities lasted for the entire decade of the flight to Pluto. At the time of the encounter, media attention was focused on the event at a level probably not seen since the Voyager missions of the 1980s, the Viking landings on Mars 1976, or the early Apollo Moon landings. More than four hundred reporters and their crews were assembled at APL.

On the eve of the encounter—for present purposes, this includes the period up to fifty days out, when LORRI on New Horizons began to get better pictures of Pluto than had ever been achieved with the HST—we thought we knew the main facts about Pluto.

For example, we knew that it is a binary object with its companion Charon. The latter is half the size of Pluto. As such, it is larger relative to its primary than any other object in the Solar System. The two are locked in a synchronous orbit with Pluto's axial rotation. In addition, at least four small satellites, probably consisting of ices, orbit Pluto farther out than Charon.

We had discovered that Pluto's surface is covered with a combination of nitrogen, methane, carbon monoxide, and ethane ices. The temperature, at least at the point in its orbit where it would meet New Horizons, is about $42°$ K (about $-231°$ C).

We realized that Pluto has a very long and complicated seasonal cycle that is bound to affect the distribution of ices on its surface and cause that distribution to change over both long and short time scales. Despite its great distance from the Sun, the ices are colored; it is yellow orange, but—as had been roughly worked out from the mutual-event observations and somewhat more definitively from HST images—some regions are highly reflective and others are almost black.

We had found that Charon's greyish surface is predominantly covered with water ice, though with occasional exposures of an ammonia compound, probably ammonia hydrate.

We knew that Pluto has an atmosphere with a possible haze layer. This atmosphere consists primarily of nitrogen gas, but with a detectable quantity of methane, and we thought it was rapidly escaping to space.

Given Pluto's great distance—and the fact that when it was first discovered in 1930, it was nothing more than a faint speck of light—we had found out a lot. Still, all that we knew could fit on a white notecard. We also knew from every previous spacecraft mission to the planets—and as one of the authors still remembered so well from the results of Mariner 4—every time we have seen a planet up close, it has surprised us.

We were about to see Pluto up close: would it disappoint or exceed expectations? Anticipation was high for everyone who had waited so long and planned so carefully as the event finally approached, nine and a half years after New Horizons departed this Earth for the frontier of the Solar System. Barring some completely unexpected disastrous mishap, we were about to live through another banner day—perhaps the last we would have the chance to experience—in the history of our exploration of the Solar System, and hence of all human exploration.

In the final few weeks before closest approach to Pluto and Charon, the team drills and practice sessions began to pay off. The earliest science data from the spacecraft were high-resolution images taken with LORRI. Pluto's rotation period is 6.4 days, long compared to the brief time that New Horizons would be close. The highest-resolution images would be taken near closest approach and would cover only the hemisphere that was in sunlight at the time. Views of the other half of Pluto would have to be taken when it was in sunlight, about 3.2 days earlier. The same situation applies to

Charon. LORRI images were taken at frequent intervals for several days on approach, and as each block of pictures was transmitted from the spacecraft the team eagerly processed them to get the first views that had been so long in coming.

The resolution of features in the images transmitted by LORRI and its wider-field companion Ralph/Multispectral Visible Imaging Camera (MVIC) is reckoned in kilometers per pixel. Higher spatial resolution translates as fewer kilometers (and eventually meters) per pixel, but it also means that near closest approach, less of the surface can be imaged at the higher resolution because the spacecraft is moving so fast and the entire disk of Pluto or Charon cannot be scanned before it's gone from view. With these constraints in mind, the choice of flyby distance had been made months in advance to maximize the resolution that could be achieved over the largest parts of Pluto's and Charon's surfaces. The flyby distance from Pluto's surface was chosen to be 12,500 km (7,767 miles), and the spacecraft velocity and timing were adjusted accordingly with the last tiny thrust for trajectory correction on June 29, as described earlier.

The highest resolution of the LORRI images of Pluto is ~79 m (~260 ft) per pixel and of Charon 157 m (515 ft) per pixel. The wider-field MVIC viewed Pluto's surface with a maximum resolution of 320 m (1049 ft) per pixel and Charon 628 m (2060 ft) per pixel. With the highest LORRI resolution, surface features on Pluto as small as Alcatraz Island in San Francisco Bay or details with the dimensions of the New York City skyline were seen. The resolving power of LORRI images is sufficient to see objects the size of the U.S. Capitol, and the clusters of people on the National Mall for the inauguration of a new president, or the masses of people bathing in the Ganges River during the annual Hindu festival Kumbh Mela.

The geology and geophysics (GGI) team was one of four subgroups of the New Horizons science team that we introduced in chapter 1. They dealt primarily with the study and interpretation of images of the surfaces of Pluto and all five of its satellites obtained with LORRI and MVIC. Going beyond the physical description of the features seen in the images, they endeavored to interpret past and current geological processes. The interpretation was especially challenging because the surface of Pluto is utterly unlike any planetary surface ever seen in the Solar System.

The composition (COMP) team was concerned with interpreting the color images and spectroscopic data obtained with MVIC and the Linear Etalon Image Spectral Array (LEISA) to reveal the compositions of various regions on the planet and satellite surfaces through their spectral signatures. In addition to identifying the compositions of the ice and non-ice components of Pluto and Charon, COMP mapped the distributions of these materials across the varied topography revealed in the high-resolution images of the geologic structures.

Scientists in the atmospheres (ATM) team used images, ultraviolet and infrared spectra, and results from the radio occultation experiment to determine the composition, temperature, and pressure of Pluto's atmosphere at all altitudes above the surface. They were particularly eager to understand the nature of the hazes that surround Pluto. They also were ready to search for an atmosphere on Charon,

In addition to the mostly steady stream of solar wind radiating outward through the Solar System, disturbances in the Sun's magnetic field cause sporadic bursts of plasma (electrons and protons) called coronal mass ejections. The interaction of this streaming mix of charged particles with Pluto and its effect on the planet's atmosphere were the subject matter of the particles and plasma group (PPG). As we noted, the plasma instruments were working all the way to Pluto and had revealed some unexpected features of Jupiter's magnetotail.

As the New Horizons spacecraft drew increasingly closer to Pluto and Charon in late June and early July, the entire science team was assembled at the APL campus. The spacecraft had been assembled and tested at APL, and all the mission operations had been conducted there since the launch. Commands to and data from the spacecraft (the ultimate long-distance calls) were sent from APL via NASA's Deep Space Network of large antennas in California, Spain, and Australia. New images from LORRI and MVIC received every day showed increasing detail of both Pluto and Charon and, together with some other data about the plasma environment at Pluto's distance of 32.9 AU (4.93 billion km, or 3.06 billion miles) from the Sun, were eagerly processed and discussed by the team.

While the great majority of the data from New Horizons, including the images, spectra, plasma measurements, and the engineering data on the spacecraft itself, were stored on board during the rapid-fire observing sequences, arrangements had been made to send back some first-look and fail-safe images almost in real time. These few images were critical in the event of a last-minute spacecraft systems failure, possibly from being hit by a dust grain. In the worst case, at least a few close-range images of Pluto's surface would survive.

The highest-resolution and most visually appealing images were also essential to fulfill promises made to the press and public media. Strategic releases of images and other information to the press and in social media, particularly in the first two weeks of July, had resulted in a crescendo of public interest and excitement that hadn't been seen surrounding space exploration for years. Contributing to the special excitement of the flyby was the decade-long and surprisingly passionate discussion of Pluto's status among the other planets of the Solar System. The Pluto question is especially resonant with young students, and is a persistent topic for debates in schools across the country, serving as a useful means for teachers to inspire student interest in science.

The wishes of NASA, APL, and senior New Horizons mission personnel were well exceeded by media attention to the flyby of Pluto and its moons, and the close-range pictures released several hours after the shutter snapped 3 billion miles away did not disappoint.

Wherever one happened to be—at APL, on Mars Hill in Flagstaff, any of the great cities and small towns of America, or anywhere on the third planet from the Sun where the interested public waited—we were once again on the verge of great events.

19

Pluto and Charon

Marvelous Worlds

Early Arrival

AFTER ITS NINE-AND-A-HALF-YEAR journey over 4.8 billion km (3 billion miles) from Earth to Pluto, the New Horizons spacecraft arrived 88 seconds early and within 45 km (28 miles) of the targeted time and position. With the timing and position so close to the nominal values, the full suite of scientific observations programmed into the spacecraft were executed just as planned months before. Now, approaching three years since the flyby, the New Horizons science team and other scientists have pored over the images and other data, and put together an unexpected and astonishingly dynamic picture of Pluto, Charon, and the four small satellites.

The closest approach to Pluto at 12,500 km (7,770 miles) from the surface occurred on July 14, 2015, at 11:49 Coordinated Universal Time (UTC), and to Charon at 28,800 km (17,900 miles) a few minutes later at 12:04 UTC. As the first-look and fail-safe images were traveling over radio waves to Earth, the spacecraft turned to aim its instruments back toward the Sun and Earth. The Sun disappeared first behind Pluto and then Charon just minutes after the spacecraft's closest approaches, giving the optical instruments, particularly the ultraviolet spectrometer, the opportunity to probe Pluto's atmosphere and to search for a thin gaseous envelope around Charon.

Powerful radio signals sent from antennas in NASA's Deep Space Network were recorded on New Horizons as Earth disappeared behind first Pluto and then Charon, also within minutes of the closest approaches. These radio occultations also provided the opportunity to acquire atmospheric data as the radio signals were extinguished for the short time the spacecraft was behind each body. Because the characterization of Pluto's atmosphere was so important, the flight plan was to take the spacecraft on precisely this trajectory. The accuracy of both the aim point in space and the timing of the flyby were critical.

In addition, during the few busy hours after closest approach to Pluto, New Horizons had opportunities to take images of Pluto's small satellites. Although the spacecraft did not pass very close to any of the four, it did obtain some good images plus spectral data for Nix and Hydra.

As soon as the most intense parts of the observing sequences were concluded and the data safely stored on board, the spacecraft turned its antenna to Earth. It then began transmitting the images, spectra, charged-particle measurements, and other crucial bits of information to the eager science and spacecraft-operations teams. With the power of its radio transmitter only 12 watts, the rate of data transmission was only two kilobits per second. By comparison, the standard digital transmission in modern internet networks is sixty-four kilobits per second. To receive the signals on Earth, 70-m (230-ft) dishes, the largest antennas of the Deep Space Network, were needed—those of the Goldstone Deep Space Communications Complex, located outside Barstow, California, the Madrid facility in Spain, and the Canberra facility in Australia. While at least one, and usually two, of these antennas would be in position to see New Horizons at all times, the use of these dishes was already heavily subscribed for communications with many spacecraft scattered throughout the Solar System, so New Horizons had to share. The data-transmission plan was to send all the science and spacecraft data in a compressed format at the highest available rate as insurance against any untoward events that might shorten the life of New Horizons, then continue transmission of the uncompressed data set afterward. This first phase of data return took about six months and was completed by October 2016.

Surprises

Inevitably, the most compelling (and press-worthy) data from New Horizons were the close-range images showing the landforms of Pluto and Charon. The robotic exploration of the Solar System has taken the virtual eyes of planetary scientists to terrains and environments of unimaginable variety, from the scorched lava fields of Mercury, to the dust-blown deserts of Mars, to the volcanically active surface of churned-up Io, to the diverse landscapes of the icy satellites circling the giant planets. With the variety of landforms seen from up close on nineteen large icy moons and half a dozen small ones,[1] Pluto was expected to offer a potpourri of things already explored, similar to vistas seen elsewhere. In particular, the best speculation was that it would resemble Neptune's large moon, Triton. Pluto and Triton had long been considered twins, and for a while, as we have seen, were even thought to have conspired with one another in a close encounter in orbit around Neptune to reverse the direction of Triton's orbital motion and send Pluto careening into the outer Solar System. After Voyager 2's Triton encounter in 1989, the argument for their being kindred worlds seemed even more compelling because of their similar sizes, densities, surface compositions, and shared cold space environment. The conjecture ended with New Horizons' close encounter with Pluto: as increasingly detailed images began to trickle in during the approach to the planet, and especially when the highest-resolution views came back during the closest-approach observing window, mission scientists and the public alike were simultaneously astonished and humbled as they made firsthand acquaintance with an extraordinary and utterly new world.

Just after the flyby, and with at least a few broad-brush images and data to look at, Principal Investigator Alan Stern and members of the New Horizons Science Team began to share their first impressions. Stern, who more than anyone else had succeeded in persuading NASA officials that this mission was worth flying, summed up in three words: "We did it!" A picture, widely published, showing Stern, Will Grundy, and other members of the team ogling in delight at an image on their computer screen, captured better than a thousand words the sheer exhilaration of exploring these remote worlds.

Even the preliminary results showed that Pluto was far different from what anyone had dreamed. They showed, in contrast to the fossilized face of Charon, a complex and, in places, unexpectedly youthful geology and varied surface composition. Parts of Pluto's surface are ancient, probably more than 4 billion years old, but no one expected the presence of young-appearing landforms evidently being fashioned by dynamical processes still taking place and attesting to the continuing role of heat in the interior of the planet.

Where does this heat come from? Long after the formation of the Solar System, the heat in the interior of a rocky body continues to be generated by the slow decay of the radioactive elements incorporated in it as it coalesced from the solar nebula. Small bodies have less radioactive content and generally cool quickly, while larger ones have more and cool more slowly. Earth's dynamic interior is driven mostly by this radiogenic heat; it is what keeps the iron core molten and the rocky mantle slowly shifting in a convective pattern that manifests itself in the crust as continental drift, volcanic activity, and sea-floor spreading. It also produces Earth's strong magnetic field.

As a planetary body smaller than Earth's Moon, Pluto was expected to have cooled since its origin to the point where any remnant heat in the interior would be too small to visibly affect the surface. In this regard, Pluto most completely defied expectations.

The Strange Geology of Pluto

The cameras on New Horizons (see appendix 2) took many images of Pluto and Charon, covering every part of their surfaces illuminated by sunlight at that time. Multiple images of the same scene taken from slightly different angles allowed the construction of three-dimensional views and the computation of a digital elevation map using the sophisticated technique of stereo photogrammetry. This information on the heights and depths of the surface structures would be critical in attempts to puzzle out how they formed and the modifications by geological forces to which they have been subjected over millions of years.

The Complex Story of Pluto's Ices

Understanding Pluto's surface features and the forces that shaped them requires an examination of the properties of ices. The Earth-like planets in the inner Solar System are forged of rock (comprising mainly silicate minerals) and metal in different proportions in different planets. Earth's mass is about 68 percent silicate rock forming the mantle and crust, and about 32 percent metal (iron and nickel) concentrated in the core. The outer Solar System's satellites, on the other hand, are composed mainly of frozen water. Triton is an exception: its surface is frozen methane and nitrogen, in addition to water ice, while much of its interior is thought to be a mix of the silicate minerals and (probably) metals that is generically termed rocky material.

While Pluto's deep interior is mostly rocky material, its mantle is largely frozen water. The geologically active surface is shaped by a combination of frozen methane, nitrogen, carbon monoxide, and water. As noted in earlier chapters, though all of these except water ice had been found from previous Earth-based telescope observations, they were corroborated by New Horizons as soon as it came close enough to Pluto for spectroscopy with the its mapping spectrometer, the Linear Etalon Image Spectral Array (LEISA). Although the spatial resolution of this instrument is lower than that of the cameras, the resulting data allows the topographic features on Pluto and Charon to be correlated with their compositions. Water ice was found in some areas of Pluto's surface from LEISA data about a week before the spacecraft made its closest approach, and the geographical distribution of the four main ices was revealed. In seeking to further decipher Pluto's geology, an understanding of the properties of these four main ices individually and in combinations, at the planet's low surface temperature of about 45° K (−228° C, −379° F), is needed. Much of the desired information about the properties of these combinations has not yet been precisely determined in the laboratory, but at least they can be estimated in broad outline.

Nitrogen evaporates from the surface into Pluto's atmosphere, whose hazy layers and blue sky were among the more sensational New Horizons findings, and re-condenses to make nitrogen ice when the temperature drops on a very long seasonal timescale. Since the nitrogen ice is soft, it flows over long periods—just as water ice does on Earth. Nitrogen and carbon monoxide are the most volatile ices, exchanging between the surface and the atmosphere as the temperature changes. Methane ice is less volatile, and at Pluto's temperature, water ice is quite literally rock: it never melts and it never evaporates. The situation is made more complicated by the fact that nitrogen and methane are soluble in one another. Thus neither one is ever found in its pure state. The proportions in which nitrogen and methane mix have a strong effect on the freezing temperature and lead to other subtle but important differences in their physical properties, differences that are only now beginning to be explored in the laboratory.

The Great Nitrogen-Ice Sea

The hemisphere of Pluto in full sunlight at the time of New Horizons' closest approach was primarily the hemisphere opposite Charon. It included Pluto's north pole and a swath of terrain along the equator. Among the several geologically distinct regions on this hemisphere, the most notable is a roughly circular plain that has no discernable impact craters. This great expanse of nitrogen ice, the lobe that is the northwestern third of the great heart-shaped light area Tombaugh Regio, is Sputnik Planitia. Sputnik Planitia covers some 870,000 km^2 (336,000 square miles, about 25 percent larger than Texas) and its surface lies 3–4 km (2–3 miles) below the highlands that surround it. It appears to mark the position of a large impact basin formed early in Pluto's history. The interior surface shows a large-scale pattern reminiscent of convection in a pan of liquid heated from below, consisting of roughly polygonal cells some 10 to 40 km (6 to 24 miles) across whose edges are defined by narrow, smooth lanes. Calculations of convection patterns expected in a layer of nitrogen ice a few kilometers thick appear to be consistent with the sizes and shapes of the features on Sputnik Planitia, with the flow of the ice down into the interior along the cell margins and upwelling in the central parts, serving to refresh the surface. Although the mechanical properties of solid nitrogen with small amounts of methane and carbon monoxide dissolved in it are not yet well known from laboratory studies, the best available data suggests that the timescale for turnover of the convection cells in Sputnik Planitia is on the order of half a million years. Geologically speaking, this

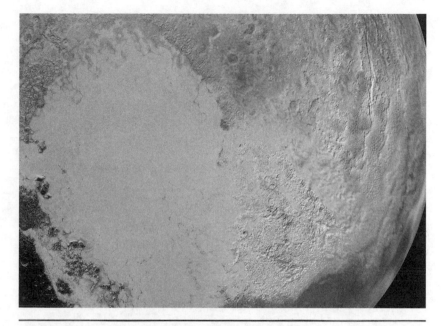

Figure 19.1 This prominent heart-shaped region consists of two major terrain types. The left half is smooth on the large scale, but is covered with pits a few hundred meters across and displays a pattern of convection cells. This region is a sea of nitrogen ice several kilometers deep in which slow-moving vertical currents refresh the surface on timescales of thousands of years. The absence of impact craters indicates that the surface is no more than about 10 million years old. The brilliantly white upland region to the right may be coated by nitrogen ice that has been transported through the atmosphere and deposited at higher elevations. This image is a composite of several individual images. NASA photo PIA19945.

is a very short timescale that agrees with the absence of detectable impact craters in this part of Pluto's surface.[2]

High-resolution images of Sputnik Planitia show that this geologically young surface is carpeted with circular or nearly circular shallow pits, typically one hundred meters to a few kilometers across. Some of these are drawn out into distinctly elongated forms; others are aligned in such a way as to suggest that some kind of flow has taken place. Many features, including the convection pattern and absence of impact craters younger than about 10 million years, can be explained by the tendency of frozen nitrogen to flow, albeit very slowly. The small pock marks all over the surface appear

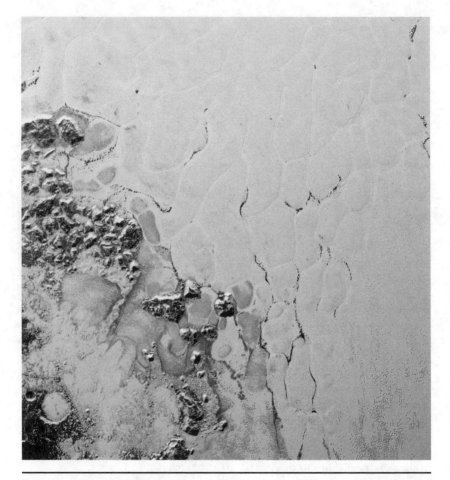

Figure 19.2 In this view of Sputnik Planitia, smooth convection cells in the solid nitrogen ice appear to be crowded against the western margin at the lower left. Irregular blocks, presumed to be frozen water, may be pushed along by the moving nitrogen ice. This scene is about 400 km (250 miles) across, NASA photo PIA20726.

to be sublimation pits formed by the direct evaporation of nitrogen (and methane) ice when mildly heated by the distant Sun. The alignments of some of the clusters of shallow pits are presumably set along the direction of the flow pattern.

Water ice is less dense than nitrogen ice, and in principle will float in a sea of frozen nitrogen. Floating mountain-size clumps of water ice may explain the broken, jagged blocks that protrude from the surface mainly at

Figure 19.3 The smooth but pockmarked surface of Sputnik Planitia borders against dark highlands called Krun Macula, which rises some 2.5 km (1.5 miles) above the surrounding plane. Krun Macula is scarred by clusters of connected, roughly circular pits that reach between 8 and 13 km (5 and 8 miles) across and up to 2.5 km (1.5 miles) deep. At the boundary with Sputnik Planitia, these pits form deep valleys more than 40 km (25 miles) long and almost 3 km (2 miles) deep, which is almost twice as deep as the Grand Canyon. The floors of the pits are covered with nitrogen ice and may have formed by the collapse of large sections of Pluto's surface. NASA photo PIA20732.

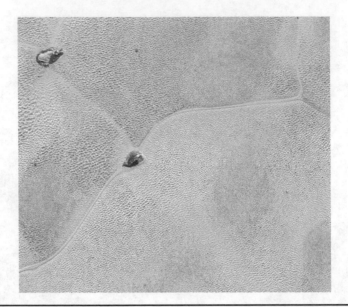

Figure 19.4 This section of a mosaic strip across Pluto includes the highest-resolution images taken by New Horizons. At this resolution of 80 m (260 ft) per pixel, details of the convecting cells of solid nitrogen in Sputnik Planitia and the abundance of sublimation pits are clearly seen. The width of this image is about 80 km (50 miles). NASA photo PIA14458.

the northwestern margin of Sputnik Planitia. These icebergs may have been broken off the edges of the basin and rafted on flowing nitrogen ice to other locations. Or they may be the shattered remains of a part of Pluto's ancient crust that have been invaded and are still being nudged, jumbled, and tilted by currents in the slow-moving nitrogen ice. Glacial flow from higher elevations into Sputnik Planitia and into the larger Tombaugh Regio is evident in several locations where the convergence of multiple tributaries to the main flow can be seen, as well as moraine-like features common in terrestrial glaciers. These nitrogen ice glaciers appear to be flowing currently.

The extraordinary Sputnik Planitia has still another story to tell: its location on Pluto is almost exactly opposite of Charon, which as we have seen remains stationary over this part of the planet because of its tidal lock with Pluto. Sputnik Planitia may be in some way responsible for this permanent locked orientation. To test this possibility, Francis Nimmo of the University

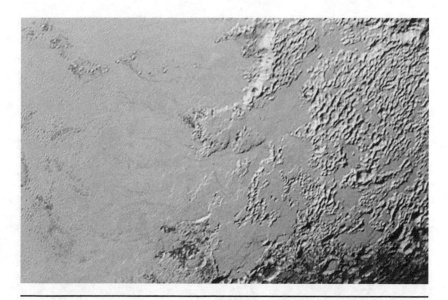

Figure 19.5 Nitrogen ice on the uplands on the right side of this 630-km (390-miles) wide image is draining (right to left) from Pluto's highlands onto Sputnik Planitia through the 3–8-km-wide (2–5-mile-wide) valleys. The origin of the ridges and pits on the right side of the image remains uncertain. NASA photo PIA 19944.

of California, Santa Cruz, has calculated the effect of a high point in Pluto's gravitational field at this location on how the two bodies revolve about their common center of gravity, or barycenter.[3] If Sputnik Planitia is an ancient impact basin where nitrogen later condensed, the positive gravity anomaly can be accounted for if Pluto has a subsurface ocean of liquid water and the planet's crust is thinner at that location than elsewhere because of the impact. A subsurface body of water close to the surface under Sputnik Planitia could, because of its effect on the gravity field, have reoriented Pluto's rotation to its present alignment. The existence of such a subsurface ocean is hardly far-fetched, since liquid water oceans under the surfaces of Jupiter's satellite Europa and Saturn's satellite Enceladus have been established to a near certainty. But the possibility of liquid water in Pluto's interior raises some fascinating possibilities about its origin.

Enough has been said to convey something of the astonishment the New Horizons science team experienced as the first of the high-resolution images were downloading from deep space. The very idea that Pluto is geologically

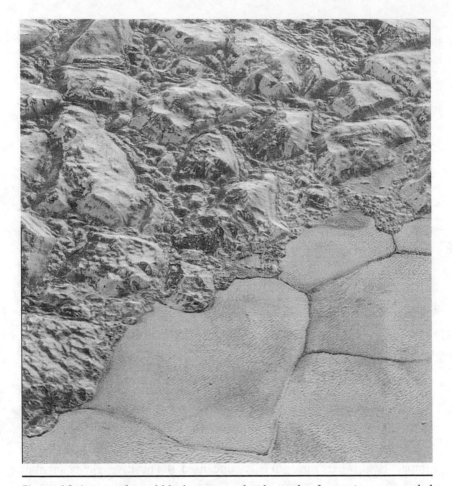

Figure 19.6 Large-faceted blocks presumed to be made of water ice are crowded against one another on the shoreline of Sputnik Planitia (bottom of frame). The tilted blocks are covered with nitrogen and methane ice but show layers of red tholin material deposited in a past epoch. This is a section of NASA photo PIA14458.

active was paradigm-shifting. The possibility had been mostly dismissed long ago, since Pluto and other small, extremely cold, and icy bodies far from the Sun were supposed to be frozen and dead. Furthermore, because the rotations of Pluto and Charon are tidally locked, no energy production was expected in their interiors, in contrast to the continually flexed and partially melted Galilean satellites of Jupiter.[4] Pluto and Charon were supposed to have long ago dissipated the heat from radioactive elements incorporated into them when they emerged from the solar nebula. Significant additional heating would have been created when a small body likely collided with Pluto, resulting in the formation of Charon.[5] But that, too, must have taken place early in their histories, and the heat dissipated long ago.

As so often as happens during spacecraft exploration of the planets, things did not turn out as expected. After New Horizons, everything about Pluto and its moon must be thought through afresh. Clearly Pluto is not entirely cold; a persistent source of heat must continue to drive Pluto's ongoing geological activity. In light of the New Horizons findings, the currently favored theory is that enough internal heating is present through the decay of natural radioactive elements to drive geological processes. If so, then other icy bodies of comparable size in the Solar System near and beyond Pluto may also be geologically active. Difficult as it would have been to fathom only a few years ago, the bodies of the outer Solar System are not, perhaps, immobile worlds in deep freeze as long seemed almost self-evident. They may not be dead worlds after all.

The Terrains and Colors of Pluto

Apart from the youthful and possibly still-active structures in Tombaugh Regio, which is perhaps the most intriguing part of the planet imaged at high resolution by New Horizons, much of the rest of Pluto's surface has a truly ancient look.

We now know that Pluto is a KBO. Of almost planetary size, it is admittedly an outlier—possibly the largest KBO to have survived the pulverizing effects of the frequent impacts that took place during the Solar System's rough-and-tumble first half-billion years. Theoretical calculations suggest that the impact of a large object with young Pluto would have produced a large cloud of fragments that would have coalesced into Charon. Pluto's scars from this collision eventually healed, and all traces of its existence were erased as Pluto collapsed gravitationally into its nearly spherical shape, with the heavy, rocky material forming the center and the lighter icy components forming the mantle and crust. Over the subsequent history of the Solar System, the KBOs continued their process of mutual grinding down.

Figure 19.7 The ancient and mostly ice-free (on the surface) expanse on Pluto informally named Cthulhu Regio extends nearly halfway around Pluto's equator, starting to the west of Sputnik Planitia, the great nitrogen ice plains. Cthulhu measures approximately 3,000 km (1,850 miles) long and 750 km (450 miles) wide, and with an area of more than 1.8 million km² (700,000 square miles), it is a little larger than Alaska. NASA photo PIA19952.

The ejection of many of the ensuing fragments over time has now left the belt depleted of large objects, but during the first billion or so years after the formation of Charon, random fragments of other KBOs must have been common in the region, regularly pummeling Pluto. The legacy of this violence is the population of ancient craters still visible on the aged parts of its surface.

A high density of craters is found in a large dark area just south of the equator: Cthulhu Regio. (The provisional name was taken from the works of the American horror writer H. P. Lovecraft.) This region is noteworthy for its dark red color and the apparent thick layering of colored material that in some places completely blankets underlying topographic features. Cthulhu also boasts mountain ranges that are capped not with snow but with methane frost. Cthulhu's craters range from about 0.5 to 250 km (0.3 to

155 miles) in diameter. The high density of craters here—and the presence of many larger ones—indicate that this area must be very old (~4 billion years). It is clearly one of the oldest and least-modified regions viewed closely by New Horizons. In other cratered regions of Pluto the crater walls and floors often expose layers of dark and light material while the rims of many of the craters are hoary with deposits of methane ice or frost. The layers and frosted crater rims bear witness to epochs in Pluto's early history when the

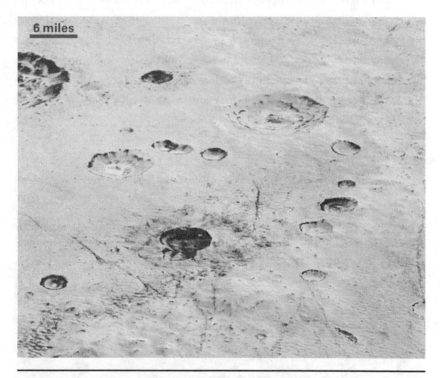

6 miles

Figure 19.8 This highest-resolution image from New Horizons reveals new details of Pluto's rugged, icy, cratered plains. Layering in the interior walls of many craters indicate changes in the geological conditions on Pluto over time. The darker crater in the lower center is apparently younger than the others because dark materials ejected from within have not been erased and can still be seen. The origin of the many dark linear features trending roughly vertically in the bottom half of the image may be tectonic. Most of the craters seen here lie within the Burney Basin, with is 250-km (144 miles) wide and whose outer rim or ring forms the line of hills or low mountains at bottom. The top of the image is to Pluto's northwest. This image was made with LORRI about 15 minutes before New Horizons' closest approach to Pluto on July 14, 2015. NASA photo PIA20200.

temperatures were much different from what they are now, allowing for the evaporation of the ices (except water) and their redeposition on the surface by the condensation of the corresponding gases from the atmosphere. In principle, working out a chronology of Pluto's past should be possible from carefully studying these layers—a task awaiting a deeper investigation of the New Horizons data set.

Cthulhu Regio is notable both for its very low reflectivity—about 9 percent, only slightly more reflective than such rocky worlds as Mercury or the Moon—as well as the deep red color that is responsible for this low reflectivity. Other, more limited areas have a similar red color, while broad expanses of terrain have coloring somewhere between light yellow and orange. Whatever pigment colors most of Pluto's surface, the deep red areas appear to have the highest concentration. It is important to emphasize that the ices that

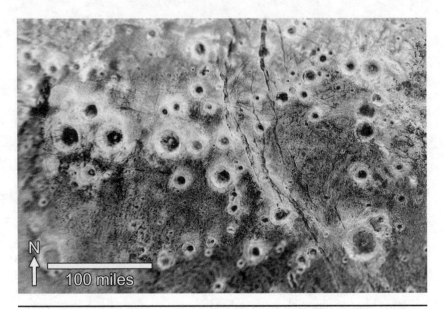

Figure 19.9 This region is in the western part of the hemisphere that New Horizons viewed. The craters are remnants of impacts on Pluto's surface; the largest is about 50 km (30 miles) across. The bright walls and rims of the craters are coated with methane frost, while the crater floors and the terrain on which they occur show the presence of frozen water. The maximum surface resolution is about 230 m (760 ft) per pixel, and the scene was imaged from a distance of 46,400 km (28,800 miles). NASA photo PIA20656.

cover Pluto's visible surface are themselves colorless. However, exposure of nitrogen, methane, and carbon monoxide to ultraviolet sunlight, charged atomic particles from the solar wind, and cosmic rays impinging on the planet's atmosphere produce chemical reactions that form the chemically complex and colorful tholins discussed previously. Pluto, which might have been expected to be composed only of bland and colorless ice, is lightly or, in some places, deeply dyed in tholins. The tholins are distributed in unexpected ways: they appear as layers in tilted water-ice blocks at the margins of Sputnik Planitia, but as deposits on the flat floors of steep-walled canyons and as layers in the canyon walls themselves. In some areas the colored material appears as little more than a light dusting on preexisting topography; in Cthulhu Regio they form thick deposits.

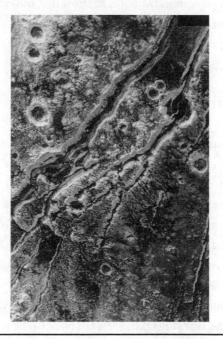

Figure 19.10 A complex network of canyons (fossae) cuts across Pluto's highlands west of Sputnik Planitia, suggestive of some kind of fluid drainage along tectonically induced lines of weakness in the planet's crust. Layers of dark material can be seen in some of the canyon walls, as well as mass wasting of that material down to the flat floor. The dark materials are thought to be tholins, but their method of formation and transport are not understood. A finer network of cracks, also suggestive of drainage, lies perpendicular to the main canyons. NASA photo PIA 20658.

Tholins not only form directly on the surface but also in Pluto's atmosphere. Ultraviolet sunlight and charged particles cause the breakdown of methane and nitrogen molecules into molecular fragments, and their recombination in turn gives rise to heavier molecules of highly varied structure. The colliding molecules eventually stick together and make aggregated aerosol particles. Depending on their sizes, some of these particles slowly sift downward from the atmosphere and reach the surface where, over time, they can build up in sufficient concentration to have a discernible color. Since Pluto's atmosphere varies in density and in the concentration of various molecules over long timescales in patterns that are not yet completely understood, the rates at which tholins form and come to rest on the surface must also be highly variable. These processes present a puzzle that is not easily disentangled, but enough is already known to show that the complex interplay of tholin production in the atmosphere and in the ices on the surface, as well as details of composition, contain important clues to Pluto's complex history.

Volcanoes of Ice

Two mountainous structures, Piccard Mons and Wright Mons in Pluto's highlands on the eastern margin of Tombaugh Regio, are of special interest. They are roughly circular in outline, about 125 km (100 miles) across, and notable for the very rough textures and great depths at their centers (the latter ranging from 6 to 10 km). Since they vaguely resemble volcanic structures on Earth and other planets, the New Horizons team has speculated that they might be cryovolcanoes. From the terminology alone one can imagine what a cryovolcanic vent might look like: icy masses, propelled upward from pressure in the interior, spew out from a vent on the surface, or blobs of icy slurry are ejected to great heights, then fall back to the surface with a *splat!* and freeze solid again. Icy masses or slush pushed out from the central vent would build up a high circular mound until all the pressure from below is gone, at which point, no longer resisted, the central part collapses into a deep, wide hole. This possibility is easy enough to imagine, and may provide a plausible explanation for the colossal gnarled ice piles with central pits like Piccard Mons and Wright Mons.[6]

Apart from the icy luster of the modern frozen-nitrogen sea that is Sputnik Planitia, Pluto largely presents a weather-beaten face that is wrinkled,
lined, and reddened by prolonged exposure to the elements. Global tectonic
forces have produced large patterns of cracks and scarps, while a great variety of terrains define geologic provinces across the hemisphere imaged by
New Horizons. We have a rather vague idea of the geological features on the
other hemisphere, which was imaged from a distance and at low resolution
on approach to the planet. That hemisphere was in sunlight some 3.2 days,
or half of a Pluto rotation, before closest approach.

Pluto's diversity of strange and curious features immediately challenges
the geologist's vocabulary. Among the terms extracted from the geologist's
traditional lexicon by the members of the New Horizons science team to
describe these features are *washboard, fretted, dissected, degraded, bladed,
pitted, fluted, dendritic,* and more. Troughs and valleys, some with flat floors,
cut through craters or terminate in craters. Water ice and red tholin deposits
coat the flat-bottomed valleys, while dark material spills from well-defined
layers in their walls down the steep slopes in a process of mass wasting.
Other ancient terrain appears wrinkled from forces of compression, and
valley systems suggest erosion by the downhill drainage of material somehow mobilized to flow in this utterly frozen world. Cliffs retreat as mass
slips from their faces, weakened by the sublimation of the ices that glue
them together.

On Earth, geological processes take place on both very long-term (e.g.,
mountain building) and very short-term (e.g., volcanic eruptions) timescales; on frozen Pluto, though geologically active, all processes seem to proceed in extreme slow motion, at least in the current epoch.

The surfaces of Pluto and Charon, like those of other planets and satellites, are a chronicle documenting past epochs in the history of the planet,
and planetary geologists have already begun trying to decipher this history,
at least in broad outline.[7] The various scenarios share one common aspect:
all involve the rigidity and ability of water ice under Pluto conditions to
physically support other ices that are more transient and mobile, particularly
nitrogen and methane, as they cycle between the atmosphere and the surface
on millennial timescales. Inevitably some nitrogen escapes into space over
time, but enough has been retained to continue to drive the dynamics of

the surface and atmosphere billions of years after Pluto formed. Moving ice, even the relatively soft nitrogen, appears to erode upland regions formed of rock-hard water ice, leaving behind the recognizable glacial terrain seen on the planet. It also affects, in ways not entirely understood, other regions by sculpting a variety of textures, some familiar looking, others frankly bizarre. Words fail, however, to adequately communicate the complexity and intricacies of Pluto, and for that reason we have no recourse but to direct the reader back to the images in the New Horizons gallery included here. Even a cursory perusal suffices to show why geologists are both perplexed and delighted at the diversity and strangeness that Pluto proffers.

Even before New Horizons, telescopic observations from Earth showed that frozen nitrogen, methane, and carbon monoxide are not distributed uniformly over Pluto's surface. However, it took the close-range spectral views afforded by LEISA to show in detail just where the concentrations of each kind of ice lie. When spectral images are overlain on the higher-resolution images from MVIC and LORRI, the correlations of composition with geology become clear. Methane gives a particularly strong signature, meaning that it is in relatively high abundance in Tombaugh Regio, including the northwestern smooth section that is Sputnik Planitia. Methane frost also favors the rims of craters, appearing to condense more easily from the atmosphere on these parts that rise a bit higher into the colder air. On the other hand, Cthulhu Regio and Krun Macula in the equatorial region show little or no exposed ice, except for the crest of a long mountain range rising from Cthulhu where condensed methane frost has brightened and highlighted the landscape. Instead, as we have seen, Cthulhu is draped in dark red tholins, which makes it stand out from the rest of the surface. In addition to these observations, the LEISA data showed for the first time the distribution of water and ethane) ices, which are not readily visible from Earth.

Clearly, the surface distribution of ices on Pluto is complex. The three main ices, nitrogen, methane, and carbon monoxide, differ in volatility (evaporation and condensation temperatures) and in their ability to mix with one another at the molecular level (which also affects volatility). The consequence of these ices' properties is that they distribute themselves across the surface in patterns related to season, latitude, altitude, local slope, and Pluto's changing distance from the Sun. They also provide different contributions to Pluto's atmosphere through seasonal cycles of evaporation and condensation at the surface.

The high-latitude regions on Pluto have relatively little nitrogen ice. Presumably, this is from the peculiar patterns of seasonal and long-term exposure to sunlight that Pluto's surface experiences. Pluto's rotational axis is tipped more than 119° from the plane of its orbit around the Sun, resulting in a strong seasonal cycle over the 248-year orbit. If this were the only factor affecting the planet's seasons, then over the full orbit each region of the surface would get about the same total dose of sunlight. However, Pluto's axis precesses rather like a top,[8] and over millions of years, precessional cycles and variations in the planet's orbit cause the north and south polar regions to get strongly contrasting periods of sunlight. For the high-latitude zones on Pluto around the north and south poles, over millions of years the precession results in long arctic-like periods with the Sun always low in the sky, alternating with long tropical-like periods with the Sun high in the sky. The equatorial region, by contrast, gets a more uniform exposure to sunlight, never the arctic or the tropical extremes. On Pluto, then, the situation is opposite of that found on Earth, where the equatorial regions are warmest and the polar regions icy. Were Earth's axis as highly tipped as Pluto, the same pattern as on Pluto would prevail. The long-term pattern of solar heating is bound to affect the distribution of Pluto's ices, and thus provides a cogent explanation of why such a small amount of nitrogen appears in the midlatitudes and north polar region: those regions are warmer than the equator where Sputnik Planitia lies.

Most planetary satellites, KBOs, and comets in the outer Solar System are well endowed with water, the most abundant ice-forming substance. Pluto's bulk density of 1.86 g/cm³ implies that much of the interior is composed of water ice. However, exposures of it on the surface are inconspicuous because there the water ice is largely masked by nitrogen, methane, and carbon monoxide. Frozen water forms the hard bedrock of Pluto. Under Pluto conditions, it never evaporates—any more than granite or schist do on Earth. But more-volatile molecules migrate between the surface and the atmosphere as the temperature fluctuates during the planet's short- and long-term seasonal cycles. LEISA spectral images show water ice in limited regions of the surface, notably in mountainous regions and, curiously, in Virgil Fossa, a deep, flat-floored channel or valley in the old terrain of Cthulhu Regio, west of Sputnik Planitia. Virgil Fossa is a gash 3–4 km (1.9–2.5 miles) deep, running for about 950 km (590 miles) and terminating in Elliot crater, which is 88 km (55 miles) in diameter. In addition to the presence of water ice here, much of the floor of the valley has a bright red-orange color that contrasts with the dark red of Cthulhu Regio. It is probable that these different colors come from two distinctly different tholins. Pulfrich crater in eastern Tombaugh Regio is another concentration of exposed water ice, but with no comparable presence of tholins.

Pluto's most prominent mountainous areas are known as Baré, al-Idrisi, Hillary, and Zheng-He Montes. They also have exposed water ice, especially in the complex and steep valleys that dissect the main mountain masses. The appearance of the ice in the valleys supports the view that the mountains themselves are made of frozen water and that nitrogen and methane ices form superficial layers that blanket the higher elevations and whose thickness may cycle with the condensation and evaporation of these more volatile materials.[9]

We have only begun to make sense of the ancient frozen records of Pluto's history. What seems clear is that the ices of Pluto, spanning a huge range of volatility from the rigid and unmoving water to the nimble nitrogen and carbon monoxide, with methane in between, have shaped and modified the surface of this dynamic object over billions of years as conditions on Pluto's surface and in its atmosphere have changed. Layers of ices and tholins, attesting to past eras, are exposed in the walls of impact craters, tilted mountain blocks, and in steep valley walls, while in Sputnik Planitia, modern glaciers continue to creep into a vast sea of slowly simmering nitrogen ice.

Spring on Pluto and Charon

The unusual circumstances of the short- and long-term seasonal cycles Pluto undergoes are reflected not only in the features of its surface but in those of its atmosphere as well.

Though discovered in the 1980s, the very fact that Pluto has an atmosphere—and even, as we now know, blue skies—remains almost hard

to believe. Moreover, at times in the long-term seasonal cycles the atmosphere is substantial enough to have wind, weather, clouds—perhaps the
most surprising result of New Horizons. Nitrogen is the main constituent of
Pluto's atmosphere (as of Earth's), and is in equilibrium with nitrogen frost
or ice on the ground. Whenever Pluto warms, the nitrogen ice sublimates
to gas, and the atmospheric pressure goes up. Conversely, whenever Pluto
cools, frost forms, and the atmospheric pressure drops. This sounds simple,
but the situation is complex and involves several other factors, only some of
which are currently understood.

The most fundamental cycle of warming and cooling is that which Pluto
experiences as the solar energy it receives varies over the course of each 248-
year Pluto orbit. Because of its extremely eccentric orbit, this cycle is far from
negligible. Indeed, at perihelion, when Pluto is only 30 AU from the Sun, it
receives a meager 1/900 as much solar radiation as Earth; at aphelion, when
it is 50 AU from the Sun, it receives only 1/2,500 as much.

Also extreme is the high inclination of its axis of rotation to the plane of
its orbit (referred to as the obliquity by astronomers), which results in very
exaggerated seasons by terrestrial standards. Pluto's obliquity is, at the present epoch, 119.6°, but the exact figure varies over millions of years, and this,
as we mentioned above, is an important factor in explaining the distribution
of ices and even of Pluto's surface features.

We might consider, for a moment, what conditions on Earth would be
like if, instead of its current inclination of 23.4°, Earth's axis had an extreme
tilt like Pluto's. Each hemisphere would then be shrouded in frigid night
for six months at a time as other sweltered under the blistering heat of a
six-month-long day. The Sun would blaze down over the day hemisphere
from as high an angle as that with which it intensely shines on our tropics
for the few hours around noon each day. Daytime temperatures would be
blistering: at the north pole they would reach 50° C (more than 120° F); at
the south pole—located in the middle of Antarctica, thus well away from
the temperature-moderating effects of the ocean and on our alt-Earth the
hottest part of the planet—they would likely reach as high as 80° C (176° F).
Meanwhile, Earth's equatorial regions, instead of being steaming tropics as
they are now, would be below freezing and covered in ice year-around.[10]
These extremes of temperature would have strong effects on the oceans as

well as the atmosphere, and whether such a planet would be habitable or not we leave to the reader's imagination.

What we consider normal conditions on Earth are therefore the product of cosmic accidents.[11] The present inclination of Earth's axis is stable, although the continents and oceans change positions over time because of continental drift. In terms of living conditions on Earth, the tilt of our its axis is only one determinant of climate; others include massive volcanic eruptions that cloud the atmosphere, rises in greenhouse gases (carbon dioxide and methane, in particular), and the occasional (but rare) impact by an asteroid or comet larger than a few kilometers. The climate of Earth, over the most recent several million years, has been moderate and relatively stable, making our planet habitable. That is not to say, however, that we are blind to the fact that the habits and needs of the nearly 8 billion people on Earth are having a steadily worsening effect on both the atmosphere and the oceans. Thanks to the presence of the Moon, our planet may be immune to large polar precession, but there is no question that Earth's climate is changing noticeably over the span of a single human lifetime, and that more frequent extremes of weather are on the rise. In contrast, while Mars in the present epoch has an axial tilt almost the same as Earth's, it is subject to a wide variation of angles over time spans of less than a million years. Large variations of climate on Mars have resulted in a very complex geological record of volcanic eruptions, erosion by water and wind, and the formation and disappearance of ices at the north and south poles.[12]

Pluto's seasons are aligned with its orbit in such a way that the most recent equinox occurred in 1989, at about the same time that Pluto reached perihelion. Though the geometry of the orbit is such that both hemispheres receive similar amounts of solar energy during a Plutonian year, the onset of summer in the northern hemisphere is much more rapid than that in the southern hemisphere. Also, as just described, any place in a polar region receives more sunlight than in the equatorial region over the course of a full year.

Much of this had been worked out in great detail before New Horizons reached Pluto which led planetary scientists to entertain multiples scenarios or educated guesses regarding what New Horizons would find when it arrived. Some expected the spacecraft would find two polar ice caps; others guessed that at any given time Pluto would have one large cap at one or the other pole. When New Horizons' imagery began to come back, most investigators still thought that the seasonal effects of the Pluto year would go a long way toward explaining the complexity of Pluto's surface, including the recent reworking of some areas.

However, in the first year since the flyby, the interpretation has become much deeper and more complicated. Recent work has considered not only Pluto's annual seasonal cycle but the influence of a 2.8-million-year oscillation caused by the perturbations of the giant planets that produces a variation in Pluto's axial tilt ranging from 103 to 127°. When the tilt is near the 103° minimum, as it last was 0.8 million years ago, some 97 percent of Pluto's surface experiences an overhead Sun, while about 78 percent experiences arctic winter (and corresponding arctic summer). As we already mentioned briefly, the effect of this degree of axial tilt is that during such eras, a broad range of Pluto's latitudes—some 75 percent of the total surface area—will experience a mixture of tropical and arctic climate regimes. However, a permanent "diurnal zone," where day-night cycles occur each 6.4-day Pluto rotation and neither arctic winter nor arctic summer is experienced for at least twenty years, prevails in latitudes within 13° on either side of the equator. As discussed further below, the existence of this diurnal zone likely accounts in part for the planetwide dark equatorial band noted on the New Horizons images. Other aspects—such as the stark contrast between bright Tombaugh Regio and dark Cthulhu Regio—are also likely because of long-term effects of these climate zones and the effect they have on volatile transport and deposition.

Charon, whose synchronous alignment with Pluto means it shares the same climate-zone structure, shows similar effects: in particular, it is noteworthy that the dark north polar cap lies entirely within the permanent "polar zone" (above 77° latitude), where the Sun never reaches the overhead point and arctic conditions have prevailed for perhaps tens of millions of years. Though all this remains a work in progress, scientists are optimistic that detailed correlations of features on Pluto with these long-term cycles may eventually become possible, though on Charon, except for the dark north polar cap, such detailed correlations are less likely simply because minimal volatile transport is apparent on its surface.[13]

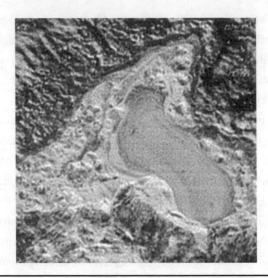

Figure 19.11 This feature may be a frozen, former lake of liquid nitrogen located in a mountain range just north of Sputnik Planitia. Liquid nitrogen may have been stable on Pluto's surface at times in the past when the atmosphere was denser and the temperature higher, and it may have flowed and ponded in various regions of the planet's surface. Nitrogen cannot exist as a liquid on Pluto under the present climatic conditions. This image shows details as small as about 130 m (430 ft). At its widest point the lake is about 30 km (20 miles) across. NASA photo PIA20543.

The Atmosphere

We already alluded briefly to the effect that seasonal cycles have had on Pluto's atmosphere. The characterization of the structure and composition of that atmosphere was, of course, one of the highest priorities of the New Horizons mission.

Images obtained with MVIC and LORRI of the night side of Pluto took advantage of the dim illumination provided by sunlight reflected from Charon's surface—a moonlight view of Pluto's surface that was otherwise in darkness at the time of closest approach just minutes before. Wide-field views of Pluto's night side, with the Sun illuminating the atmosphere, show a beautiful blue-tinted halo surrounding the planet that reveals nearly two dozen layers of haze encircling the dark hemisphere. (Who would have ever imagined that Pluto has blue skies!) Sunlight reflected from this haze to the ground gave a dim illumination to parts of Pluto's surface, allowing twilight views of terrain that could not otherwise have been seen.

A new view of Pluto's atmosphere emerged from the combined results from the New Horizons' Alice ultraviolet spectrometer, the radio occultation experiment, the solar occultation experiment, and the images taken with MVIC and LORRI.[14] New Horizons also investigated Pluto's atmosphere using radio beams. A powerful radio beam sent from Earth was monitored by the spacecraft's REX (Radio Silence Experiment) instrument as it flew behind Pluto as seen from Earth (ingress), remained out of contact with Earth for eleven minutes, and then recovered the signal as Earth reappeared on the opposite edge of Pluto (egress). Analysis of the radio beam's profile showed that Pluto's atmosphere has a temperature inversion layer near the surface—in other words, instead of the atmospheric temperature steadily decreasing with altitude, a layer of colder air was near the surface

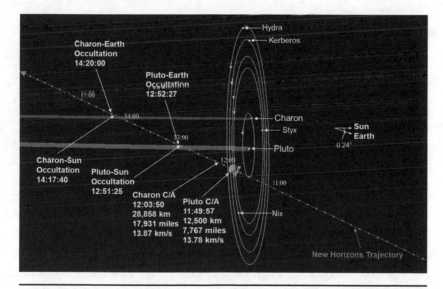

Figure 19.12 The trajectory of the spacecraft through the Pluto-Charon system. Seen from Earth, New Horizons passed behind Pluto, and a few minutes after emerging from behind the planet passed behind Charon. Each time the spacecraft crossed over the edge of either body, there were opportunities to measure the extent of the atmosphere, or in the case of Charon, determine an upper limit to how much atmosphere there might be. None was detected. NASA image [from the NH Pluto Flyby Press Kit].

that influences how the radio beam passes through it. The effect was nearly symmetric on both the ingress and egress sides of Pluto, which indicates that the atmosphere is likely uniform around the planet. This result suggests that winds in the lower atmosphere are weak at best.

A similar experiment, taking advantage of the fact that—from the spacecraft's point of view—the Sun set behind Pluto and then reappeared eleven minutes later, utilized Alice to record the spectrum of the atmosphere. In addition to methane, whose existence had already been established, acetylene, ethylene, and ethane were found. Theoreticians had expected these hydrocarbon molecules to form in the atmosphere from methane molecules fragmented by ultraviolet light, and on freezing, to fall to the surface as ice. Of the three, only ethane ice has been identified so far on the surface. Hydrogen cyanide was also discovered—this time not from New Horizons but from Earth-based observations made shortly before the spacecraft arrived at Pluto. It also is a product of the dissociation of the abundant methane and nitrogen molecules.[15]

The Alice data show a low abundance of nitrogen in the upper atmosphere and an unexpectedly low temperature there. Although the reason for this isn't clear, one consequence is that the escape of nitrogen gas into space is about ten thousand times less than had been predicted. The low rate of escape also has implications for the size of the atmosphere, which in turn influences the nature of the atmosphere's interaction with the streams of charged particles coming from the Sun, discussed below. Of course, the atmosphere's current temperature and nitrogen abundance may be different from those in the remote past, but at least at the present epoch the atmospheric nitrogen abundance is consistent with sublimation from the great reservoir of nitrogen ice in Sputnik Planitia at the measured surface temperature.

After the close flybys of Pluto and Charon, when New Horizons looked back at the night side of Pluto with the Sun nearly behind the planet, a beautiful and complex halo of hazes leapt into view. The halo was too tenuous to be seen in direct sunlight on the approach to Pluto, but was stunning when backlit, that is, when haze particles scatter sunlight forward toward the observer. MVIC and LORRI resolved some distinct layers reaching to more than 200 km (120 miles) above the surface, with each layer having a

Figure 19.13 Just fifteen minutes after its closest approach to Pluto on July 14, 2015, New Horizons looked back toward the Sun and captured this near-sunset view of the rugged, icy mountains and flat ice plains extending to Pluto's horizon. The smooth expanse of Sputnik Planitia (right) is flanked to the west (left) by rugged mountains up to 3,500 m (11,000 ft) high. To the right, east of Sputnik Planitia, rougher terrain is cut by apparent glaciers. The backlighting highlights more than a dozen layers of haze in Pluto's tenuous but distended atmosphere. The image was taken 18,000 km (11,000 miles) from Pluto; the scene is 1,250 km (780 miles) wide. NASA photo PIA19948.

typical thickness of a few kilometers. In color images from MVIC, the hazes have a distinct blue color, indicating that the particles composing them are very small, around 10 nm (0.00001 mm). Other properties of the haze point to larger particles, suggesting that the tiniest particles aggregate in clumps.

The tiny particles making up the haze are presumably tholins produced by exposure of methane, nitrogen, and carbon monoxide gases to the intense far ultraviolet light from the Sun and surrounding space and to charged particles entering the atmosphere from the solar wind. Laboratory experiments with mixtures of methane and nitrogen gas readily produce tholin particles of various colors and compositions, depending on the gas mixture, pressure, temperature, and other factors. The addition of oxygen atoms from carbon monoxide further enriches the mix of complex chemicals, and all this eventually leads to the production of solid (but probably fluffy) aerosol particles that can reach about one-half to one micrometer in size. The exact nature of the haze at different levels no doubt depends on local conditions. However, in general the haze particles appear to be organized into roughly equally spaced layers by gravity waves that arise as the air slowly blows across surface topography, such as mountains. Such orographic processes can only work, obviously, if the atmosphere is moving.

Haze particles do not remain suspended in the atmosphere indefinitely, but as they are removed, they are continually being replenished as methane gas evaporates from ice on the surface. The rates of falling to the surface and being replenished must ultimately come into some sort of equilibrium, which presumably changes over time as long- and short-term oscillations in Pluto's temperature play out. In the present epoch, the fallout of particles, which depends on their size, appears to occur in a matter of days to weeks. Although the particles are very small, their accumulation over thousands to millions of years is likely to affect the color of the surface, but the situation is not entirely simple—if it were, the surface of Pluto should have a uniform color, which clearly is not the case. In our present state of knowledge, the story is "to be continued."

Pluto's Plasma and Particle Environment

Though one sometimes thinks of the "vacuum of space," the term is a misnomer. The vast volumes of space between the planets and the stars are populated, albeit very sparsely, with dust particles, atoms and molecules, and streams of plasma, which are basically electrons and protons. Like other stars, the Sun is a continuous source of plasma ejected outward from active regions in the photosphere and corona. The Sun's magnetic field guides this solar wind deep into the planetary region of the Solar System, where it interacts with the planets' magnetic fields, an interaction that further influences the paths that the moving particles take. Earth's magnetic field shields the surface by deflecting the solar wind into the thin atmosphere at altitudes above 80 km (50 miles) in the regions of the north and south magnetic poles, which produce the glorious displays of the Aurora Borealis and Aurora Australis. Astronauts aboard the International Space Station can look down on the aurora as they orbit Earth between 330 and 435 km (205 and 270 miles) above the surface.

Cosmic rays—atom fragments produced by energetic events elsewhere in the Galaxy—also enter the Solar System from all directions, and when they interact with the planets they can, because of their very high energies, break apart molecules in their atmospheres. Because of their high energies, even after their flights through the atmosphere they are still able to penetrate several meters into planetary surfaces.

Pluto and Charon exist in this stew of atomic particles and bits of dust, but the specific nature of their interactions is important to understand in terms of the evolution of Pluto's atmosphere and the effects of charged-particle bombardment on its (and Charon's) solid surface. The characteristics of the solar wind and streams of interplanetary particles at Pluto's great distance are not well known. While numerous spacecraft have explored the space environment of Earth and all the major planets, only a few, notably the two Voyagers, have probed the outer region beyond about 10 AU. The environment in the outer regions is known to be highly variable on time scales of days and weeks, depending on the activity of the Sun and occurrences known as interplanetary compression events. As the Solar System orbits around the center of the Galaxy, it encounters clouds of interstellar gas and dust. The effect of an interstellar cloud is to push against the heliosphere, the "bubble" of gas and plasma blown out by the Sun, bringing the region of interaction of the solar and interstellar gas closer to the planets. In extreme but rare cases, the interaction region, containing atomic particles with very high energies, may press in as far as Pluto, or even Neptune and Uranus, likely affecting their atmospheres, or in Pluto's case, the surface as well. Just what New Horizons would find at Pluto was difficult to predict, but it was of the greatest interest to the science team.

Monitoring the particles in open space during the spacecraft's outward journey, the instruments of the Pluto Energetic Particle Spectrometer Science Investigation (PEPSSI) and Solar Wind Around Pluto (SWAP) experiments had already found unexpected effects in the solar wind and the particle environment 5 AU from the Sun during its unique traverse of Jupiter's long magnetic tail in 2007.[16] On the spacecraft's arrival in the vicinity of Pluto, the science team was eager to understand the reciprocal interactions between the highly variable solar wind and Pluto's atmosphere. Fortunately, they had had some experience predicting what might be involved by studying solar wind interactions with Mars and with comets.

Mars has a thin atmosphere that doesn't extend far out into space, while comets approaching the Sun are in a constant state of rapid evaporation of their ices, producing a very large, if temporary, gaseous envelope ("atmosphere"). This comet envelope, or coma, also includes dust, and its extension for millions of kilometers opposite the direction of the Sun forms the comet's tail. Pluto's atmosphere of rapidly escaping nitrogen and the planet's small size (hence weak gravity) suggested that its atmosphere would also be very large (extending beyond the orbit of Charon), and it seemed likely that its solar-wind interaction would be more like that of a comet than of Mars. The part of Pluto's atmosphere extending in the direction of the Sun would be compressed by the flow of the solar wind, while the solar wind in turn would pick up ionized atoms and molecules from the atmosphere, slowing the wind's flow in a process called mass loading. Depending on the properties of Pluto's atmosphere, the interaction region where mass loading occurs could be as distant as about 20 Pluto radii, Rp (2 million km or 1.2 million miles), or as close as about 7 Rp (13,000 km or 8,100 miles) from the surface. The details of the physics and size of Pluto's region of interaction with the solar wind plasma, as well as the characteristics of the space environment on the downstream side of Pluto, were intensively studied by PEPSSI and SWAP.

When the spacecraft flew by, the interaction region was found to be closer to Pluto's surface than predicted, at only about 6 Rp (11,000 km or 6,800 miles). This is a consequence of the unexpectedly slow rate of escape of Pluto's nitrogen atmosphere, discussed above, due to the atmosphere being much colder than expected. Since the speed and distances of molecules in a planet's atmosphere correlate with temperature, a cold atmosphere like Pluto's will hug a planet more closely than if it were warm.

Neither PEPSSI nor SWAP made measurements during the several hours of New Horizons' closest approach to Pluto and Charon because the spacecraft had to be oriented to make observations with the cameras and spectrometers, but later, under more favorable orientation conditions, the particle observations were resumed and continued as Pluto and Charon receded into the distance. The "disturbance" to the space environment caused by the

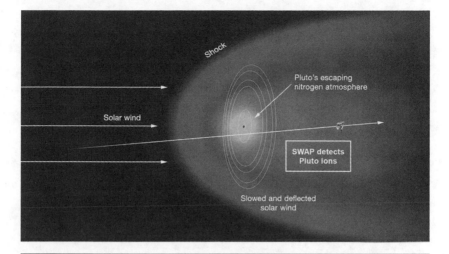

Figure 19.14 Artist's concept of the interaction of the solar wind with Pluto's predominantly nitrogen atmosphere. Some of the molecules that form the atmosphere have enough energy to overcome Pluto's weak gravity and escape into space, where they are ionized by solar ultraviolet radiation. As the solar wind encounters the obstacle formed by the ions, it is slowed and diverted, possibly forming a shock wave upstream of Pluto. The ions are "picked up" by the solar wind and carried in its flow past the planet to form an ion or plasma tail. The SWAP instrument on New Horizons made the first measurements of this region of low-energy atmospheric ions shortly after closest approach on July 14. Such measurements enabled the SWAP team to determine the rate at which Pluto loses its atmosphere. The trajectory of the spacecraft is also shown. NASA image PIA19719.

obstacle that Pluto presents to the flow of the solar wind was measured to extend all the way out to about 400 Rp (750,000 km or 465,000 miles), or almost twice the distance between our Earth and the Moon.[17]

We mentioned above that one of the results the New Horizons science team was most interested in was whether Pluto's interaction with the solar wind would resemble Mars or a comet. The SWAP results say both. The cometary aspects described above are certainly present, but data show that heavy ions of methane, which is the gas that dominates the loss of the atmosphere, are folded in with the light hydrogen ions in the solar wind to shape Pluto's downstream tail of charged particles in a manner similar to that found on Mars (and also Venus). Thus Pluto's interaction with the solar wind is not entirely like that of a comet or a planet, but a hybrid—something that has not been seen elsewhere in the Solar System. This result is not just relevant to conditions in our own Solar System, it also may help us understand the interaction of plasma winds in planetary systems hosted by other stars.

Dust

The New Horizons spacecraft carried a unique instrument, the first-ever instrument to be student-built and managed on any spacecraft (a significant achievement in its own right). The Venetia Burney Student Dust Counter (SDC)—named for the eleven-year-old girl who had proposed the name for Pluto—was developed by advanced students at the University of Colorado who were tasked with proposing a simple device that could measure the impacts of dust particles encountered on the long flight to Pluto.

Particles of dust, mostly made of silicate minerals, are ejected from asteroids during grinding collisions that scatter the pieces in all directions, and from comets as they propel outward the dust particles from the streams of gas produced by the comet's evaporating ices. As has been known for some time, comets and some asteroids produce dust trails throughout the inner Solar System, which persist in space for hundreds of years. Indeed, several meteor showers occur every year as Earth passes through them; in many cases, the specific objects responsible for a given meteor shower have been identified.

In addition, particles of galactic, or interstellar, dust also drift through the Solar System, but they are rare and difficult to distinguish from interplanetary dust. The Cassini spacecraft that orbited Saturn between 2004 and 2015 has detected and analyzed three dozen dust particles identified as having come from beyond the Solar System.[18] These particulate interstellar sojourners belong to the outflow of dust and gas from stars at various phases of their development and demise.

Outside the asteroid belt and the region where comets begin to evaporate (mostly inward of 5 AU), dust produced by collisions between KBOs at Pluto's distance and much farther out slowly spirals in toward the Sun. Measurements of the amount of dust and its distribution at different distances from the Sun can be used to estimate the rate of collisions between the KBOs. The SDC on New Horizons registered the impacts of interplanetary dust particles on the way to Pluto by means of an array of detectors about the size of an 18-inch by 12-inch cake pan. Its sensitivity was for any particles greater than a few picograms in mass, where a picogram is one trillionth (10^{-12}) of a gram.

As noted in chapter 18, the potential existence of a large amount of dust in Pluto's environment was regarded as a great hazard to New Horizons. Fortunately, that potential hazard was downgraded as observations of the space environment close to Pluto were made during the last weeks of the approach, allowing spacecraft managers to approve the original planned trajectory for the encounter. The probability of a spacecraft-debilitating collision with dust particles was found to be exceedingly low, an expectation that was borne out during the Pluto-Charon flyby, when only a single dust particle was registered by the SDC. Based on the known volume of space traversed during the time the measurements were being made, and the detection of one dust particle, the dust density was estimated to be only about 1.2 particles per cubic kilometer of space.

The Geology of Charon

So far we've said little about Charon, which was largely overshadowed by its larger companion. In contrast with Pluto's complex and largely unexpected geology, Charon presented an aspect that at least at first glance looked rather familiar to planetary scientists: it looked strikingly like our own Moon as seen in a small telescope. Cratered plains laced with lunar-like narrow channels cover a substantial part of Charon's visible surface. But the apparent analogies between Charon and the Moon are somewhat misleading: in terms of bulk composition, the Moon is made mostly of rock while Charon has a rocky core encased in a thick mantle of ice. Despite the morphological similarities between the two objects, Charon is more directly comparable to icy planetary satellites elsewhere in the outer Solar System.

The frozen water that makes up the crust of Charon is as rigid as rock, and lacks the patina of nitrogen and methane ices that covers most of Pluto. For this and other reasons, the geologic structures of Charon are quite different from those of its larger companion. While much of Pluto's tableau of geological textures and structures result from surface-atmosphere interactions and ices that are mobilized when warmed, Charon's geology comes about by the more familiar and better-understood processes of impact, large-scale surface flow, fracturing, and blanketing by the ejecta of crater-producing impacts.

A band of ridges and canyons fractured and displaced by tectonic forces cuts across the entire hemisphere of Charon imaged at closest approach. North of this zone, which is more than 200 km (120 miles) wide in places, the terrain is rugged and cratered; to the south it consists of smooth and fractured plains. The north polar region is geologically complex and tinted orange-red, recalling colors seen on Pluto. The tinted region has been provisionally called Mordor Macula after Mordor in J. R. R. Tolkien's *Lord of the Rings*, and the largest impact crater, which lies close to Charon's pole, is also partly orange-tinted. This crater is about 230 km (145 miles) across and 6 km (3.7 miles) deep. Many of Charon's geological structures show significant topographic relief, on the order of 20 km (12 miles) in some places. This demonstrates the rock-like strength and rigidity of frozen water at a temperature of about 40° K (−233° C) as well as the enormity of forces that created this topography. Some of that force probably came from the interior as liquid water froze and expanded early in Charon's history.

The density of large craters in Charon's rugged terrain shows that this region is very old, retaining the scars of impacts that occurred about 4 billion years ago when the Kuiper Belt was actively grinding itself up. In contrast, the smoother plains in the region south of the tectonic belt, called Vulcan Planum, are peppered with well-defined craters and cut by narrow troughs like the narrow channels on the lava plains of Earth's Moon. The texture of Vulcan Planum suggests that a formerly more rugged terrain has been resurfaced, possibly by water ice somehow fluidized and spread over the preexisting topography. Cryovolcanism again appears in this context as a possible mechanism for producing the smooth and relatively young surface in this part of Charon, but—and again in contrast with what is seen on Pluto—the resurfacing event was itself ancient, occurring about 4 billion years ago, at a time when Charon still possessed enough internal heat to melt and mobilize water.

Figure 19.15 Charon's surface is distinguished in part by a complex system of chasms about 1,800 km (1,100 miles) long, possibly caused by pressure from the interior of the satellite as formerly liquid water expanded upon freezing. The image shows surface features as small as about 395 m (1,290 ft), and was taken with LORRI about an hour and forty minutes before the spacecraft's closest approach to Charon on July 14, 2015. NASA photo PIA21864.

In general, the surface of Charon is composed of water ice, and as on so many moons of Jupiter, Saturn, and Uranus, the crystal structure of the ice is hexagonal. This structure is evident by the characteristic profile of the infrared spectrum produced. However, at Charon's low temperature the amorphous, noncrystalline form is also expected. If heated, amorphous ice can acquire the hexagonal form, but the regular structure is readily damaged or destroyed by charged atomic particles coming in from space. Thus the persistence of the crystalline structure in many places in the outer Solar System is something of a mystery. The explanation is probably that many (or all) icy bodies have occasionally undergone warming events, preventing the amorphous form from persisting over eons.

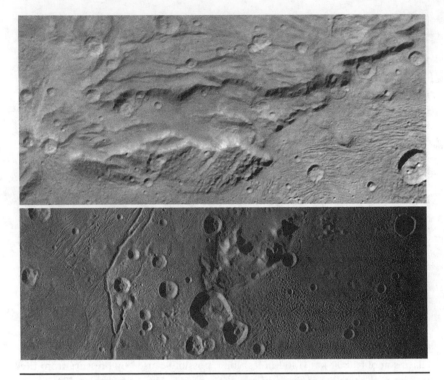

Figure 19.16 These two enlargements of sections of Charon's surface show surface features as small as about 395 m (1,290 ft). The top frame shows a section of the chasm shown in figure 19.15 that is about 385 km (240 miles) long and 175 km (110 miles) wide. The Vulcan Planum view in the bottom panel includes the "moated mountain" Clarke Mons just above the center of the image. As well as featuring impact craters and sinuous troughs, the water ice–rich plains display a range of surface textures, from smooth and grooved at left to pitted and hummocky at right. The full image was taken with LORRI about an hour and forty minutes before the spacecraft's closest approach to Charon on July 14, 2015. NASA photos PIA20467 and PIA 20535.

The water ice covering the surface of Charon is laced with a twist of ammonia. First seen in spectra obtained with large Earth-based telescopes, the signature of ammonia also appears in LEISA data. While this relatively minor component is seen all over the surface, it is concentrated in a few locations, notably in and around a fairly recent crater called Organa.[19] Curiously, a nearby crater of comparable fresh appearance shows no such concentration. At present, what form the ammonia takes is not clear. It may be as a hydrate structure in which ammonia is surrounded by water molecules, or possibly ammonium hydroxide (NH_4OH) in which the water and ammonia molecules have formed a new compound as $NH_3+H_2O \rightarrow NH_4OH$. The spectral data from telescopes and from New Horizons have not yet differentiated between these possibilities, but additional work in the laboratory should clarify the details.

Another puzzle is the origin and longevity of ammonia molecules at the surface of Charon. Is ammonia native to Charon, inherited from the components of one or more KBOs from which it formed, or has it formed by chemical processes in more recent times? Complicating the matter is the short lifetime of ammonia molecules when exposed to ultraviolet light from the Sun and other sources: ammonia is quickly destroyed, and its presence on an ancient surface strongly suggests that it is being replaced from the interior of Charon.

Yet another puzzle is the origin of the orange-red tint of the north polar region, which presents a striking contrast to the nearly uniform neutral gray of nearly all the rest of the surface. On Pluto, such warm colors are presumed to be due to tholins produced by ultraviolet light and charged-particle processing of methane and nitrogen ices on the surface and similar gases in the atmosphere. But Charon has no atmosphere, and no detectable methane or nitrogen ices. A proposed solution to this conundrum relies on the fact that Pluto's methane-rich atmosphere leaks off from the planet at a high rate, estimated at 5×10^{25} molecules per second. (Nitrogen is currently lost much more slowly, at a rate of about 1×10^{23} molecules per second, though the rate might have been greater in the past.)[20] On migrating to Charon, this leaked methane gas could freeze on the very cold north (and south) polar zones when the Sun doesn't shine for long periods and where ultraviolet light and charged particle radiation from space would do their usual work of making tholins, bringing a dash of color to an otherwise gray, drab little world when the pole finally emerges into sunlight again.[21]

Pluto's Small Satellites

When New Horizons was selected by NASA for flight in 2001, Pluto was known to have only one satellite—Charon, found by James Christy in 1978.

Using the HST, two additional satellites, Nix and Hydra, were found in 2005. Two additional smaller moons were found, again with HST, while New Horizons was en route to Pluto, Kerberos in 2011 and Styx in 2012. The presence of these four small bodies plus the possibility of several more led mission engineers and scientists to worry about the possible existence of a significant quantity of dust near Pluto. This dust could, they feared, present a far-from-negligible risk of disabling the spacecraft. Fortunately, as we have already related, this possibility was dismissed through further observations and dynamical calculations.

HST observations of all four small satellites were made at intervals during New Horizons' outward flight to refine their orbits, measure their brightness, and revise estimates of their sizes, shapes, and rotation periods. These moons are exceedingly faint because of their small sizes and great distance, and observations with the HST were made even more difficult by the high

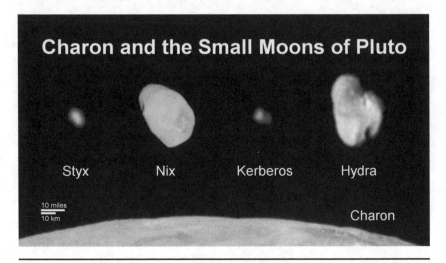

Figure 19.17 This composite image shows all four of Pluto's small moons, and a sliver of the large moon, Charon, as resolved by LORRI. All the moons are displayed with a common intensity stretch and spatial scale (see scale bar). Charon is by far the largest of Pluto's moons, with a diameter of 1,212 km (751 miles). Nix and Hydra have comparable sizes, approximately 40 km (25 miles) across in their longest dimension, while Kerberos and Styx are much smaller and have comparable sizes, roughly 10–12 km (6–7 miles) across in their longest dimension. All four small moons have highly elongated shapes, a characteristic thought to be typical of small KBOs. NASA photo PIA20033.

number of background stars while Pluto was moving slowly through dense parts of the Milky Way in Serpentis and Ophiuchus. The observers persisted because advance information would enable observations by the spacecraft that could give details only discoverable by close-range examination of as many satellites as possible within the other limits imposed on the spacecraft's trajectory.

Unfortunately, those other limits, which prioritized scientifically important observations of Pluto and Charon, meant that the spacecraft never got very close to any of the four. Images of Styx and Kerberos showed their shapes, but not much surface detail. Higher-resolution images of Nix and Hydra showed their shapes and also several craters presumably formed by impacts with smaller pieces of KBOs. The largest crater on Nix is tinted red, possibly because the impacting body had a different composition and left some of itself deposited on the surface, though the colored material may have been excavated from the moon's interior.

LEISA obtained near-infrared spectra of both Nix and Hydra. Both clearly show the characteristic spectral bands of water ice and, more surprisingly, the same ammonia absorption band first seen on Charon. These findings suggest the likelihood that Charon and at least Nix and Hydra and probably all four small satellites—are kindred objects, formed from the debris produced by a giant impact on a very young Pluto by another KBO of similar composition. In this scenario, they would all possess a similar inventory of ice that is highly reflective and mostly water laced with methane. So far, so good—though the problem of how ammonia compounds exposed on the surfaces of these bodies could survive over billions of years remains unresolved.

Additional support for the giant-impact theory of the origin of the satellites comes from their orbital and rotational characteristics (see table below).[22] They are all irregular in shape, spin rapidly, and revolve in nearly circular orbits about the Pluto-Charon barycenter in the same orbital plane. As is typical of such objects, the rotations of all four are chaotic, which means that the orientation of their rotational axes are apt to change abruptly in what seems like a random way. This tumbling is caused by the changing strength and orientation of the gravitational field they experience relative to Pluto and Charon as those two bodies revolve around their barycenter in 6.4 days.

Additionally, the small satellites' irregular shapes enhance the apparent randomness of their changes in orientation. Despite the apparent chaotic nature of their motions, these tumbling little moons remain steadfast at their fixed distances from Pluto and rotate at uniform rates—the orientation of their rotational axes is the only thing that rapidly changes.[23]

Object	Size (km)	Orbital period (days)	Rotation period (days)
Styx	16 × 9 × 8	20.16155 ± 0.00027	3.24 ± 0.07
Nix	50 × 35 × 33	24.85463 ± 0.00003	1.829 ± 0.009
Kerberos	19 × 10 × 9	32.16756 ± 0.00014	5.31 ± 0.10
Hydra	65 × 45 × 25	38.20177 ± 0.00003	0.4295 ± 0.0008

The numbers and sizes of craters visible on Nix and Hydra are consistent with ages greater than about 4 billion years. Again, this is compatible with the big-impact theory in which they are leftovers from the same catastrophic event that formed Charon early in Pluto's history.[24]

At present, this is the state of knowledge from scientists' analysis of the data returned from New Horizons. Though additional interpretations and discoveries can be expected, at least the broad outlines are clear. Thanks to the spacecraft's nearly flawless performance, and the extraordinary skills and efforts of the spacecraft team, navigation team, and science team, all the science experiments proposed during the development of the mission were accomplished. We now stand in possession of a stunning array of results demonstrating the many unexpected characteristics of Pluto, Charon, and the small satellites that could hardly have been imagined when the spacecraft set out from Earth in January 2006. Importantly, we now have a detailed and precious baseline of scientific understanding of two large KBOs against which future studies of other bodies in the third zone of the Solar System can be conducted as our exploration reaches beyond Pluto and Charon to the next new horizon beckoning to human curiosity.

20

On to the Kuiper Belt

AS THE LATE ASTRONOMER and philosopher Carl Sagan has written,

> In all the history of mankind, there will be only one generation that will be
> the first to explore the Solar System, one generation for which, in childhood,
> the planets are distant and indistinct disks moving through the night sky, and
> for which, in old age, the planets are places, diverse new worlds in the course
> of exploration.[1]

We are that generation.

Exploration and courageous ventures into unknown territories are core
visceral drivers of human behavior. They account for the migration of our
species out of Africa, and for the spread of languages, cultures, and empires
across lands and seas. Beginning in the sixteenth century, Russian culture
and language extended eastward across the Eurasian continent over a span of
some four hundred years, while America's rapid westward expansion across
North America began in earnest when Meriwether Lewis and William Clark
were dispatched to explore as far as the Pacific by President Thomas Jefferson in 1804. Polynesian cultures filled the vast Pacific basin using the
stars for navigation and venturing in boats made only of wood, while Norse-
men extended their sphere of influence in the northern latitudes in voyages

reaching as far as Newfoundland, their adventures vividly related in the Icelandic sagas. Each continent and every oceanic island cluster has had its own rich and multidimensional history of discovery and migration. Each culture has its own story to tell.

Though the visual telescopic surveillance of other worlds began with Galileo in 1609–10, the actual physical exploration of these distant long dreamed-of destinations by means of spacecraft began only in the early 1960s. Enabled by the development of rocket technology, which was first invented to serve the interests of warfare and mass destruction, it has at least partly outgrown those ignoble roots and become a servant of human curiosity and the quest for knowledge about the larger Universe. Though a strong thread of primitive tribalism and international rivalry was woven through the first few decades of this endeavor, the exploration of the Solar System is ultimately something that transcends terrestrial boundaries. It is by and for all humankind.[2]

Its most recent culmination has been New Horizons' exploration of Pluto and Charon, which rank at the time of writing—March 2017—as first and fifth largest among KBOs. (Pluto, at 2,372 km in diameter, is slightly larger than Eris at 2,326 km; Charon, at 1,212 km, is surpassed by those two and by Makemake at 1,430 km and Haumea at 1,240 km.) A still-healthy New Horizons continues onward thorough the outer Solar System, on a three-and-a-half-year, billion-mile add-on journey to a more representative Kuiper Belt object, 2014 MU_{69}. Discovered only in 2014 (and given its official number 486958 by the Minor Planet Center in 2017), it will be the first object explored by a spacecraft that was not even known when the spacecraft was launched. We saw in chapter 18 that MU_{69} was one of only three KBOs found in the search area corresponding to a zone that could be reached by New Horizons, and as we describe below, it may be quite a remarkable object indeed. Another token of the rapidity of progress in the field is that, only fifteen years before New Horizons was launched, the existence of a population of planetary bodies in the zone of the Solar System now called the Kuiper Belt was nothing more than an abstraction. Abstraction became reality only when David Jewitt and Jane Luu, bothered by the seeming emptiness of the vast region beyond Neptune, discovered the first KBO to be recognized as such, 1992 QB_1. It was only later that it was realized that Pluto and Charon, though exceptional in many ways, also belonged with the KBOs. The rapid pace of discovery has now added a thousand more objects near and beyond Pluto, with no end to the discoveries in sight. Consequently, the Kuiper Belt has now taken its place as the third zone of the Solar System.

The first zone is defined by the terrestrial planets and the asteroid belt, all distinguished in composition by combinations of metal and rocky material. The second consists of the giant planets, each with a dense, cloudy atmosphere dominated by hydrogen. When Pluto was the only known body beyond Neptune, it seemed a curious, rather bizarre oddity, difficult to categorize and fit into standard schemes of classification. In some ways it still does. However, it is no longer a singular curiosity. With a thousand known KBOs and an estimated hundred thousand more larger than 100 km (60 miles) yet to be discovered, the third zone has emerged as the largest and most populous region of the Solar System, embracing not only Pluto and all the other discovered and predicted members of the family but also a practically infinite number of smaller lumps of ice and dust that can only be seen when they take a fateful orbital turn to approach the Sun and become comets.

Though the observational history of the Kuiper Belt begins with Jewett and Luu's 1992QB$_1$ (or, if one prefers, with the discovery by Clyde Tombaugh in 1930 of the object to which the IAU has now assigned the official asteroid number 134340), the third zone was anticipated by at least some astronomers as they pondered the nature of the Solar System on the largest scale. At least two questions inspired modern thinking about the evolution and current state of the Solar System: First, does the planetary system really end at about 40 AU with Pluto? Second, where do comets come from? Among the numerous early theories of the origin of the Solar System in a rotating and flattening turbulent nebula of gas and dust, none provided a compelling reason for a restriction on its absolute size. Meanwhile comets—despite the occasional majesty of those that passed through the inner Solar System— were usually regarded, in the larger scheme of things, as second tier. They were interesting, but minor and inconsequential, mere will-o'-the-wisps whose tails were so exiguous than even when Earth passed through them no detectable result was produced. Quantitative theories of solar-system structure, though much more sophisticated than in the "Bode's law" era, remained hard-pressed to explain the details. On the other hand, conceptual frameworks could be constructed, even in the face of, or perhaps because of, the paucity of observed facts.

The first half of the twentieth century was marked by many theories and qualitative concepts for the origin of the Solar System to incorporate astronomical discoveries of the asteroids, numerous satellites of the giant planets, many comets and, of course, Pluto. Notably, the Irish-born Kenneth Edgeworth, who won a scholarship to study in England, was decorated for his military service with the Royal Engineers during World War I, and returned to Ireland on his retirement in the 1930s, published in 1943 the paper "The Evolution of Our Planetary System" in the *Journal of the British Astronomical Association*.[3] A rank amateur in every sense of the word, Edgeworth took a broad view in trying to understand, qualitatively, how the Solar System might have been born and evolved. This first paper contained a mere sketch of his ideas, but in 1949 he followed up with a more expansive and semi-quantitative treatise that explicitly considered the origin of the comets. "It is not unreasonable to suppose," he wrote, "that this outer region [beyond Neptune] is now occupied by a large number of comparatively small clusters [of condensed material], and that it is in fact a vast reservoir of potential comets. From time to time one of these clusters is displaced from its position, enters the inner regions of the solar system, and becomes a visible comet."[4]

Published in an amateur journal, and without the rigor of a paper vetted by qualified referees and published to professional standards, Edgeworth's views had little influence at the time. Nevertheless, he deserves credit for having at least the kernel of the right idea. Meanwhile, as described in chapter 7, the German astronomer Carl von Weizsäcker's paper on the origin of the Solar System, originally published in 1943 but not immediately appreciated outside of Germany because of the war, was now being enthusiastically taken up and modified by Kuiper and his colleague at Yerkes Observatory, the theoretical astrophysicist Subrahmanyan Chandrasekhar. Also during this period of burgeoning research, a growing number of astronomers were coming to appreciate the importance of comets to an understanding of the origin of the Solar System. Though most astronomers still believed in the interstellar origin of comets, the paradigm was shifting. As early as 1932, the Estonian-born Ernst Öpik had shown that comets would remain bound to the Sun at distances as far as 1 million AU, and he had mooted the possibility of a comet cloud bound to the Sun. Another significant contribution was made in 1948, when Adrianus J. J. Van Woerkom, in his PhD dissertation at the University of Leiden (his thesis adviser was the famous Jan Oort), examined Jupiter's role in changing the orbits of comets as they passed through the Solar System. Van Woerkom showed that in only a million years or so Jupiter would eject all the long-period comets into interstellar space. Thus they had to be replenished from somewhere, and Van Woerkom, agreeing with Öpik's idea that comets would remain bound to the Sun even at enormous distances, suggested as a possible source a distant comet cloud moving permanently with the Sun. Two years later, Oort himself published a landmark paper in which he strikingly demonstrated that the observed distribution of long-period comet orbits could be explained by positing the existence of a large halo of small icy bodies surrounding the Solar System at distances of 50,000 to 150,000 AU, now called the Oort cloud.[5] While the Oort cloud has not been seen directly with even the largest telescopes, the comets it spawns continue to enter the inner Solar System, often to be absorbed by the Sun or ejected after a single pass on a hyperbolic orbit that sends the one-time comet deep into the Galaxy, never to return. (Oort clouds are thought to encircle other planetary systems orbiting other stars, providing episodic infusions of icy bodies that evaporate from the heat near the star. In doing so, they give a temporary, but detectable, spectroscopic signature.)

The year 1950, in which Oort's paper appeared, was a banner year for comets. The same year saw the publication of the first of a series of papers by Fred L. Whipple presenting his comet model. As we saw earlier, as a graduate student at the University of California, Berkeley in 1931, Whipple had helped calculate the first orbit for Pluto. Later that same year, he headed to Harvard, where he spent the remainder of a long and productive career. (He died in 2004 at age ninety-seven.) In explaining the observed behavior of short-period comet Encke as it approached the Sun, Whipple famously proposed that comets were "dirty snowballs," consisting of a mixture of ice and dust.[6] Both the arresting phrase and the model that inspired it have remained viable as more and more comets have been observed, including from spacecraft.[7] (Many comet researchers, however, prefer the phrase "icy dirtballs," as the observational evidence suggests that the non-ice component is dominant, and that the dirt is held together by a strong but relatively sparse matrix of ice.)

Kuiper, as usual, followed developments with keen interest. His overriding ambition had always been to understand the origin of the Solar System, and in 1951, he finally published his own important paper on the subject. He saw the importance of comets, but in contrast to Oort, his interest lay not with the long-period but with the short-period comets—those with orbital periods about two hundred years or fewer and lying very close to the orbital planes of the planets. As noted in chapter 7, Kuiper was working with a semiquantitative model of the origin of the Solar System based on von Weizsäcker's ideas (as modified by Chandrasekhar), in which large-scale instabilities in the solar nebula became the sites of major-planet formation while gas and dust not incorporated into the planets and their satellites were swept by the solar wind into the outer parts of the solar nebula beyond Neptune. In this view, the leftover material became the feedstock of comet formation in the plane of the planetary orbits. Note that Kuiper did not explicitly predict the existence of a population of large, icy bodies in the region beyond Neptune; rather, in noting Pluto's similarity to Triton, he accepted the view of Cambridge mathematical astronomer Raymond Lyttleton (and others) that Pluto was an escaped satellite of Neptune. Apparently influenced by Kuiper, Edgeworth, at least by 1961, was also subscribing to the escaped-satellite theory. Thus he too missed his chance to ascribe to Pluto original membership in the reservoir of icy bodies he had posited beyond Neptune.[8]

Still accepting the relatively large mass of Pluto because of the flawed analysis of the residuals in the motions of Uranus and Neptune, Kuiper further proposed that as Pluto combed the cometary region of the outer Solar System, it would toss many of these icy bodies inward toward the giant planets. Though the giant planets in turn would expel most of them outward into the remote regions of the Oort cloud, a few would be captured and become short-period orbits.

None of this answers the question of why, since Kuiper never actually predicted the existence of a population of large icy bodies in the region beyond Neptune, his name has been ascribed to the Kuiper Belt. The explanation presumably has something to do with the fact that Kuiper was such a remarkably eclectic thinker whose ideas were always many-sourced and fluid. Like most voracious researchers, he absorbed and embellished the ideas of others whenever they were useful to him. His concepts and supporting calculations of the origin of the Solar System were influenced to varying, and to some extent unknown, degrees by the work published and unpublished of many other astronomers, including von Weizsäcker, Chandrasekhar, Van Woerkom, Oort—and also H. P. Berlage and Dirk ter Haar—all of whom were working on related problems at the same time.[9] Some of them were even at Yerkes or the University of Chicago, either as staff (Chandrasekhar) or as visiting fellows (Oort and von Weizsäcker) during this period. It is at least possible, though not necessarily likely, that he was aware of Edgeworth's 1943 paper as well as a brief and mostly unrelated 1944 paper in the British journal *Nature*.[10] As for Edgeworth's longer and more substantial paper in the *Monthly Notices of the Royal Astronomical Society* for 1949, since that issue of the journal was not distributed until March 1950, by the time it appeared, the writing of Kuiper's 1951 paper on the origin of the Solar System was likely finished. Published in a Yerkes Observatory fiftieth-anniversary volume edited by former Yerkes student J. Allen Hynek,[11] it contains no reference to Edgeworth; nor does any of Kuiper's other papers on the origin of the Solar System. While astronomer John Davies, in his book *Beyond Pluto*, in calling into question the originality of some of Kuiper's ideas, notes his propensity to reference the work of other scientists sparsely and highly selectively, one cannot say with any certainty whether Edgeworth had any influence on him.[12]

The question of priority aside, we can certainly agree that it was Edgeworth who most clearly predicted the likelihood of a population of bodies, including potential comets, lying in the plane of the planets, as well as much larger bodies made largely of ices. In reading Edgeworth's papers, one sees that his ideas were rather impressionistic and unsubstantiated by anything more than intuition and his own incomplete and elementary calculations. His obscurity—and Kuiper's prominence, then and later—explains why it is Kuiper's name, not Edgeworth's, that was adopted as the designation for the Solar System's third zone. Indeed, though Kuiper's name began as a casual (and in some ways careless) attachment, it has now become a barnacle. Though the third zone is occasionally referred to pedantically as the Edgeworth-Kuiper Belt or as the Kuiper-Edgeworth Belt, these awkward names are clearly bound to go the way of "Georgium Sidus" or "Ceres

Ferdinandea." For better or worse, we are stuck with "Kuiper Belt." (For anyone seriously bothered by that, there is always the option of using the generic
term "trans-Neptunian objects," or TNOs, encompassing Pluto, QB_1, and the
rest of the bodies found in the third zone.)

Whatever we call the vast assemblage of icy bodies beyond Neptune, MIT
planetary scientist Richard P. Binzel waxes justly poetic in noting, "What was
previously our solar system's *regio incognita* has become the new frontier."[13]
Discoveries since $1992QB_1$ have defined the population and structure of the
Kuiper Belt and have revealed several objects that clearly have orbital characteristics different from those of Pluto, QB_1, and their cohorts. Emerging
information on their compositions and their propensity to have attendant
satellites demands a deeper level of understanding of all their properties.
To address a broad distribution of known bodies and their clustering in
groups with similar characteristics, as in biology, a finer classification of the
phenotypes is needed.

Just as a newborn human baby invites relatives and friends for a close,
admiring look, and with each passing day grows in beauty, complexity, and
captivating fascination, the newborn baby of solar-system research dazzles
astronomers of all persuasions. Among planetary scientists, David Jewitt
notes, "The Kuiper Belt is amazing in its ability to link together research
areas that previously seemed unlinked. Different populations of small bodies
used to be described separately and given different labels, like strange animals in an exotic zoo. Now, we see their connections more clearly."[14] Meanwhile, astronomers have begun to explore the dusty disks observed around
other stars of similar mass and temperature as our Sun in light of the Kuiper
Belt and its evolution over the age of the Solar System. This dusty debris
around stars of similar age to the Sun must be refreshed to be maintained
over several billion years, suggesting that the slow grinding of solid bodies
trapped in their own Kuiper belts provides a continuing source of small
particles of rock and ice that form the flattened disks observed with the
largest telescopes. As Jewitt says, "Planetary scientists and astronomers have
something to talk about."[15]

With a known population of already more than a thousand, the exotic zoo
of trans-Neptunian objects sorts itself into five major categories on the basis
of their orbital characteristics. The Kuiper Belt encompasses the bodies that

orbit the Sun close to the plane of Earth's orbit (the ecliptic plane), and range in average heliocentric distance from about 30 to about 55 AU. Pluto fits into this category, although its orbit is inclined some 17.1° to the ecliptic plane. Beyond the Kuiper Belt is the second category, the scattered disk population, having orbits of high eccentricity and large maximum heliocentric distances. These bodies are dynamically unstable, and through gravitational interactions with Neptune can be ejected from the Solar System or drawn in even closer than Neptune. The third category, detached trans-Neptunian objects, have highly eccentric orbits that never bring them close enough to Neptune for a significant gravitational interaction, and while they orbit the Sun with all the other bodies, they are always very distant. The fourth general category is the centaur population, former trans-Neptunian objects that have orbits relatively close to the Sun and cross the orbits of one or more major planets. The centaurs are dynamically unstable and are destined for a brief sojourn of about ten thousand years or fewer in the planetary region of the Solar System before being ejected or colliding with one of the planets. The third zone of the Solar System includes all the objects beyond Neptune, embracing at its extremity the Oort cloud, a fifth category of TNOs.

The existence of another massive planet is at least possible. The tentative indications, as in the cases of Adams and Le Verrier regarding Neptune and Lowell regarding Planet X, lie in its putative gravitational effects on other bodies. First proposed in 2014 by KBO researchers Chad Trujillo and Scott S. Sheppard from similarities in the orbits of distant KBOs Sedna and 2012 VP$_{113}$, and then further elaborated two years later by Konstantin Batygin and Michael E. Brown of Caltech, who calculated that a planet of perhaps ten Earth masses would explain the orbital similarities of six distant objects, this hypothetical planet recalls those proposed long ago by Forbes, Gaillot, and Pickering. It follows a highly elliptical orbit, with perihelion at about 200 AU and aphelion at 1,200 AU; its orbital period is estimated at ten thousand to twenty thousand years. With an orbital inclination to the ecliptic at 30°, it might explain a curious feature in the Solar System's arrangement: even though, according to the nebular hypothesis, the planets formed in the same plane, the axis of rotation of the Sun itself is tilted by 6° to the mean plane of the planets' orbits. A highly inclined massive planet might have twisted the planets out of alignment. As for how it ended up in such a remote outpost in the Solar System, it could have been ejected by Jupiter during the early era of the Solar System's evolution, or formed in a distant circular orbit later perturbed during an encounter with another star. So far, the evidence for its existence of the planet is at best circumstantial. But if it exists, it will clearly be very faint, with a magnitude greater than twenty-two—at least six hundred times fainter than Pluto. Astronomers have already begun searching for it with large telescopes, but so far the results are reminiscent of the unsuccessful searches for Planet X. A point against it is that a careful analysis of the motion of the Cassini spacecraft in orbit around Saturn failed to detect any effects of such a planet on the orbital motion of Saturn. Though not enough to disprove the existence of the planet, it does place significant constraints on where it might be lurking, and suggests it may presently be near aphelion, which of course will make it that much harder to find.[16]

To Boldly Go

Now that we have defined the "geography" of the third zone, we can chart the course of the New Horizons spacecraft as it plunges beyond Pluto into that vast and frigid frontier.

Recall that the spacecraft passed Pluto and Charon on July 14, 2015, at a speed of about 50,400 kph (31,300 mph). After concluding all its observing sequences, which included images and spectra of the other four small satellites and the looks backward toward the Sun and Earth as the bright ring of Pluto's atmosphere came into view, the spacecraft was spun up again for its ongoing journey. In October and November 2015, the onboard hydrazine thrusters fired four times to make very small corrections to the trajectory toward 2014 MU_{69}.

Immediately after the July 14, 2015, flyby of Pluto, observations of MU_{69} intensified to improve our knowledge of its exact position, size, shape, and the reflectance properties of its surface. All these factors are critical to the success of the planned flyby on January 1, 2019.

Then, in July 2017, just two years after the Pluto flyby, a series of stellar occultations by MU_{69} was predicted to occur. We described in chapter 14 the power of stellar occultations for the detection of Pluto's thin atmosphere, and these events are routinely used to determine the sizes and shapes of asteroids and other small Solar System bodies. The successful observation of a stellar occultation by a tiny, distant KBO would depend on knowing its position in the sky to very high precision, and in view of its small size would be difficult in the best of circumstances. New Horizons team members, led by Marc W. Buie, made the most accurate possible predictions of the time of the occultation and the position of the shadow of MU_{69} on Earth's surface. The shadow was predicted to cross part of the southern Pacific Ocean and then cross over Chile and Argentina. Buie's team of some sixty observers with twenty-four portable telescopes set up a "picket fence" of observing stations across the predicted shadow. Around midnight on July 17, in the clear skies of a remote region of Chubut and Santa Cruz, Argentina, five of the telescopes detected the winking out of the star as tiny MU_{69} crossed in front of it. None of the small telescopes could see MU_{69} itself, but as it passed in front of the star, its presence and its motion were evident.

The occultation event showed at once that MU_{69} is small, but it revealed an astonishing shape. As they analyzed the data, Buie and his team found that the MU_{69} is either a highly elongated object about 30 km (20 miles) long, or quite possibly a binary pair of bodies that are both revolving about a common center of gravity. If it is a pair of bodies, each is about 15–20 km (9–12 miles) in size. They could actually be in contact with one another. This extraordinary discovery adds drama and excitement for the possibility to examine two objects during the impending New Horizons flyby on the first day of 2019. If it is indeed a binary object, the New Horizons flyby will be the first close investigation of such a complex body of asteroidal size.

The three-and-a-half-year journey to MU_{69} is anything but monotonous—en route, New Horizons will repeatedly measure the brightness of more than a dozen KBOs to determine their rotation periods and shapes and to look for satellites. The spacecraft's remote vantage point affords unique opportunities to measure the brightness of KBOs at angles not possible from Earth. Measurements taken over a range of phase angles will give crucial information on the albedo and fine structure of the surfaces[17]—a coarse granular surface of great roughness reflects sunlight differently from a smooth surface. The albedo can be estimated from the brightness measurements over a range of phase angles, and the albedo in turn gives a good determination of the size of the body.

Just as the PEPSSI instruments measured the flux of solar-wind particles and atomic particles coming from outside the Solar System en route to Pluto, the measurements will continue along the billion-mile trajectory toward MU_{69}. The charged-particle environment far from the Sun has been measured by only a few spacecraft, and New Horizons will provide unique information on this region at a time of unusually low solar activity. Similarly, the SDC will continue to detect fine particles chipped off KBOs and permeating the vast outer Solar System, giving more information on the rate of demolition of these bodies as they continue to grind one another down by violent collisions.

After more than three years of flight past Pluto, the new target will come into view of LORRI, providing an opportunity to make any needed corrections to the trajectory to arrive at MU_{69} on time and at the desired distance above the surface. The optimum flyby distance has been set at 3,500 km (2,175 miles) to get high resolution of the surface from the cameras and spectrometers while maximizing the length of time over which the observations can be made. The timing of the encounter is critical, as it was at Pluto, because the observing sequences must be loaded onto the spacecraft's computer well in advance of closest approach for the commands to be executed automatically at the correct times. The timing is even more critical for the MU_{69} flyby because the target is, at only 20–40 km (12–24 miles) across, much smaller than Pluto and Charon, and thus at optimum viewing geometry for a very short time.

This epoch-making journey will cross another new horizon as the spacecraft passes the small icy body on January 1, 2019 (218 years to the day after Piazzi discovered Ceres). At that moment, it will unseat Pluto from its briefly held distinction as the most distant targeted object ever visited by a spacecraft.

Not only is MU_{69} the most distant targeted object, it is surely the most primitive. In planetary science, primitive means that a body has undergone a minimal amount of physical and chemical alteration since it condensed and aggregated from the feedstock materials in the solar nebula about 4.6 billion years ago. Objects that grew through this process to about a hundred kilometers and larger in size underwent significant heating from the gravitational energy that brought together their component bits of dust, ice, and gas. As the gravitational heat dissipated, more heat was generated by radioactive elements mixed in with their other component materials.[18] Objects smaller than a hundred kilometers in size, such as MU_{69}, are thought to be too small to have heated significantly from either gravity or the decay of natural radioactive elements. It formed cold, and has remained far from the Sun ever since—sequestered in the Solar System's deep freeze at a distance ranging from 42.7 AU at perihelion to 45 AU at aphelion. Its orbital period is 293 years, and its inclination to the ecliptic about 2°. (By coincidence, except for the inclination, its orbit is very similar to that Lowell posited for Planet X.)

MU_{69} is in the category of classical KBOs because its orbit is nearly circular and only slightly inclined to the ecliptic. This ensures that it has been in this stable arrangement since the origin of the Solar System. As a frozen leftover from the ancient era of planet formation, MU_{69} must preserve not only the original recipe stirred up in the solar nebula, but also hold important clues to the processes by which ices, dust, and gases stuck together to make a solid body sufficiently durable to survive since the beginning.

Even though MU_{69} most likely escaped significant internal heating, its surface has doubtless endured billions of years of bombardment by high-velocity fragments knocked off other KBOs by mutual collisions, as well as other tiny bits of ambient dust, some presumably interstellar.[19] These impacts would affect the outermost surface layers, gradually vaporizing ices and slowly chipping away at a microscopic level any rocky material that might be exposed. What the surface of MU_{69} might look like is impossible to know in advance—it could be icy and moderately reflective, or a dark, porous, sponge-like layer. It could well be a combination of those two possibilities. If we have learned anything from the history of spacecraft exploration of Solar System objects, however, perhaps most likely of all is that it will present surprises that we haven't seen or even imagined.

While New Horizons will give us an unprecedented view of an original and primitive small member of the Kuiper Belt population as it speeds by, it will barely have enough time to get acquainted. Until the spacecraft is quite close, in the several days prior to the close flyby its cameras will give only indistinct images MU_{69}, but will show if it is a true binary object or a single, highly elongated body. During the few hours of closest approach and flyby, images and spectra of the surface will reveal details of the shapes and compositions of geological features on the surface. The highest-quality data will be taken and recorded over a span of just a few hours, and then transmitted to Earth in the days after closest approach as MU_{69} recedes into the distance. Unless MU_{69} reveals fresh patches of surface exposed by a recent collision with another KBO, New Horizons may view only an ancient, battered landscape, providing no opportunity to see deeper into the interior where so much of the early history of the Solar System lies hidden.

This scarred surface will, however, have a compelling story to tell of the history of impacts with fragments of other KBOs. The rate of grinding of the Kuiper Belt population by mutual collisions is key to estimating the total original mass of the solar nebula and the solid objects that aggregated within it in the earliest millennia of Solar System history. This information is crucial because it informs theories of large-scale dynamical events affecting not only the activity in the third zone but events in the planetary region as well. Orbital turmoil beyond Neptune dislodged trans-Neptunian objects and propelled them inward where they collided with the planets, possibly resetting their geological clocks by the immense forces unleashed in the resulting explosive impacts. A major event in Solar System called the Late Heavy Bombardment (LHB) is conjectured by some planetary scientists who seek to explain the statistics of craters on the Moon and the terrestrial planets. The LHB, which may or may not have actually occurred, is conceptualized as an intense rain of comets and asteroid-size rocky bodies about 4 to 3.8 billion years ago, which left characteristic patterns of craters on the surfaces of the terrestrial planets after they had largely solidified some half a billion years after their origin. It isn't likely that pictures of MU_{69} alone can answer the LHB proponents and skeptics definitively. However, if its surface is truly primitive, the impact record it displays will surely add solid information to the discussion.

The imaging spectrometers on New Horizons will probe the composition of the surface of MU_{69}, just as they did for Pluto and Charon, searching for the characteristic signatures of ices and minerals. Telescopic observations show that most or all of the largest trans-Neptunian bodies have icy surfaces, some with only water ice, some with methane and ethane and likely some frozen nitrogen mixed in as well. The smaller bodies, including MU_{69}, are much too faint to allow ices to be distinguished from minerals with current Earth-based telescopes. This makes discerning the surface composition of MU_{69} even more important. LEISA will accomplish this goal from observations in the near-infrared wavelengths, and at the closest approach of a few thousand kilometers, will provide spatially resolved spectral images that can detect regional concentrations of such different kinds of ices or minerals as may be present. Alice, the ultraviolet imaging spectrometer, will concentrate on searching for gas escaping from MU_{69}. It may be a vain quest, however, since ices on the surface and in the shallow interior are expected to have dissipated long ago.

Leaving MU_{69} behind, until it vanishes back into the space from which it emerged, New Horizons will continue unimpeded on its lonely course farther outward from the Sun, covering some 1,210,000 km (752,000 miles) a day and 442,000,000 km (275,000,000 miles) a year. With the images, spectra, and other data taken at the MU_{69} encounter safely transmitted back to Earth, the spacecraft will continue to make a few more observations of KBOs and measurements of particles from the Sun and other sources in the Galaxy, and the SDC will continue to register the occasional dust particle. With the increasing distance of the spacecraft from Earth, the radio signals from the 12-watt transmitter will be increasingly difficult to detect, even with the largest antennas on Earth. At the same time, the electrical energy available to the spacecraft from the radioisotope thermal generator powered by slowly decaying radioactive plutonium will ebb to where it will no longer be able to support the onboard computers and other hardware. The limited amount of hydrazine in the spacecraft's thrusters will be depleted, leaving New Horizons unable to orient itself, either for astronomical observations or to point the radio antenna to Earth to report back home.

Even after New Horizons goes silent, its performance will remain a significant milestone in our species' history of exploration. It will have added its chapter to the story of humankind's first reconnaissance not only of Earth's Moon, the five planets known since classical times, and Uranus, Neptune, and Pluto, but of numerous comets and asteroids. Its own peculiar achievement has been to open exploration long confined to the first two zones of the Solar System to the almost incomprehensibly vast third zone, the Kuiper Belt.

Gone but not forgotten, the spacecraft will continue to wend its long and lonely way through the Galaxy. Carrying on board a few mementos from its home planet as well as a small container of the ashes of the discoverer of Pluto, and long-since faded into the background of the night sky, New Horizons will, we can only hope, achieve a yet more signal success than it has done so far: that its epic journey will inspire the efforts of future generations of our species to continue the unfinished work of uncovering the further secrets of the Solar System and the Universe.

It may well be that this quest, whose genesis came in our own time, is only just beginning.

Appendix 1

New Horizons Science Team

Principal Investigator: S. Alan Stern, Southwest Research Institute
Project Scientist: Harold A. Weaver, Johns Hopkins University Applied Physics Laboratory
Deputy Project Scientists: Kimberly Ennico-Smith, NASA Ames Research Center; Catherine B. Olkin, Southwest Research Institute; Leslie A. Young, Southwest Research Institute
Director of the Office of the Principal Investigator: Catherine B. Olkin

Coinvestigators

Fran Bagenal, University of Colorado
Richard P. Binzel, Massachusetts Institute of Technology
Marc W. Buie, Southwest Research Institute
Bonnie J. Buratti, NASA Jet Propulsion Laboratory
Andrew F. Cheng, Johns Hopkins University Applied Physics Laboratory
Dale P. Cruikshank, NASA Ames Research Center
Heather A. Elliott, Southwest Research Institute
Kimberly Ennico-Smith, NASA Ames Research Center
G. Randall Gladstone, Southwest Research Institute

Will M. Grundy, Lowell Observatory

Matt E. Hill, Johns Hopkins University Applied Physics Laboratory

David P. Hinson, SETI Institute

Mihaly Horanyi, University of Colorado

Donald E. Jennings, NASA Goddard Space Flight Center

Ivan R. Linscott, Stanford University

Jeffrey M. Moore, NASA Ames Research Center

David J. McComas, Southwest Research Institute

William B. McKinnon, Washington University in St. Louis

Ralph L. McNutt Jr., Johns Hopkins University Applied Physics Laboratory

Catherine B. Olkin, Southwest Research Institute

William Joel Parker, Southwest Research Institute

Harold J. Reitsema, Independent Consultant

Dennis C. Reuter, NASA Goddard Space Flight Center

Paul Schenk, Lunar and Planetary Institute

John R. Spencer, Southwest Research Institute

Darrell F. Strobel, Johns Hopkins University

Mark R. Showalter, SETI Institute

Michael E. Summers, George Mason University

G. Leonard Tyler, Stanford University

Harold A. Weaver, Johns Hopkins University Applied Physics Laboratory

Leslie A. Young, Southwest Research Institute

Appendix 2

The New Horizons Spacecraft*

Size: About the size of a baby grand piano; 0.7 m (27 inches) tall, 2.1 m
   (83 inches) long, and 2.7 m (108 inches) at its widest. The dish antenna is
   2.1 m (83 inches) in diameter.

Weight: At launch, with 77 kg (170 lb) of hydrazine and propellant, the spacecraft weighed 478 kg (1,054 lb), including the science-instrument payload
   weighing 30 kg (66 lb).

Instruments: The science payload of New Horizons consists of seven instruments. These include three optical instruments, two plasma-measuring
   devices, a radio science receiver/radiometer, and a student-built dust
   counter. These highly miniaturized instruments draw only 28 watts of power.

Alice is an ultraviolet imaging spectrometer that probed the composition of Pluto's atmosphere. It makes spectral images in the wavelength range of about
   500 to 1,800 Angstroms (0.05–0.18 μm), where many atoms and molecules
   expected in Pluto's atmosphere have their spectral signatures.

Ralph is built around a single telescope that shared the light of Pluto and other
   target objects between two main instruments, *MVIC* (Multispectral Visible

   *Information in this appendix is from "New Horizons Pluto Flyby," National
Aeronautics and Space Administration, press kit, July 2015, 34–39, http://pluto.jhuapl
.edu/News-Center/Resources/Press-Kits/NHPlutoFlybyPressKitJuly2015.pdf.

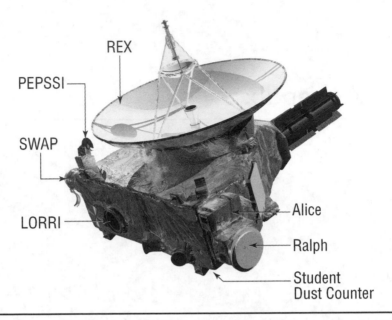

Figure A2.1 The New Horizons spacecraft with the principal instruments identified.

Imaging Camera) and *LEISA* (Linear Etalon Imaging Spectral Array). MVIC
takes high-resolution images through color filters that define the colors of
the target objects to highlight geographical regions of different compositions. LEISA makes images of the surfaces of Pluto, Charon, and other target
objects at many wavelengths across the near-infrared spectrum, 1.25–
2.5 μm. Combining MVIC and LEISA data defines both the geology and
the composition of regions on the surfaces of Pluto and the other targets.
LORRI (Long Range Reconnaissance Imager) is the third optical instrument,
consisting of a telephoto high-resolution camera that allows high-quality
imaging of the target objects at large distances.

REX, the Radio Science Experiment, is a miniature electronic device that
received a powerful radio beam from Earth as New Horizons passed behind
Pluto and then Charon, probing the properties of Pluto's atmosphere and
searching for an atmosphere of Charon.

SWAP (Solar Wind at Pluto) measured charged atomic particles emanating
from the Sun and the solar wind to evaluate the interaction of this particle
stream with the planet.

PEPSSI (Pluto Energetic Particle Spectrometer Science Investigation) also measured charged atomic particles and molecules escaping from Pluto's atmosphere and their interaction with the solar wind.

SDS (Venetia Burney Student Dust Counter) was built by students at the University of Colorado and detects microscopic dust grains that are made by collisions between asteroids and that are ejected from comets as the ice in those bodies evaporates into space. It is named in honor of Venetia Burney, an eleven-year-old English girl who is credited with suggesting the name "Pluto" for the ninth planet when it was discovered in 1930.

Notes

Chapter 2

1. See Francis Burney, *Journals and Letters, selected with an introduction by Peter Sabor and Lars E. Troide* (London: Penguin, 2001), xiii.
2. Patrick Moore, *80 Not Out: The Autobiography* (London: Contender, 2003), 121.
3. W. Davis Jerome, quoted in the Mozart Orchestra, Richard Woodhams, and Davis Jerome, *Sir William Herschel—Music by the Father of Modern Astronomy* (Newport, RI: Newport Classic, 1995), CD jacket.
4. Constance A. Lubbock, ed., *The Herschel Chronicle: The Life-Story of William Herschel and his Sister Caroline Herschel* (Cambridge: Cambridge University Press, 1933), 59.
5. Caroline Herschel, *Caroline Herschel's Autobiographies*, edited by Michael Hoskin (Cambridge: Science History Publications, 2003), 53.
6. J. L. E. Dreyer, "Herschel's Life and Work," in William Herschel, *The Scientific Papers of Sir William Herschel* (London: Royal Society and Royal Astronomical Society, 1912), 1:xxv.
7. Michael Hoskin, "Vocations in Conflict: William Herschel in Bath, 1766–1782," *History of Science* 41 (2003): 315–33.
8. Astronomers use a logarithmic scale of brightness of objects called stellar magnitudes. On this scale, the faintest star or planet visible to the naked eye in a completely dark sky is magnitude 6, and each magnitude step is 2.5 times fainter; 5 magnitudes corresponds to 100 times fainter. On this scale, Pluto at magnitude 14 is about 250 times fainter than the faintest star visible to the naked eye.

9.   Herschel, *Caroline Herschel's Autobiographies*, 57.

10. William Watson Jr. to William Herschel, December 18, 1781.

11. From Shakespeare's *Merchant of Venice*, 5:1, where Lorenzo says,

> How sweet the moonlight sleeps upon this bank!
> Here will we sit and let the sounds of music
> Creep in our ears: soft stillness and the night
> Become the touches of sweet harmony.
> Sit, Jessica. Look how the floor of heaven
> Is thick inlaid with patines of bright gold.

12. From Agnes Mary Clerke, *The Herschels and Modern Astronomy* (London: Cassell and Company, 1901), 100: "Sitting next Herschel one day at dinner, Henry Cavendish, the great chemist, a remarkably taciturn man, broke the silence with the abrupt question—'Is it true, Dr. Herschel, that you see the stars round?' 'Round as a button,' replied the Doctor; and no more was said until Cavendish, near the close of the repast, repeated interrogatively, 'Round as a button?' 'Round as a button,' Herschel briskly reiterated, and the conversation closed."

13. Sir William Watson, communication to the Royal Society, March 22, 1781; quoted in Michael Hoskin, *Discoverers of the Universe: William and Caroline Herschel* (Princeton, New Jersey: Princeton University Press, 2011), 50.

14. Pierre-Simon Laplace. "Une mémoire sur la determination des orbit des comètes." Following Charles Coulston Gillispie, *Pierre-Simon Laplace, 1749–1827: A Life in Exact Science* (Princeton, New Jersey: Princeton University Press, 1997), 289. This work, which Laplace read to the Academy on March 21, 1781, was a draft of the analytical part (Articles I-VII) of the work "Mémoire sur la détermination des orbites des comètes," published in *Mémoires de l'Académie royale des sciences de Paris, année 1780; 1784*, 13–72. (The publication was always two to four years in arrears. Thus the volume for the year 1780 appeared in 1784 and contained memoirs submitted in at any time between the nominal and publication dates, thus 1780 through 1783.) Laplace's attempt to apply the method described in the preceding memoir to Herschel's "comet," "Une application de sa méthode a la comete qui paroît actuellement," was read to the Academy on May 2, 1781.

15. William Herschel, *Collected Scientific Papers of William Herschel*, edited and with an introduction by J. L. E. Dreyer (London: Royal Society and Royal Astronomical Society, 1912), 1.

16. Richard Baum, *The Haunted Observatory: Curiosities from the Astronomer's Cabinet* (Amherst, NY: Prometheus, 2007), 19.

17. For "space-penetrating power," see William Herschel, "On the Power of Penetrating into Space by Telescopes; with a Comparative Determination of the

Extent of that Power in Natural Vision, and in Telescopes of Various Sizes and Constructions; Illustrated by Select Observations," *Philosophical Transactions of the Royal Society of London* 90 (1800): 49–85.

18. William Herschel, "On the Construction of the Heavens," *Philosophical Transactions of the Royal Society of London* 75 (1785): 213–66. For an account of this phase of Herschel's career, see William Sheehan and Christopher J. Conselice, *Galactic Encounters: Our Majestic and Evolving Star-System, from the Big Bang to Time's End* (New York: Springer, 2015), 42–53.

19. Dennis Rawlins, "Vindication of William Herschel's Claimed 1801/4/17 Discovery of Uranus's Satellite Umbriel," *Astronomy & Space* 3 (1973) 29, 26–40. A photographic examination by Charles Kowal at Mount Palomar in 1973 of the April 17, 1801, place in the sky where Herschel had recorded the satellite showed that there was no star in the position remotely comparable to Umbriel.

20. For the reader interested in pursuing the mathematical aspects of perturbation theory further, several monographs can be recommended, such as C. H. H. Cheyne, *An Elementary Treatment on Planetary Theory* (London, Macmillan, 1883) and Forest Ray Moulton, *An Introduction to Celestial Mechanics*, 2nd ed. (1914; repr., New York: Dover, 1970).

21. Johann Elert Bode, *Berliner Astronomisches Jahrbuch* (1784), Seite 219, quoted in A. F. O'D. Alexander, 1965. *The Planet Uranus: A History of Observation, Theory, and Discovery* (London: Faber and Faber), 217.

22. Pierre-Simon Laplace, *The System of the World*, translated by J. Pond (London: Richard Phillips, 1809), 2:58.

23. Ibid., 372.

Chapter 3

1. Johann Elert Bode, *Einleitung zur Kenntnis des Gestirnten Himmels* [Introduction to the Knowledge of the Starry Heavens], 2nd ed. (Hamburg, Germany: Dieterich Harmsen, 1772), 462.

2. Clifford J. Cunningham, *The Discovery of the First Asteroid: Historical Studies in Asteroid Research* (New York: Springer, 2016), 28.

3. Basil Hall, *Patchwork* (London: Edward Moxon, 1841), 292–93.

4. Giuseppe Piazzi to Barnaba Oriani, January 24, 1801. Quoted in G. Cacciatore and G. V. Schiaparelli, eds., *Corrispondenza Astronomica fra Giuseppe Piazzi e Barnaba Oriani* (Milan: University of Hoepli, 1874), 49.

5. Franz Xaver von Zach, "Über einen zwischen Mars und Jupiter längst vermutheten, nun wahrscheinlich entdeckten neuen Hauptplaneten unseres Sonnensystems," *Monatliche Correspondenz zur Beförderung der Erd und Himmelskunde* 3 (1801): 592–623.

6. Gauss's method of least squares is often incorrectly assumed to be the method he later published in *Theoria motus corporum coelestium in sectionibus conicis solem ambientium* (Hamburg, 1809). The least-squares method was in fact first published by Adrien-Marie Legendre, though Gauss probably possessed it first. See Stephen M. Stigler, "Gauss and the Invention of Least Squares," *Annals of Statistics* 9, no. 3 (1981): 465–74, https://doi.org/10.1214/aos/1176345451.

7. Regarding Piazzi, after discovering Ceres he continued to work diligently on his star catalog. It was published 1803 and included 6,748 stars. Von Zach called the achievement "epochal," and the Institut de France acknowledged it by awarding Piazzi the Lalande Prize. It says much of Piazzi's makeup that he no sooner published than he began to ponder the possibility of coming out with a new edition. However, he was forestalled by fate; with his eyesight failing by 1807 he had no choice but to entrust the observations to his assistant, Niccolò Cacciatore. The second edition of the great star catalog, containing 7,646 stars, appeared in 1814. In 1817, King Ferdinand called Piazzi to Naples to oversee the building of the Capodimonte Observatory, and for a time the Theatine monk-astronomer served as director-general of both the Naples and Palermo observatories; the day-to-day work of running the Palermo Observatory was taken up by Cacciatore. At about that time, Piazzi was also named president of the Royal Academy of Sciences in Naples. Though Piazzi missed Palermo, he was at least spared the attack of the observatory during the Sicilian Revolution in 1820, when Cacciatore was imprisoned. Piazzi visited the Palermo Observatory for the last time in 1825, writing on his return to Naples, "Perhaps I will never see it again." Clifford Cunningham, "Giuseppe Piazzi," in *Biographical Encyclopedia of Astronomers*, edited by Thomas Hockey (New York: Springer, 2007), 2:903. His words proved prophetic; the following year he fell ill during an outbreak of cholera and died, at age 80, on July 22, 1826.

8. Lubbock, *The Herschel Chronicle*, 273.

9. Laplace, *The System of the World*, 35.

10. Alexis Bouvard, "Extrait des Registres des observations astronomiques faites par Lemonnier, à l'Observatoire des Capucins, rue Saint Honoréé, à Paris," *Connaissance des Tems* [sic], *ou des mouvemens célestes a l'usage des astronomes et des navigatuers, pour l'an 1821* (Paris: Bureau of Longitudes, 1819), 339–40. The official astronomical yearly publication in France has been published without interruption since 1679, when it was known as *La Connoissance des Temps ou calendrier et éphémérides du lever & coucher du Soleil, de la Lune & des autres planètes* until just after the French Revolution, when the title appeared as *Connoissance des Temps*, and for several years afterward as *Connaissance des tems*. Since the 1820s, it has been—and remains—the *Connaissance des Temps*.

11. According to E. M. Antoniadi, Bouvard told this to Arago; see *L'Astronomie* 50 (1936), 252.

12. "Il a ete prive de l'honneur d'une belle decouverte." Bouvard, "Extrait des Registres des observations astronomiques faites par Lemonnier," 339.

13. Alexis Bouvard, *Tables astronomiques publiées par le Bureau des Longitudes de France contenant les Tables de Jupiter, de Saturne et d'Uranus contruites d'après la théorie de la Mécanique céleste* (Paris: le Bureau des Longitudes, 1812), xiii.

14. Ibid.

15. Ibid., xiv.

Chapter 4

1. Roger Hutchins, *British University Observatories 1779–1939* (Aldershot, Hampshire, England: Ashgate, 2008), 88.

2. Ibid., 89.

3. George Biddell Airy, "Report on the Progress of Astronomy During the Present Century," *Report on the First and Second Meetings of the British Association for the Advancement of Science* (London, 1833), 189.

4. Bessel was drawn to the idiosyncratic notion of the Göttingen physicist Johann Tobias Mayer, son of the great astronomer Tobias Mayer, according to which a specific attraction existed that depended on the nature of the masses—an idea that Bessel did not finally give up until he had squandered many hours on experiments with pendulums, using all kinds of bodies, even meteoritic iron, which in the end gave the same results.

5. Hussey had been born in 1792 in Lamberhurst, Kent, the only son of the Reverend John Hussey and Catherine Jennings. The Husseys were of an old Anglo-Norman family, and were substantial local landowners—the Reverend John Hussey's brother was Edward Hussey, of Scotney Castle. After Thomas's father died in Allahabad, India, when he was seven, Catherine sent him to Eton, where something of a scandal ensued. An Irish barrister and confidence man, J. P. Maccabe, arrived on the scene, claiming to be Catherine's nephew. According to a deed of gift Maccabe claimed was written by Catherine, Thomas had supposedly rendered her miserable, "took a considerable sum of money from me, and went away, I knew not whither. He enlisted: my hopes were blasted of ever seeing him either respectable in life, or a man of education." Meanwhile, Maccabe had absconded to Dublin with Catherine's inheritance and with Thomas, and had sent the latter to Trinity College. Eventually, a case against Maccabe went to the Irish Court of Chancery, where he was judged to have obtained the money by "undue influence, fraud, and misrepresentation," a judgment that was upheld by the House of Lords in 1831. Maccabe was forced to

repay a large sum to Catherine, thus securing Thomas's own financial situation. Thomas married Anna Maria Reed, later a noted illustrator and mycologist, and was awarded his degree as Doctor of Divinity from Trinity College in 1835.

6. This letter is in Airy's "Neptune file," long lost, but recovered in Chile in 1999 and now at Cambridge University Library; it was published verbatim by Airy as item no. 1 in "Account of Some Circumstances Historically Connected with the Discovery of the Planet Exterior to Uranus," *Memoirs Royal Astronomical Society* (1847): 385–414, which contains generally faithful extracts of the more important items in this file.

7. George Biddell Airy to Thomas J. Hussey, November 23, 1834, "Neptune file."

8. T. J. Hussey to G. B. Airy, undated letter in the "Neptune file," evidently written in response to Airy's letter to Hussey, October 23, 1846.

9. On the sale of Hussey's observatory to Durham University, see Allan Chapman, *The Victorian Amateur Astronomer: Independent Astronomical Research in Britain, 1820–1920* (Chichester and New York: Wiley-Praxis, 1998), 20. In the 1840s, in addition to working on a revised edition of the Bible, with "a brief hermeneutic and exegetical commentary," Hussey edited a magazine, *The Surplice*, to which his wife was an occasional contributor. Up to that time she had presumably been kept busy looking after their children (they had six in all, but only two survived into adulthood). She now began making a name for herself as a mycologist and illustrator. In 1847 she published, under the name Mrs. Thomas John Hussey, the beautiful (and expensive) *Illustrations of British Mycology: Containing Figures and Descriptions of the Funguses of Interest and Novelty Indigenous to Britain, First Series*, in 1847, which today commands very high prices on the used-book market; even individual color plates (of which there are ninety) sell for hundreds of pounds each. She died in 1853, with Hussey following the next year. A second volume of the *Illustrations*, containing fifty colored plates, appeared posthumously in 1855. Both sides of the family shared scientific interests. Anna Maria's sister Fanny Reed was also an illustrator; her brother, George Varenne Reed, served as tutor for Charles Darwin's sons before taking over as rector at Hayes, Kent; while the Reverend James Hussey was a member of the Botanical Society. One wonders if the Husseys might rather have resembled Dorothea Brooke and the Reverend Edward Casaubaum in George Eliot's *Middlemarch*.

10. Quoted in Morton Grosser, *The Discovery of Neptune* (Cambridge: Harvard University Press, 1962), 50–51.

11. The definitive account of the Cacciatore and Wartmann objects is found in Baum, *The Haunted Observatory*, 267–78.

12. The German astronomer Theodor Oppolzer finally explained the Wartmann affair in 1880, pointing out that on average the positions given by Wartmann

were six minutes of arc greater in right ascension and twenty-six seconds farther north in declination than the corresponding places of Uranus. Clearly, Wartmann was guilty of a gross error. "Whether Wartmann erred in plotting a map or in reading positions from a map, or whether precession was incorrectly applied is difficult to say," Oppolzer writes. But he leans toward the latter explanation, adding, "Nevertheless, it is remarkable that subtraction of the approximate precession in a century removes the difficult." See *Astronomische Nachrichten* 97 (1880): 253.

13. Benjamin A. Gould, *Report of the History of the Discovery of Neptune* (Washington, DC: Smithsonian Institution, 1850), 1.

14. Heinrich Christian Schumacher, in his preface to Bessel's *Populäre Vorlesungen* (Hamburg, 1848), promised that he would "later" publish Flemming's reductions, made, as he says, at the behest of Bessel, "with the most rigorous care." He was good to his word. See H. C. Schumacher, *Astronomische Nachrichten*, band 30 (1850): Nr. 705, "Reduction der in Greenwich, Paris und Königsberg gemachten Beobb. des Uranus, von 1781 bis 1837," 129; Nr. 707, 707, 708, "Reduction der in Greenwich, Paris und Königsberg gemachten Beobachtunen des Uranus, von 1781–1837, 145–182"; Beil. zu Nr. 708, "Reduction der in Greenwich etc. gemachten Beobb. des Uranus (Beschluss)," 193. Anyone who examines this work in detail cannot but be impressed by the thoroughness and skill with which it was carried out, and have no doubt that Bessel and Flemming would, but for the circumstances that tragically intervened—Flemming's death and Bessel's illness—have soon produced a position for the unknown planet and moved quickly, with the excellent instruments at Königsberg, to find it.

15. The Moon's average distance from Earth is 384,400 km (238,855 miles), which is 0.00257 AU.

16. George Biddell Airy, *Schreiben des Herrn Airy, Astronomer Royal, an den Herausgeber, Astronomische Nachrichten* 15 (1838), columns 217–20. Column 219 contains the tabular errors of the radius vector of Uranus.

Airy's previous publications on the radius vector problem include his textbook written for the *Penny Cyclopaedia, Gravitation, an Elementary Explanation of the Principal Perturbations of the Solar System* (London: Charles Knight, 1834), where he describes how the radius vector of a planet is affected by conjunction with a planet exterior to itself—at the time, the classic case was, of course, that involving Jupiter as it passed by Saturn, which gave rise to the "Great Inequality." He writes, "The effect then of a force in the direction of the planet's motion, which increases the planet's velocity, is to increase the size of its orbit; and the bigger the orbit is, the longer is the time of revolution" (38). Put more simply, what is going on is a transfer of angular momentum, which pushes the planet into a higher orbit (one with greater potential energy).

As late as December 17, 1846, Airy was still insisting to Richard Sheepshanks,

> Concerning the radius vector of Uranus.—The error was certain as to sign. It was determined with reasonable accuracy as to magnitude. . . . Now, suppose that Adams' elements which gave longitude-corrections had given a wrong sign for the correction of the radius vector, What would his theory have been worth? The alternation of signs of errors +– in longitude does not exclude any other hypothesis than that of an exterior planet. If the law of force differed slightly from that of inverse square of the distance (of which two years there was great probability) and if tables were calculated strictly on the law of inverse square of distance (as was done in existing tables), then the discordances in longitude would have the alternate signs +–. Le Verrier evidently attached great importance to the radius vector. . . . The radius vector, as you say, was to be used as an indirect verification: but its error demanded explanation quite as imperatively as the other [i.e., the discordances in longitude].

He ended, understandably, "I am quite tired of radius vectors."

17. A. F. O'D. Alexander, *The Planet Uranus: A History of Observation, Theory, and Discovery* (London: Faber and Faber, 1965), 97.

18. Alexis Bouvard to G. B. Airy, May 27, 1844.

19. Joseph Bertrand, "Élogie historique de Urbain-Jean-Joseph Le Verrier," *Annales de l'Observatoire de Paris, Mémoires* 15, 3–22 (1880): 5.

20. Ibid., 6.

21. Brian Sheen, "Neptune: Towards Understanding How Adams's Upbringing Enabled Him to Predict, Then Miss the Discovery of the Century," *Antiquarian Astronomer* 7 (March 2013): 71.

22. Henry Spencer Toy, *The Adamses of Lidcott* (Penzance, England: Wordens of Cornwall, 1969), 3. Toy cites John Allen, *History of Liskeard* (1856), 554, as authority for this remark. In full, the passage reads, "He [Adams] went with his father to Sticklepath on a visit to Mr. G. Pearse, who, having a son a few years older, questioned the boys on their knowledge of algebra, and finding there was something superior in young Adams called in a neighbouring schoolmaster, who examined him closely and was surprised at his mathematical genius. When Mr. Pearse brought him back, he told Mr. Adams that his son would be a great man, adding, 'If he were my boy, I would sell the hat off my head rather than not give him a good education.'"

23. This quotation was supplied by Hilda Roseveare, John Couch Adams's niece; quoted in Toy, *The Adamses of Lidcott*, 10.

24. The first Mechanics' Institute had been established in Edinburgh in October 1821 as the School of Arts of Edinburgh (later Heriot-Watt University) with the pur-

pose of addressing "societal needs by incorporating fundamental scientific thinking and research into engineering solutions." By the mid-nineteenth century more than seven hundred such institutes were in the United Kingdom and overseas.

25. J. C. Adams to his parents, October 17, 1835; Adams papers, Truro Library.

26. J. C. Adams to Thomas Adams, May 13, 1836; Adams papers, Truro Library.

27. George Adams, "Reminiscences of our Family," 1892, reprinted in H. M. Harrison, *Voyager in Time and Space: The Life of John Couch Adams, Cambridge Astronomer* (Lewes, Sussex: The Book Guild, 1994), 18.

28. Quoted in Morton Grosser, *The Discovery of Neptune* (Cambridge: Harvard University Press, 1962), 74. Campbell came to realize that there were few if any in Adams's league, and that "if I could keep near him in the examinations I should do very well." He did well enough to become a fellow in mathematics at St. John's before leaving for Jamaica to oversee plantations owned by his family. His son, A. Y. G. Campbell, was also a good mathematician; he won a scholarship to Trinity College and wrote a creditable undergraduate dissertation on the composition of stars, which won for him a Smith's Prize. Sir Robert Ball, the Lowndean Professor of Astronomy at Cambridge, called him one of the most brilliant mathematicians of his time. However, instead of remaining at Cambridge, A. Y. G. Campbell sat the 1895 examinations for the Indian Civil Service and placed second—behind Alan Turing. He had a distinguished career in the Indian Civil Service, serving as chief secretary of Madras from 1925 to 1935.

29. James Challis, "Account of Observations of the Cambridge Observatory for detecting the Planet exterior to Uranus," *Monthly Notices of the Royal Astronomical Society* 7, no. 9 (November 13, 1846): 145.

30. The text, which was not published until 2013, reads as follows:

Sir: I must apologize for having called at the observatory the other day at so unreasonable an hour, the reason was that I had only arrived in town that morning & it was necessary for me to be in Cambridge the same day, so that I had no other opportunity. The paper I then left contained merely a statement of the results of my calculation; I will now, if you will allow me, trouble you with a short sketch of the method used in obtaining them. My attention was first directed to the anomalies in the motion of Uranus by reading, some time since, your valuable Report on Astronomy. If the action of known planets really proved insufficient to account for the perturbations of Uranus, it appeared to me that by far the most probable hypothesis which could be formed for that purpose, would be that of the existence of an undiscovered planet beyond. If this were the case, I conceived it might be possible to find from an examination of the observed perturbations, the approximate position of the new planet, so as to assist

Astronomers in discovering it. The solution of this problem, however, would be clearly impracticable at present, without making some assumption as to the mean distance. Fortunately, Bode's law supplies us with a value which, at any rate, has a claim to be first tried. Accordingly, using the differences between Obs[n]. and the Tables given in Bouvard's Equations of condition as far as they extend, and subsequently those supplied by the Cambridge and Greenwich observ[ns]. together with those in the various numbers of the Astron. Nachrichten, I obtained values of the mass and position of the assumed planet, so as to satisfy all the observations very nearly. The series however, not being sufficiently continuous at one or two points, Professor Challis had the kindness to request you to communicate the results of a few of the Greenwich obs[ns]. to supply the deficiency. For your kindness in sending the whole of the Greenwich ob[sns]. of Uranus, I beg you will accept my warmest thanks. On receiving them, I determined to make them the base of a new investigation, taking into account several quantities which I had previously neglected. The later observations used were the same as before. (J. C. Adams, unfinished draft letter to George Biddell Airy, November 13, 1845, Cornwall Record Office: Adams family papers, AM 330. Quoted in Sheen, "Neptune," 5.)

31. G. B. Airy to the Reverend Adam Sedgwick, December 4, 1846.
32. James Challis to *The Athenaeum*, December 17, 1846.
33. George Biddell Airy, "Account of some circumstances historically connected with the discovery of the Planet exterior to Uranus." *Monthly Notices of the Royal Astronomical Society* 7 (1846), 121–44.
34. "Excusez-moi, Monsieur, d'insister sur ce point." Urbain Le Verrier to G. B. Airy, June 28, 1846. Airy Neptune file, Royal Greenwich Observatory archives, Cambridge University Library.
35. "Si je pouvai ésperer que vous surez assez de confiance dans mon travail pour chercher cette planète dans le ciel, je m'empresserais, Monsieur, de vous envoyer sa position exacte, dès que je l'aurai obtenue." Ibid.
36. William James, *Principles of Psychology* (New York: Henry Holt and Company, 1890), 2:530.
37. Robert W. Smith, "The Cambridge Network in Action: The Discovery of Neptune," *Isis* 80 (1989): 395–422.
38. Edwin Holmes, "The Planet Neptune," *Journal of the British Astronomical Association* 18 (1907): 36–37.
39. Johann Galle to Urbain Le Verrier, September 25, 1846. The diameter of the disk as measured by Galle and Encke was somewhat too large; on the date in question it was only 2.5″. Undoubtedly philosophers of science will note here the tendency of expectation to influence what is observed, as had earlier been

the case with William Herschel, whose series of measures indicated an increase in Uranus's diameter in accordance with the expectation it was an approaching comet, even though it was shrinking slightly, as noted by R. H. Austin, "Uranus Observed," *British Journal for the History of Science* 3 (1967): 275–84.

40. Quoted in James Lequeux, *Le Verrier: Magnificent and Detestable Astronomer* (New York: Springer, 2015), 50.

41. Ibid.

Chapter 5

1. H. H. Turner, obituary notice of Johann Gottfried Galle, *Monthly Notices of the Royal Astronomical Society* 71 (1911): 280.

2. John Herschel, "Sir J. F. W. Herschel's Address on the Subject of the Award of the Testimonials," *Monthly Notices of the Royal Astronomical Society* 8 (1848): 102–119:111.

3. Kepler's third law of planetary motion states that the square of the orbital period one body around another is proportional to the cube of the distance between the centers of the two bodies. On Lassell's suspected observation of the ring and early searches for a satellite, see Richard Baum, "Neptune 1846–7," in *The Planets: Some Myths and Realities* (Newton Abbot, UK: David & Charles, 1973): 120–46. The name Triton was proposed by the French astronomer Camille Flammarion, who was also responsible for naming the fifth satellite of Jupiter, Amalthea.

4. John Herschel to Friedrich Wilhelm Struve, December 27, 1846; Royal Society.

5. The significance of Galileo's observations was recognized only in 1980. It turns out that in December 1612 and January 1613, Galileo caught Neptune in the same field as Jupiter. Neptune was then near a stationary point (changing the direction of its motion from retrograde to direct), when Galileo noted its position relative to the satellites and another star on a sketch on January 27, 1613, and again the following night, when he entered his suspicion that the "two stars" appeared slightly further apart. However, he failed to check into the matter. See Stillman Drake and Charles T. Kowal, "Galileo's Sighting of Neptune," *Scientific American* 243 (1980): 74–79.

6. "Report of Peirce's Communication of Investigation in the Action of Neptune to Uranus," *Proceedings of the American Academy of Arts and Sciences* 1 (1846–48): 65.

7. See our discussion of resonances in chapter 15.

8. "Report of Peirce's Communication," 67.

9. See, for instance, Richard Baum and William Sheehan, *In Search of Planet Vulcan: The Ghost in Newton's Clockwork Universe* (New York: Plenum, 1997). This remains the definitive account.

10. According to David W. Hughes and Brian G. Marsden, "Planet, Asteroid, Minor Planet: A Case Study in Astronomical Nomenclature," *Journal of Astronomical History and Heritage* 10, no. 21–30 (2007): 28.

11. Quoted in Baum, *The Haunted Observatory*, 41.

12. The "strange interlude" of Ferguson's star is delightfully told by Baum, *Haunted Observatory*, 41–66.

13. Ibid., 65.

14. Simon Newcomb, *An Investigation of the Orbit of Uranus, with general tables of its motion* (Washington, DC: Smithsonian Contributions to Knowledge, 1873), 170.

15. Simon Newcomb, *The Reminiscences of an Astronomer* (Boston: Houghton Mifflin, 1903), 141.

16. Newcomb's preference for theory over observation figured in his encounter with a young self-taught astronomer at the meeting of the American Association for the Advancement of Science, which was held at the capitol in Nashville in the summer of 1877. The young man was E. E. Barnard, who had recently saved up his earnings at the photograph gallery where he was employed to acquire a 5-inch refracting telescope, and succeeding, through friends, in gaining an audience with the great astronomer from Washington, DC, asked him how a young man with a small telescope might make himself useful in astronomy. Newcomb's answer: "Lay aside at once that telescope and master mathematics, for you will never be what you seek to become without this mastery." Fortunately, Barnard did not take Newcomb's discouraging response to heart, and went on to become the greatest observational astronomer of his era. See William Sheehan, *The Immortal Fire Within: The Life and Times of Edward Emerson Barnard* (Cambridge: Cambridge University Press, 1995), 23–25.

17. Todd's career, with an emphasis on his transit-of-Venus involvements, is described in William Sheehan and A. Misch, "Ménage à trois: David Peck Todd, Mabel Loomis Todd, Austin Dickinson, and the 1882 transit of Venus," *Journal for the History of Astronomy* 35 (2004): 123–34.

18. Like Todd's, Forbes's genius lay in more practical directions. Within a few years of publishing his planet predictions, Forbes left Anderson's University to become manager of the British Electric Light Company, manufacturers of carbon filaments and arc lamps. He patented the use of carbon in electricity-generating machines—an invention still universally used today. However, he sold the American rights to Westinghouse Electric, and seems never to have received British royalties for his invention. It was later said of him, "There can be no doubt that he presented to the world an idea of great engineering and commercial value, the importance of which he does not seem to have fully grasped at the time." G. L. Addenbroke, quoted in "George Forbes (scientist),"

Wikipedia, last updated October 14, 2017, https://en.wikipedia.org/wiki /George_Forbes_(scientist). He died, a virtual recluse, in increasing poverty, at "the Shed," the wooden structure with an observatory on the upper story that he built near Pitlochry, Scotland, in 1936.

19. Camille Flammarion, *Popular Astronomy: A General Description of the Heavens*, translated by John Ellard Gore (London: Chatto & Windus, 1894), 471.

Chapter 6

1. Agnes M. Clerke, *Popular History of Astronomy During the Nineteenth Century*, 3rd ed. (London: Adam & Charles Black, 1893), 392, 393.

2. Quoted in *Lowell National Historical Park Handbook* (Lowell, MA: National Park Service, 2015).

3. See Don Lago, "The canals of Lowell," *Griffith Observer* 49, no. 8 (August 1985): 2–11. For a scholarly history of the canals and engineering works of Lowell, see Patrick M. Malone, *Waterpower in Lowell: Engineering and Industry in Nineteenth-Century America* (Baltimore: Johns Hopkins University Press, 2009).

4 John Amory Lowell's first marriage was to his cousin Susan Cabot Lowell, daughter of Francis Cabot Lowell; his second marriage, which produced Augustus, was to her cousin, Elizabeth Cabot Putnam. On the Lowell family, see Ferris Greenslet, *The Lowells and their Seven Worlds* (Boston: Houghton Mifflin, 1946).

5. Percival Lowell, "Augustus Lowell," *Proceedings of the American Academy of Arts and Sciences* 37 (August. 1902): 652.

6. P. Lowell to Barrett Wendell, January 1, 1877; Houghton Library, Harvard University.

7. Percival Lowell, "The Lowell Observatory and Its Work," text of address to the Boston Scientific Society, May 22, 1894; *Boston Commonwealth*, May 26, 1894.

8. For a general discussion of Lowell's Mars theories, see, for example, William Graves Hoyt, *Lowell and Mars* (Tucson: University of Arizona Press, 1976); Michael J. Crowe, *The Extraterrestrial Life Debate: 1750–1900* (Cambridge: Cambridge University Press, 1986), especially chapter 10, "The Battle over the Planet of War"; William Sheehan, *Planets & Perception: Telescopic Views and Interpretations, 1609–1909* (Tucson: University of Arizona Press, 1988); William Sheehan, *The Planet Mars: A History of Observation and Discovery* (Tucson: University of Arizona Press, 1996); and K. Maria D. Lane, *Geographies of Mars* (Chicago: University Chicago Press, 2011). The best biography of Lowell to date is David Strauss, *Percival Lowell: The Culture and Science of a Boston Brahmin* (Cambridge: Harvard University Press, 2001).

9. E. E. Barnard to Simon Newcomb, September 11, 1894; Simon Newcomb papers, Library of Congress.

10. A marvelous appreciation of the telescope, on the occasion of its restoration, is Kevin Schindler, *The Far End of the Journey: Lowell Observatory's 24-inch Clark Telescope* (Flagstaff, AZ: Lowell Observatory, 2016).

11. See Strauss, *Percival Lowell*, 173.

12. Percival Lowell to Elizabeth Lowell, April 12, 1901; Lowell family papers, Houghton Library, Harvard University.

13. For details, see Hoyt, *Lowell and Mars*. See also William Sheehan, *Planets and Perception: Telescopic Views and Interpretations, 1609–1909* (Tucson: University of Arizona Press, 1988).

14. Clyde Tombaugh, "The Discovery of Pluto: Generally Unknown Aspects of the Story," *Astronomy Beat* no. 23 (May 18, 2009): 2.

15. Johann Gottfried Galle, *Verzeichniss der Elemente der bisher berechneten Cometenbahnen, nebst Anmerkungen und Literatur-Nachweisen* (Leipzig, Germany: W. Engelmann, 1984).

16. Percival Lowell, *Trans-Neptunian Planet* (unpublished manuscript, ca. 1909); Lowell Observatory, Planet X calculations, Box 21. Note: Comet 1862 III, Swift-Tuttle, was shown by Giovanni Schiaparelli to be the source of the debris giving rise to the Perseid meteor shower, so this represents a kind of double entry. According to the most recent orbit, its perihelion is at 0.9595 AU and its aphelion is at 51.225 AU. It was recovered at its most recent return in 1992. Then Comet 1889 III was long lost, but was recovered in 2006. Its perihelion is at 1.077 au and its aphelion is at 47.232 AU.

17. Percival Lowell, *The Solar System* (Boston: Houghton Mifflin; Riverside Press, 1903). 113–15.

18. P. Lowell to Carl Otto Lampland, October 13. 1904; LOA.

19. Robert Grant Aitken, "Note on the Comets Discovered at the Lowell Observatory," *Publications of the Astronomical Society of the Pacific* 58 (1906): 83–84.

20. William Graves Hoyt, *Planets X and Pluto* (Tucson: University of Arizona Press, 1980), 91.

21. Ibid., 95.

22. Simon Newcomb, *An Investigation of the Orbit of Uranus with General Tables of Its Motion* (Washington, DC: Smithsonian Contributions to Knowledge, 1873), 111.

23. Percival Lowell, *Memoir on a Trans-Neptunian Planet* (Flagstaff, AZ: Lowell Observatory, 1915), 4.

24. P. Lowell to William H. Pickering, November 16, 1908; Lowell Observatory Archives (LOA).

25. P. Lowell to William F. Carrigan, November 19, 1908; LOA.

26. William H. Pickering, "A Search for a Planet Beyond Neptune," *Annals of the Astronomical Observatory of Harvard College* 61 (1909): 2:113–62.

27. Hoyt, *Planets X and Pluto*, 104.

28. P. Lowell to W. F. Carrigan, May 11, 1909; LOA.

29. Percival Lowell, untitled draft; probably written in early 1909; LOA.

30. Quoted in Abbott Lawrence Lowell, *Biography of Percival Lowell* (New York: Macmillan, 1935), 149.

31. Edward Charles Pickering, *Diary of Trip to International Solar Conference*, 1910; Pickering papers, Harvard University Archives. Subsequently—perhaps because of Mrs. Lowell's intervention—Lowell became somewhat more conciliatory. He came to Pickering's room at the station and said very cordially before the latter set off for the Grand Canyon, "You see the impression you have made on Mrs. Lowell, as she wants to go to Pasadena." In any event, neither Mr. or Mrs. Lowell did, though V. M. Slipher was allowed to attend the International Solar Conference, where he met with Edward Fath, who encouraged him in spectrographic work on the nebulae. See Sheehan and Conselice, *Galactic Encounters*, chapter 9.

32. P. Lowell to J. Trowbridge, December 9, 1914; LOA.

33. P. Lowell, *Memoir on a Trans-Neptunian Planet*, 7.

34. Hoyt, *Planets X and Pluto*, 95.

35. Ernest W. Brown, "On a Criterion for the Prediction of an Unknown Planet," *Monthly Notices of the Royal Astronomical Society* 92 (1931), 94.

36. See, for instance, Elizabeth Williams to Wrexie Louise Leonard, November 4, 1909, where she begins, "I do not wish to disobey—so please not mention to Dr. Lowell what I am about to write unless you are sure that it is best. I think I have discovered an error which runs throughout some tables in the 'Notes' on 'Planets and their Satellite Systems.'" The "Notes" were for Lowell's book *The Evolution of Worlds* (New York: Macmillan, 1909). Miss Leonard must have been sure that it was best, for the errors are corrected in the published version.

37. "The Millionaire Calculating Machine," advertisement, *The Executive Economist* 1, no. 4 (April 1911): back cover.

38. A. L. Lowell, *Biography of Percival Lowell*, 186–87.

39. P. Lowell, telegram to C. O. Lampland, March 13, 1911; LOA.

40. W. L. Leonard to Vesto Melvin Slipher, April 3, 1911; LOA.

41. P. Lowell to C. O. Lampland, April 27, 1911; LOA.

42. The blink comparator was a refinement of the earlier "stereo comparator." Both were invented by Carl Pulfrich (1858–1927), an engineer with the Carl Zeiss firm in Germany. According to William Sheehan and Thomas A. Dobbins, *Epic Moon* (Richmond, VA: Willmann-Bell, 2001), the motive for these inventions

was originally to conveniently compare images of the Moon in the search for lunar changes. Thus they write,

> The search for lunar changes . . . led to the development of two novel instruments for detecting small differences between successive photographs of the same scene. Invented by an engineer with the Carl Zeiss firm named Pulfrich, the "stereo comparator" simultaneously views two exposures with binocular vision. Any feature which differs on the photographs will appear to stand out of the plane of the resulting composite image and is thus immediately brought to the attention of the observer. In the case of the more refined "blink comparator," introduced in 1904, the two exposures are observed alternately in very rapid succession rather than simultaneously. Objects with identical positions in both exposures remain stationary, but those that have altered their position appear to jump back and forth. Objects that have changed in brightness appear to pulsate. Pulfrich's inventions made it possible to quickly recognize variable stars, novae, comets, and asteroids on photographs of star fields. Indeed, Clyde Tombaugh's discovery of Pluto in 1930 would have been all but impossible without the aid of a Zeiss blink comparator. But when Pulfrich exhibited his stereo comparator at an international astronomical convention at Göttingen in 1902, he touted it first and foremost as the perfect means by which to critically compare lunar photographs in order that "the often-debated question whether changes are still occurring on the Moon might be answered relatively easily and definitely." (278)

Pulfrich followed up with the blink comparator, which uses an electromagnet to flip a small mirror back and forth to redirect the light path from one plate to another, in 1904. He died in 1927, drowning when his canoe capsized in the Baltic Sea. He has now been honored with a crater on Pluto.

43. P. Lowell to V. M. Slipher, July 8, 1911; LOA.

44. P. Lowell to C. O. Lampland, August 24, 1912; LOA.

45. W. L. Leonard to C. O. Lampland, September 4, 1912; LOA.

46. P. Lowell to C. O. Lampland, September 12, 1912; LOA.

47. W. L. Leonard to Earl Slipher, October 30, 1912; LOA.

48. W. L. Leonard to C. O. Lampland, December 11, 1912; Leonard to V. M. Slipher, December 12, 1912; LOA.

49. P. Lowell to C. O. Lampland, February 21, 1913; LOA.

50. Hoyt, *Planets X and Pluto*, 122.

51. P. Lowell to C. O. Lampland, telegram, July 10, 1913; P. Lowell to C. O. Lampland, telegram, August 21, 1913; LOA.

52. P. Lowell to C. O. Lampland, telegram, May 5, 1914; LOA.

53. P. Lowell to V. M. Slipher, August 11, 1914; LOA.

54. P. Lowell to C. O. Lampland, December 21, 1914; LOA.

55. A. L. Lowell, *Biography of Percival Lowell*, 188.

56. P. Lowell, *Memoir on a Trans-Neptunian Planet*, 105.

57. Ibid., 101.

58. Ibid., 103.

59. Ibid., 104.

60. C. O. Lampland to P. Lowell, August 15, 1915; LOA.

61. C. O. Lampland to P. Lowell, September 15, 1915; LOA.

62. Hoyt, *Planets X and Pluto*, 132.

63. A. L. Lowell, *Biography of Percival Lowell*, 192.

64. Ibid., 193. His brother adds, "Fortunately he did not know how near it was."

65. Ibid.

66. William Lowell Putnam, *Explorers of Mars Hill* (Kennebunkport, ME: Phoenix, 1994), 97.

Chapter 7

1. Putnam, *Explorers of Mars Hill*, 84.

2. Wrexie, dismissed by Constance immediately after Percival's death, went back East, and ended sadly. Her financial condition seemed to be comfortable until the market crash in 1929, when all this changed; she then went to New York to live with a niece, and resided for many years in Roxbury's Trinity Church Home for the Aged. In 1937, at age seventy, she moved to the Medfield State Hospital and died soon afterward.

3. Putnam, *Explorers of Mars Hill*, 101. In all, Constance and her lawyers managed to bleed the value of the Lowell estate to about a third of what it was when Percival died.

4. V. M. Slipher to Robert S. Richardson, July 24, 1924; LOA. The young astronomer was Robert S. Richardson. The Lowell archives are silent as to whether he was ever considered for the Lowell job, though as a trained astronomer, he would likely have been regarded as overqualified for the drudge-like position Slipher had in mind. Richardson went on to become a solar astronomer at Mount Wilson Observatory and a writer of popular books on astronomy, especially Mars. Astronomy, of course, has never been lucrative. Hence the large contributions of "Grand Amateurs," individuals who acquired their means—by other means—and pursued their astronomical interests avocationally. However, because of the terms under which the observatory finally settled Constance's draining litigation against Percival Lowell's trust, the Lowell Observatory was famously broke at the time. By 1925, a dollar was not worth quite what it had been in 1917; roughly, it

was worth about thirteen times a dollar in 2016. To put all this in perspective, on January 1, 1926, when Constance was receiving an annuity from Percy's estate of $60,000 a year, the right to occupy his house at 11 West Cedar Street (and around the corner to include her place at 102–104 Mount Vernon Street) in Boston, and spending $40,000 on Percy's granite mausoleum, V. M. Slipher's salary was raised by just 5 percent to $4,365.90. As for the other staff astronomers, Lampland's was $3,960 and Earl Slipher's was $3,120. In terms of rough equivalencⅽes, Constance's annuity would have been worth almost $800,000 while the astronomers would have been making in the range of $40,000 to $50,000 per year in today's dollars. Of course, these were "boom" times, the roaring twenties, with the Great Depression lurking just around the corner. The financial straits through which V. M. Slipher as director had to navigate the observatory during the difficult years after Lowell's death may account in part for Slipher's increasing involvement in business and real-estate transactions and relative neglect of astronomy in the last decades of his career. (Among other things, he was an investor in the Monte Vista Hotel in Flagstaff.) It may be gratifying to note that by the end of his life he had become a very wealthy man.

5. V. M. Slipher to Roger Lowell Putnam, December 31, 1928; LOA

6. Clyde Tombaugh, personal interview with William Sheehan, April 27, 1990.

7. Robert S. Richardson, *Exploring Mars* (New York: McGraw-Hill, 1954), 227–28.

8. Ibid., 228–29.

9. Clyde Tombaugh and Patrick Moore, *Out of the Darkness: The Planet Pluto* (Harrisburg, PA: Stackpole, 1980), 117.

10. Ibid., 119.

11. Ibid., 122.

12. Tombaugh, "The Discovery of Pluto," 4.

13. Abbott Lawrence Lowell to R. L. Putnam, March 1, 1930; LOA.

14. V. M. Slipher to Harvard College Observatory, telegram, March 12, 1930; LOA.

15. Frank E. Seagrave to V. M. Slipher, March 14, 1930; LOA. Dave Huestis, of the Skyscrapers Astronomy Club in Providence, Rhode Island, first brought this interesting correspondence to co-author Sheehan's attention.

16. F. E. Seagrave to V. M. Slipher, April 15, 1930; LOA.

17. V. M. Slipher to F. E. Seagrave, April 17, 1930; LOA. Seagrave wrote back to Slipher on April 30, "I thank you for the positions of planet 'X' that you sent to me. I did not receive them until yesterday as . . . people neglected to forward your letter to me here. Too bad that we have not a larger arc to base orbit elements on. I will see what I can do." Seagrave was seventy at the time and, probably discouraged by the small arc of the observations, apparently never did work out an orbit.

18. *Harvard College Observatory Announcement Card* 112 (April 7, 1930); and Ernest C. Bower and Fred L. Whipple, "Preliminary Elements and Ephemeris of the Lowell Observatory Object," *Lick Observatory Bulletin* 421 (1930).

19. Hoyt, *Planets X and Pluto*, 207.

20. V. M. Slipher to John A. Miller, April 19, 1930; LOA.

21. V. M. Slipher to R. L. Putnam, April 19, 1930; LOA.

22. Ibid.

23. Constance Lowell to V. M. Slipher, March 9, 1930; LOA.

24. During Lowell's absence from the observatory in the late 1890s, several members of the staff at the time were asked to provide their impressions of See's character to the Lowell trustee, William Lowell Putnam II. They included W. A. Cogshall (who thought him "half sane"), Daniel A. Drew (who called him "a mental and moral degenerate"), and Andrew E. Douglass, who wrote, "Personally I have never had such an aversion to a man or beast or reptile or anything disgusting as I have had to him. The moment he leaves town will be one of vast and intense relief and I never want to see him again under any circumstances. If he comes back, I will have him kicked out of town." A. E. Douglass to W. L. Putnam, June 28, 1898; LOA.

25. The asteroid 93 Minerva. It was discovered (and named) by James Craig Watson of the University of Michigan on August 24, 1867.

26. C. Lowell to V. M. Slipher, March 14, 1930; LOA.

27. H. P. Lovecraft to Elizabeth Toldridge, April 1, 1930. At the time, Lovecraft was working on his short story, "The Whisperer in Darkness" (*Weird Tales*, August 1931), which depicts the planet Yuggoth and the creatures known as Mi-go—large, pink, crustacean-like fungi that transport themselves between worlds. Clearly, Yuggoth was inspired by Pluto; he says, "Yuggoth . . . is a strange dark orb at the very rim of our solar system."

28. C. Lowell to V. M. Slipher, January 1, 1931; LOA.

29. Disney's affable dog first appeared in the 1930 cartoon features "The Chain Gang" and "The Picnic," and the 1931 feature "The Moose Hunt," albeit with different names. According to Disney-lore sources on the internet, Walt settled on Pluto the Pup in honor of the new planet, and most probably with an eye to alliteration afforded by the name.

30. R. L. Putnam to V. M. Slipher, March 28, 1930; LOA.

31. William H. Pickering, "Planet P, Its Orbit, Position, and Magnitude. Planets S and T," *Popular Astronomy* 39 (1931): 385.

32. Ernest Brown, "On the Prediction of Trans-Neptunian Planets from the Perturbations of Uranus," *Proceedings of the National Academy of Sciences* 16 (1930): 364–71; and Ernest Brown, "On a Criterion for the Prediction of an Unknown Planet," *Monthly Notices of the Royal Astronomical Society* 92 (1931): 80–101.

33. William H. Pickering, "The Discovery of Pluto," *Monthly Notices of the Royal Astronomical Society* 91 (1931): 812–17.

34. Putnam, *The Explorers of Mars Hill*, 188. Pluto was within 0.314° of Jupiter on November 9, 1930, and within only 0.076° on May 26, 1931. Had Nicholson

made such a serendipitous discovery before Tombaugh found Planet X, astronomy would have followed a dramatically different course. The same would have been the case had Galileo, continuing in late December 1612 and January 1613 to track the four satellites of Jupiter he had discovered with his small telescopes in January 1610, discovered Neptune, which was in the same field as Jupiter. Galileo marked it down as a star, and even suspected its movement—but did not follow up at the time.

In 1938, while pursuing a systematic search, Nicholson discovered three more satellites of Jupiter on plates taken with the 100-inch telescope.

35. Ernest C. Bower, "On the Orbit and Mass of Pluto with an Ephemeris for 1931–32," *Lick Observatory Bulletin* 437 (1931).

36. See, for instance, Andrew Claude de la Cherois Crommelin, "Pluto, the Lowell Planet," *Journal of the British Astronomical Association* 41 (1930): 265.

37. Dinsmore Alter was an astronomer at Kansas University from 1917 until 1935, when he received a Guggenheim Fellowship that allowed him to study in Britain for two years. He had just left before Clyde Tombaugh, who had only a high school education when he was hired as an assistant at Lowell Observatory, came to Kansas University to complete his bachelor's and master's degrees. For Alter's work on Pluto, see Dinsmore Alter, George W. Bunton, and Paul E. Roques, "The Diameter of Pluto," Publications of the Astronomical Society of the Pacific 63 (1951): 174–76 and Dinsmore Alter, "The Story of Pluto," *Journal of the Royal Astronomical Society of Canada* 46 (1952): 1–10.

38. Walter Baade, "The Photographic Magnitude and Color Index of Pluto," *Publications of the Astronomical Society of the Pacific* 46 (1934): 218–21.

39. Lowell's need for sharp definition and intolerance of ambiguity is notable, of course, in his passion for mathematics; but it is apparent even in his drawings of planets, where everything, even the notoriously diffuse and nebulous Venus markings, is depicted as sharp—clearly bounded and well defined. This was no doubt an aspect of his obsessive personality type; but perhaps it owed something to the legacy of the family puritanism that—as he himself said in the case of his father—was "double-distilled." Hattie Bundy, sister of Bill and McGeorge Bundy, once said of her mother, Katharine Lawrence Putnam Bundy, Percival's niece, "Mother's sense of righteousness was very deep. . . . How well I remember our fights over the dining room table. . . . For her things were black and white. It's an outlook that descends directly from the Puritans and we all have it." Quoted in Kai Bird, *The Color of Truth: McGeorge Bundy and William Bundy: Brothers in Arms* (New York: Simon and Schuster, 1998), 36.

40. William James, "Is Life Worth Living?" in *Pragmatism and Other Writings*, edited and with an introduction by Giles Gunn (New York: Penguin Books, 2000), 238.

41. Quoted in Hoyt, *Planets X and Pluto*, 207.

42. The discussion here follows Clyde Tombaugh, "The Trans-Neptunian Planet Search," in *Planets and Satellites*, vol. 3, edited by Gerard P. Kuiper and Barbara M. Middlehurst, 12–30 (Chicago: University Chicago Press, 1961).

43. Ibid., 20.

44. Ibid.

45. Ibid., 26.

Chapter 8

1. Joseph N. Tatarewicz, *Space Technology and Planetary Astronomy* (Bloomington and Indianapolis: Indiana University Press, 1990).

2. Sheehan, *The Immortal Fire Within*, 247.

3. Norriss S. Hetherington, *Science and Objectivity: Episodes in the History of Astronomy* (Ames: Iowa State University Press, 1988), 62.

4. Henry Norris Russell, "Percival Lowell and his Work," *The Outlook* 114 (1916): 781–82.

5. Gerard P. Kuiper and Barbara M. Middlehurst, eds., *Planets and Satellites* (Chicago; London: University Chicago Press, 1961), vi.

6. Charles Percy Snow, *Variety of Men* (New York: Scribner's, 1967), 3.

7. Charles Percy Snow, *The Physicists: A Generation that Changed the World* (London: Macmillan, 1981), 79

8. Bart J. Bok, text of remarks delivered at the memorial service for G. P. Kuiper, December 31, 1973.

9. Sheehan, *The Immortal Fire Within*.

10. As noted in the previous chapter, for a brief while in the 1930s Pluto was thought by some to possibly be a white dwarf, which was a way of reconciling its apparently small size with a mass large enough to perturb Uranus.

11. "Near Death" is the title of chapter 3 of Osterbrock's *Yerkes Observatory, 1892–1950* (Chicago: University of Chicago Press, 1997).

12. Dale P. Cruikshank, "Gerard Peter Kuiper, 1905–1973," *Biographical Memoirs*, vol. 62 (Washington, DC: National Academy of Sciences, 1993), 259–95; includes an extensive bibliography of Kuiper's publications.

13. Coauthor Cruikshank, who accompanied Kuiper on many observing runs on telescopes at various observatories, including McDonald Observatory, was present when the infrared spectroscopic observations of Betelgeuse (Alpha Orionis) were made (probably in 1962 or 1963), and jotted down the following first-person recollections more than half a century later:

    On one of the cold McDonald winter nights, I was assisting Kuiper, and we were using the PbS spectrometer to record the spectrum of Betelgeuse at higher resolution than had ever been done (by Kuiper or anyone else). As we stood for hours on the platform taking turns guiding the telescope the spectrometer was scanning the region around 2.3 micrometers, and as it slowly scanned through the wavelengths the spectrum was being drawn out on the strip-chart recorder—no digital data in those days. The pen slowly traced out a series of absorption bands, strongly and clearly enough to ensure that they were real, and not just noise. Although we didn't know what the cause of the bands was at the time, Kuiper and I were both transfixed at the sight of the beautiful regular pattern emerging. Kuiper quietly said at one point, "This has never been seen before." I can hear him saying it as if it were yesterday. Recalling this incident nearly brings a tear to my eye because it was profound and reflects no doubt dozens of previous first-times that Kuiper had experienced in his researches. And it was a forerunner of experiences that I have been privileged to have as an astronomer in the years since then. [The spectrum showed for the first time the regular pattern of rotational lines from carbon monoxide molecules in the cool atmosphere of the star.]

14. Ronald E. Doel, *Solar System Astronomy in America* (Cambridge: Cambridge University Press, 1996), 56. Herzberg went on to win the Nobel Prize in Chemistry in 1971.

15. Ibid., 55.

16. Gerard P. Kuiper, *Atmospheres of the Earth and Planets* (Chicago: University Chicago Press, 1949), plate 13. The "better spectra" to which Kuiper refers remain unpublished to this day. These plates are very tiny and difficult to handle when cutting, hypersensitizing, exposing, and process in darkroom chemicals. Coauthor Cruikshank always wondered whether the best Triton plate might have shown the 8900-Å methane band, despite its being so much fainter than the same band on Titan. Having the original plates on loan from the Lunar and Planetary Laboratory in Tucson, he examined the Triton plate. The results were inconclusive because, unfortunately, someone had laid a fingerprint right at the wavelength of the methane band!

17. Gerard P. Kuiper, "On the Origin of the Solar System," *Proceedings of the National Academy of Sciences* 37, no. 1 (1951): 1–14; "On the Origin of the Solar System," in *Astrophysics: A Topical Symposium*, edited by J. Allen Hynek, 357–424 (New York: McGraw-Hill, 1951); "On the Evolution of the Protoplanets," *Proceedings of the National Academy of Sciences* 37, no. 7 (1951): 383–93; "On the Origin of the Irregular Satellites," *Proceedings of the National Academy of Sciences* 37, no. 11 (1951): 717–21; "Note on the Origin of the Asteroids," *Proceedings of the National Academy of Sciences* 39, no. 12 (1953): 1159–61; "Satellites,

Comets, and Interplanetary Material," *Proceedings of the National Academy of Sciences* 39, no. 12 (1953): 1153–58; "On the Formation of the Planets," *Journal of the Royal Astronomical Society of Canada* 50 (1956): 57–68, 105–21, 158–76.

18. Gerard P. Kuiper, "Pluto's Diameter," *Sky & Telescope* 10 (1950): 50.

19. Baade started his career at the Hamburg Observatory, and made an early set of positional measures of Pluto. See Walter Baade, *Beobachtungen des Pluto an Spiegelteleskop der Hamburger Sternwarte* (Mitteilungen Hamburger Sternwarte Bergedof, 1931), vii, 44. After emigrating from Germany to the United States, he made an early attempt at photographic photometry (and colorimetry) of Pluto and Triton using the Mount Wilson 60-inch telescope. See Walter Baade, "The Photographic Magnitude and Color Index of Pluto," *Publications of the Astronomical Society of the Pacific* 46 (1934): 218–21.

20. Robert L. Marcialis, "The First 50 Years of Pluto-Charon Research," in *Pluto and Charon*, edited by in S. Alan Stern, and David J. Tholen, 27–84 (Tucson: University of Arizona Press, 1997), 43.

21. Seth B. Nicholson and Nicholas U. Mayall, "Positions, Orbit and Mass of Pluto," *Astrophysical Journal* 73 (1931): 1–12.

22. Robert L. Marcialis and W. J. Merline (unpublished; cited in Marcialis, "The First 50 Years of Pluto-Charon Research").

23. Gerard P. Kuiper, "The Diameter of Pluto," *Publications of the Astronomical Society of the Pacific* 62 (1950): 136.

24. Marcialis, "The First 50 Years of Pluto-Charon Research," 46.

25. Kuiper, "On the Origin of the Solar System," 13.

Chapter 9

1. Harold Clayton Urey, *The Planets* (Yale University Press, New Haven, 1952).

2. Charles A. Wood, *The Modern Moon—A Personal View* (Cambridge: Sky Publishing, 2003).

3. "Harold C. Urey—Biographical," Nobel Media, 2014, https://www.nobelprize .org/nobel_prizes/chemistry/laureates/1934/urey-bio.html. Also published in *Nobel Lectures, Chemistry 1922–1941* (Amsterdam: Elsevier, 1966).

4. Richard Rhodes, *The Making of the Atomic Bomb* (New York: Simon & Schuster, 1986).

5. Carl Sagan to Dale P. Cruikshank, December 1, 1982.

6. John F. Kennedy, speech, Rice University, September 12, 1962. Kennedy spoke to a crowd of 35,000 people. Video and transcript of this historic speech can be accessed at "1962-09-12 Rice University," John F. Kennedy Presidential Library and Museum, https://www.jfklibrary.org/Asset-Viewer/MkATdOcdU06X5u NHbmqm1Q.aspx.

7. Richard S. Lewis, "At Sea on the Moon," *Bulletin of the Atomic Scientists* 20 (November 1964): 35–37.

8. Fred L. Whipple, "A Comet Model 1. The Acceleration of Comet Encke," *Astrophysical Journal* 111 (1950): 375–94.

9. Gerard P. Kuiper, "Infrared Observations of Planets and Satellites," *Astrophysical Journal* 62 (1957): 245–46.

10. Vasili Ivanovich Moroz, "Infrared Spectrophotometry of the Moon and the Galilean Satellites of Jupiter," *Astronomicheskii Zhurnal* 42 (1965): 1287 (English translation of 1964 paper).

11. Carl B. Pilcher, Stephen T. Ridgway, and Thomas B. McCord, "Galilean Satellites: Identification of Water Frost," *Science* 178 (1972): 1087–89.

12. Gerard P. Kuiper, "Comments on the Galilean Satellites," *Communications of the Lunar and Planetary Laboratory* 10 (1973): 28–39.

13. The *Communications* ceased publication shortly after Kuiper's death in December 1973.

14. Moroz's original paper was published in 1964 in Russian, and he reviewed the subject in his book, *Physics of the Planets* (also in Russian) in 1967. The book was translated into English in 1968, but not widely distributed. We add parenthetically that in Russian, the word "moroz" means a chill or a frost in meteorological terms. The Russian equivalent of Santa Claus is Dyed Moroz, or Grandfather Frost. How fitting!

15. John S. Lewis, "Low Temperature Condensation in the Solar Nebula," *Icarus* 16 (1972): 241–52.

16. Sarah E. Dodson-Robinson et al., "Ice Lines, Planetesimal Composition and Solid Surface Density in the Solar Nebula," *Icarus* 200 (2009): 672–93.

17. See, for example, Daniel Clery, "Forbidden Planets," *Science* 353 (2006): 438–41.

18. Kuiper, "On the Origin of the Solar System," in *Astrophysics*.

19. Gerard P. Kuiper, "The Diameter of Pluto," 133–37.

Chapter 10

1. Harry Y. McSween Jr. and Gary R. Huss, *Cosmochemistry* (Cambridge: Cambridge University Press, 2010). The age of calcium-aluminum inclusions (CAIs) in meteorites determined from the decay of radioactive elements is set at 4568.2 +/– 0.5 million years. There are some uncertainties in this absolute age, but all other Solar System materials dated with the abundances of radioisotopes are reckoned relative to the adopted CAI age.

2. John W. Valley et al., "Hadean Age for a Post-Magma-Ocean Zircon Confirmed by Atom-Probe Tomography," *Nature Geoscience* 7 (2014): 219–23.

3. John S. Lewis, "Low-Temperature Condensation from the Solar Nebula," *Icarus* 16 (1972): 241–52; "Chemistry of the Outer Solar System," *Space Science Reviews* 14 (1973): 401–11.

4. Jeffrey L. Coughlin et al., "Planetary Candidates Observed by Kepler. VII. The First Fully Uniform Catalog Based on The Entire 48-Month Dataset (Q1-Q17 DR24)," *Astrophysical Journal Supplemental Series* 224, no. 1 (2016): 12. As of July 2016, about 20 percent of the stars with planets are sunlike.

5. Kuiper, "The Diameter of Pluto."

6. U. Fink et al., "Infrared spectra of the satellites of Saturn: Identification of water ice on Iapetus, Rhea, Dione, and Tethys," *Astrophysical Journal Letters* 207 (1976): L63–L67.

7. A telescope on Mauna Kea was yet another of Kuiper's legacies. He had been the first to recognize Mauna Kea's promise as a site for infrared observations, and by 1967 had set up a small telescope on Pu'u Poli'ahu, a cinder cone peak that was the second-highest mountain on the island; he avoided the highest mountain because it was regarded by the native Hawaiians as holy ground and thus he did not think it likely to be available as site for a telescope. Kuiper was satisfied with the results of the telescope on Pu'u Poli'ahu, and tried to interest NASA in funding a larger, permanent facility with a large telescope, housing, and other needed structures. NASA was supportive, but decided to open the project to competitive bidding. Instead of awarding it to Kuiper, it awarded it to John Jefferies, a solar physicist researcher of the University of Hawaii and director of its Institute for Astronomy. Deeply disappointed, Kuiper abandoned the site-testing program he had conducted on Mauna Kea and moved on to other projects, eventually building, with NASA support, a 61-inch telescope for planetary studies on Mount Lemmon, just above Tucson. Nevertheless, despite Kuiper's personal disappointment, he was certainly far-sighted about the location, and Mauna Kea Observatory quickly proved its preeminence in atmospheric conditions for infrared astronomy in the northern hemisphere.

Chapter 11

1. The Quaternary glaciation is defined as having begun 2.588 million years ago, and it extends to the present time. It marks the onset of a cyclic pattern of alternating widespread continental ice sheets and subsequent melting because of global climate changes caused by long-term changes in Earth's orbit and its tilt on its rotational axis.

2. EPICA = European Project for Ice Coring in Antarctica. The core that records nearly 800,000 years of ice deposition is from a drilling site that "was chosen to

obtain the longest undisturbed chronicle of environmental change, in order to characterize climate variability over several glacial cycles, and to study potential climate forcings and their relationship to events in other regions. The core goes back 740,000 years and reveals eight previous glacial cycles." Laurent Augustin et al., "Eight Glacial Cycles from an Antarctic Ice Core," *Nature* 429 (June 10, 2004): 623–28, https://doi.org/10.1038/nature02599.

3. Paul Voosen, "2.7-Million-Year-Old Ice Opens Window on Past," *Science* 357, no. issue 6352 (2017): 630–31.

4. Jane Qui "Tibet's Primeval Ice," *Science* 351 (2016): 436–39.

5. The only other edible rock is salt, which occurs naturally as the mineral halite. For humans, that is. Many microorganisms happily make their livings by munching on minerals in the rocks in caves, mines, and other harsh environments. These are the lithotrophs, or "rock eaters."

6. The true distances between the planets were not known until a reasonably accurate absolute distance of Earth from the Sun was determined through observations of the transits of Venus across the Sun's disk in 1761 and 1769. See William Sheehan and John Westfall, *The Transits of Venus* (Amherst, New York: Prometheus, 2004), chapters 7–9.

7. Isaac Newton, *Opticks, or a Treatise of the Reflexions, Refractions, Inflexions and Colours of Light* (London, printed for Sam. Smith and Benj. Walford, printers to the Royal Society, 1704). This was one of the first major works in science published first in English and later translated into Latin.

8. We use the terms "evaporate" and "sublimate" interchangeably. They both refer to the change from the solid to the gaseous state without an intermediate liquid phase.

9. Quoted in Harlow Shapley and Helen E. Howarth, *A Source Book in Astronomy* (New York: McGraw-Hill, 1929).

10. John Heilbron, *Galileo* (Oxford: Oxford University Press, 2010), 235. The utility of the Galilean satellites was not limited to the practical geographical and nautical issue of longitude. To clever minds, they provided the means to measure perhaps the most important physical quantity of all—the speed of light. In 1676, Ole Romer used timing of the entries and exits from Jupiter's shadow to demonstrate that the speed of light is finite and to measure that speed to within about 26 percent of its true value.

11. James B. Pollack and Ray T. Reynolds, "Implications of Jupiter's Early Contraction History for the Composition of the Galilean Satellites," *Icarus* 21 (1974): 248–53.

12. The two Pioneer spacecraft (10 and 11) preceded Voyager, passing by Jupiter in 1973 and 1974, respectively. Pioneer 11 went on to Saturn, making its closest approach in 1979.

13. Scott Sheppard's website (https://home.dtm.ciw.edu/users/sheppard/), which keeps track of all the discoveries.

14. J. Hunter Waite et al., "Cassini Finds Molecular Hydrogen in the Enceladus Plume: Evidence for Hydrothermal Processes," *Science* 356, no. 6334 (2017): 155–59.

15. C. R. Glein, J. A. Baross, J. H. Waite, "The pH of Enceladus' Ocean," *Geochimica et Cosmochimica Acta* 162 (2015): 202–19.

16. Anne J. Verbiscer, Michael F. Skrutskie, and Douglas P. Hamilton, "Saturn's Largest Ring," *Nature* 461 (October 22, 2009): 1098–100.

17. Gianrico Filacchione et al., "Saturn's Icy Satellites Investigated by Cassini-VIMS. IV. Daytime Temperature Maps," *Icarus* 271 (2016): 292–313.

18. Dale P. Cruikshank et al., "Carbon Dioxide on the Satellites of Saturn: Results from the Cassini VIMS Investigation and Revisions to the VIMS Wavelength Scale," *Icarus* 206 (2010): 561–72.

19. Gerard P. Kuiper, "Infrared Observations of Planets and Satellites," *Astrophysical Journal* 62 (1957): 245.

20. Carl B. Pilcher et al., "Saturn's Rings: Identification of Water Frost," *Science* 167 (1970): 1372–73.

21. The Cassini mission to Saturn was launched in 1997, and went into orbit around Saturn in 2004. On September 15, 2017, with the mission successfully completed, the spacecraft was sent plunging into the planet.

22. Named by John Herschel for characters from the works of Shakespeare and Pope.

23. Whipple, "A Comet Model 1."

24. The naming convention for comets is complex. Assigning names or other designations is the purview of the Central Bureau for Astronomical Telegrams (CBAT), which, for naming astronomical bodies, functions under the guidance of the IAU. The comets mentioned here are periodic comets, hence the notation 67P/Churyumov-Gerasimenko indicates that is periodic comet number 67 in the CBAT catalog and it bears the names of the two astronomers (Klim I. Churumov and Svetlana I. Gerasimenko) who discovered it on September 20, 1969.

25. Dale P. Cruikshank, "Stardust Memories—A Perspective on Comet Hale-Bopp," *Science* 275, no. 5308 (1997): 1895–96.

26. Astronomers yield to an occasional streak of vanity. It may amuse readers to know that the authors of this book have asteroids named after them: 3531 Cruikshank and 16037 Sheehan, designations conferred and sanctioned by the IAU. Both of these asteroids are comfortably and unthreateningly orbiting between Mars and Jupiter.

27. Martin A. Slade, Bryan J. Butler, and Duane O. Muhleman, "Mercury Radar Imaging: Evidence for Polar Ice," *Science* 258 (1992): 635–40.

28. We retain those ancient names today: Mare Tranquillitatus, Mare Serenitatis, Mare Imbrium, and so on. There is something vaguely comforting about these old Latin names, especially perhaps to scientists who find them to be a link to the past, when scientists were known as "natural philosophers" and communicated entirely in Latin. And we will never forget Neil Armstrong reporting directly from the Apollo 11 spacecraft after landing in Mare Tranquillitatus on the Moon, "Tranquility base here. The Eagle has landed."

Chapter 12

1. R. L. Duncombe, P. K. Seidelmann, and W. J. Klepczynski, "Dynamical Astronomy of the Solar System," *Annual Review of Astronomy and Astrophysics* 11 (1973): 135–54; "The Masses of the Planets, Satellites, and Asteroids," *Fundamentals of Cosmic Physics* 1 (1973): 119–65; Michael E. Ash, Irwin I. Shapirio, and William B. Smith, "The System of Planetary Masses," *Science* 174 (1971): 551–56.

2. Dale P. Cruikshank, Carl B. Pilcher, and David Morrison, "Pluto: Evidence for Methane Frost," *Science* 194 (1976): 835–37.

3. Ibid., 836.

4. Larry A. Lebofsky, George H. Rieke, and M. J. Lebofsky, "Surface Composition of Pluto," *Icarus* 37 (1979): 554–58.

5. B. T. Soifer, G. Neugebauer, and K. Matthews, "The 1.5–2.5 Microns Spectrum of Pluto," *Astrophysical Journal* 85 (1980): 166–67.

6. Dale P. Cruikshank and Peter M. Silvaggio, "The Surface and Atmosphere of Pluto," *Icarus* 41 (1980): 96–102.

7. In 1978, Cruikshank, Terry Jones, and Carl Pilcher published new spectra of Io extending farther into the infrared than had been achieved before ("Absorption Bands in the Spectrum of Io," *Astrophysical Journal Letters* 225: L89–L92). The new spectra showed previously unseen and unidentified absorption bands around 4 μm wavelength. Almost simultaneously, James B. Pollack and colleagues discovered the same bands from the KAO. The following year, Voyager 1 engineers and scientists reported their discovery of active volcanoes on Io from images taken in March 1979, as the spacecraft flew by. This was followed by a few additional papers, also in 1979, in which the connection between the spectral bands and the volcanoes was drawn, leading to the identification of sulfur dioxide frost or ice on Io's surface.

8. Craig B. Agnor and Douglas P. Hamilton, "Neptune's Capture of Its Moon Triton in a Binary-Planet Gravitational Encounter," *Nature* 441 (May 11, 2006): 192–94.

9. This brightness difference changes over Pluto's year, and by the end of 2015, as seen from Earth, Pluto had dimmed by moving farther from the Sun than in the 1970s, while Triton stayed constant. Thus in mid-2015 Triton was nearly 55 percent brighter than Pluto, rather than only 40 percent as in the mid-1970s.

10. A relatively high surface temperature was calculated on the assumption that the albedo of the surface was in the range 0.2–0.4 on the basis of a separate study by Dale P. Cruikshank et al., "The Diameter and Reflectance of Triton," *Icarus* 40 (1979): 104–14). The higher albedo, later found to be the case, gives a lower temperature and therefore implies much less gas in the atmosphere.

11. Dale P. Cruikshank and Peter M. Silvaggio, "Triton: A Satellite with an Atmosphere," *Astrophysical Journal* 233 (1979): 1016–20.

12. Dale P. Cruikshank and Jerome Apt, "Methane on Triton: Physical State and Distribution," *Icarus* 58 (1984): 306–11.

13. The north–south convention in Cruikshank and Apt's 1984 paper is technically correct, but a different convention was adopted later—when Voyager 2 viewed the south polar region when it flew by Triton in 1989.

14. Cruikshank and Apt, "Methane on Triton," 309.

15. Dale P. Cruikshank, Robert Hamilton Brown, and Roger N. Clark, "Nitrogen on Triton," *Icarus* 58 (1984): 293–305.

16. Jonathan I. Lunine and David J. Stevenson, "Physical State of Volatiles on the Surface of Triton," *Nature* 317 (September 19, 1985): 238–40.

17. Janusz Eluszkiewicz, "On the Microphysical State of the Surface of Triton," *Journal of Geophysical Research* 96 (1991): 19217–29.

18. While Triton's surface is not now covered by a sea of liquid nitrogen, many regions of the surface are smooth and reminiscent of the flow of a liquid now vanished. Triton's history may have had episodes during which liquid nitrogen appeared on some parts of the surface, producing the unique landscape that is now frozen. Planetary scientists have become aware of another interesting property of solid nitrogen—it isn't very rigid, and on timescales of thousands of years can flow, as in glaciers, or fill in topographic lows across a planetary surface. This realization has come from the study of frozen nitrogen on Pluto, as we shall see in chapter 19.

Chapter 13

1. Dale P. Cruikshank, "Mauna Kea: A Guide to the Upper Slopes and the Observatories" (Honolulu: University Hawaii Institute for Astronomy, 1986).

2. The story of the recognition of Mauna Kea as an astronomical site and subsequent developments that led to the establishment of several observatories

there is told in M. Michael Waldrop, "Mauna Kea: Halfway to Space," *Science* 214 (1981): 1010–13, 1110–14.

3. Tobias C. Owen et al., "The Surface Ices and the Atmospheric Composition of Pluto," *Science* 261 (1993): 745–48. Detailed modeling of Pluto's spectrum with laboratory data is given in S. Douté et al., "Evidence for Methane Segregation at the Surface of Pluto," *Icarus* 142 (1999): 421–44.

4. K. A. Tryka et al., "Spectroscopic Determination of the Phase Composition and Temperature of Nitrogen on Triton," *Science* 261 (1993): 751–54.

5. E. Quirico, et al., "Composition, Physical State and Distribution of Ices at the Surface of Triton," *Icarus* 139 (1999): 159–78. See also Dale P. Cruikshank et al., "Water Ice on Triton," *Icarus* 147 (2000): 309–16.

6. KBO (50000) Quaoar, discovered in 2002, is roughly half the size of Pluto, and it has a moon named Weywot.

7. KBOs on which methane bands are shifted, implying the presence of nitrogen. M. A. Barucci et al., "(50000) Quaoar: Surface Composition Variability," *Astronomy & Astrophysics* 584 (2015): 107–13.

8. Charles R. Fisher et al., "Methane Ice Worms: *Hesiocaeca methanicola* Colonizing Fossil Fuel Reserves," *Naturwissenschaften* 87 (1997): 184–87.

Chapter 14

1. Kuiper, "The Diameter of Pluto," 133–37. Kuiper notes in this paper that he had searched for a Plutonian atmosphere by ultraviolet spectroscopy, but detected no evidence for an atmosphere with surface pressure greater than about 10 percent that of Earth's atmosphere.

2. The temperature of Pluto is not known from direct measurement, but is dependent on the albedo according to $T_{ss} \approx 394(1 - A)^{0.25}/R^{0.5}$, where R is the distance to the Sun in astronomical units, A is the spherical albedo (which we will assume is the same as the geometric albedo for this calculation), and T_{ss} is at the subsolar point, the place on the surface where the Sun is exactly overhead. In this computation the average surface temperature is $T_{ss}/2^{0.5}$. This calculation (for R = 40 and A = 0.6) gives $T_{ss} = 49.5°$ K, and the average temperature is then 35.0° K.

3. Dale P. Cruikshank and Peter M. Silvaggio, "The Surface and Atmosphere of Pluto," *Icarus* 41 (1980): 96–102.

4. If both dry ice (frozen carbon dioxide) and frozen water are held at room temperature, the water ice will thaw and turn to liquid before beginning to evaporate while the dry ice will go directly from the solid state to a gas. These two materials have different phase behavior in room conditions.

5. Michael H. Hart, "A Possible Atmosphere for Pluto," *Icarus* 21 (1974): 242–47; Georgy S. Golitsyn, "A Possible Atmosphere on Pluto," *Soviet Astronomy Letters* 1 (1975): 19–20.

6. L. Trafton, "Does Pluto have a Substantial Atmosphere?" *Icarus* 44 (1980): 53–61; "Pluto's Atmospheric Bulk near Perihelion," *Advances in Space Research* 1 (1981): 93–97.

7. Planets are normally detected by sunlight reflected from their surfaces and atmospheres, and the brightness of the Sun is maximum at about 560 nm (0.56 μm). Since the 1920s and before, chemists have focused their laboratory spectroscopic studies on a longer-wavelength region of the spectrum (2.5 μm and longer) where most molecules have their simplest (and more easily interpreted) spectral bands. Detecting radiation from planets at the long wavelengths favored by chemists has always been very difficult for a variety of reasons, thus leading to a mismatch between the laboratory data on molecules (such as methane and nitrogen) and the data that an astronomer can obtain at the telescope.

8. Since an airborne observatory usually departs from and returns to the same airport, an all-night observing flight of the KAO and its successor SOFIA (Stratospheric Observatory for Infrared Astronomy) is sometimes referred to as "the red-eye to nowhere".

9. Over its twenty-one-year history of making critical infrared astronomical observations from 12 km (40,000 ft) altitude, above most of the interfering water vapor in the atmosphere, many scientists and their students used the Kuiper Airborne Observatory, and an entire generation of astronomers developed around the unique capabilities afforded by this facility. The KAO was decommissioned in 1995, before its useful lifetime had come to an end, to save operating costs that would be applied to the creation of an even more capable airborne platform for astronomy, the Stratospheric Observatory for Infrared Astronomy (SOFIA), a Boeing 747SP aircraft that would carry a much larger telescope of 2.7 m aperture and be flying by 2001. Funding delays and shortfalls, and unfortunate management decisions, resulted in the delay of the first astronomical observations until December 2010. SOFIA reached full operational capability only in 2014. The nineteen-year hiatus in airborne astronomy between the last flights of the KAO in 1995 and the onset of regular flights of SOFIA in 2014 retarded the progress of infrared astronomy on an international scale, and some unknown number of students who would have used the airborne platform for their research elected to work on other problems. The history of SOFIA and details of the aircraft and telescope are outlined in Edwin F. Erickson and Allan W. Meyer's *NASA's Kuiper Airborne*

Observatory, 1971–1995: An Operations Retrospective with a View to SOFIA, NASA SP-2013-216025, 2013, 53–64.

10. This and the remainder of the poem "KAO" are in NASA SP (Special Publication) 2013–216025.

11. Robert L. Millis, "Occultation Studies of the Solar System," NASA research and technology resume, Grant NGS-7603 (1988). Emphasis in original.

12. Robert L. Millis et al., "Pluto's Radius and Atmosphere—Results from the Entire 9 June 1988 Occultation Data Det," *Icarus* 105 (1993): 282.

13. S. Alan Stern et al., "The Pluto System: Initial Results from Its Exploration by New Horizons," *Science* 350, no. 6258 (2015). Some of this information is taken from New Horizons project internal documents.

14. In Earth's atmosphere methane (natural gas) is a very strong greenhouse gas, more powerful even than carbon monoxide. Release of methane into the atmosphere from commercial gas wells, melting permafrost, cattle feed lots, termites, and other sources is of great interest in the field of climate science because of its contribution to global warming. See Jonathan I. Lunine, *Earth: Evolution of a Habitable World* (Cambridge: Cambridge University Press, 1999).

15. Leslie A. Young et al., "Detection of Gaseous Methane on Pluto," *Icarus* 127 (1997): 258–62.

16. Tobias C. Owen et al., "The Surface Ices and the Atmospheric Composition of Pluto," *Science* 261 (1993): 745–48.

17. For example, MAVEN (Mars Atmosphere and Volatile EvolutioN Mission), launched by NASA on November 18, 2013.

18. NASA's Galileo spacecraft, launched on October 13, 1989, and orbiting Jupiter for eight years, dispatched an atmospheric-entry probe that fell into Jupiter's atmosphere on December 7, 1995, measuring composition, temperature, and pressure for about one hour as it descended to a level where the temperature and pressure reached about 153° C (426° K) and more than 23 bar, respectively. It stopped working at those high temperatures and pressures.

19. The Huygens probe from Cassini.

20. Vladimir A. Krasnopolsky and Dale P. Cruikshank, "Photochemistry of Pluto's Atmosphere and Ionosphere Near Perihelion," *Journal of Geophysical Research* 104, no. 21 (1999): 979–21, 996.

21. Michael E. Summers, Darrell F. Strobel, and G. Randall Gladstone, "Chemical Models of Pluto's Atmosphere," in *Pluto and Charon*, edited by S. Alan Stern and David J. Tholen, 391–434 (Tucson: University of Arizona Press, 1997). Xun Zhu, Darrell F. Strobel, Justin T. Erwin, "The Density and Thermal Structure of Pluto's Atmosphere and Associated Escape Processes and Rates," *Icarus* 228 (2014): 301–14.

22. Some calculations suggested that Pluto is losing 10^{28} molecules per second, evaporating into space. At this rate, a surface layer of nitrogen ice about 80 m thick would be lost every ten years.

23. The hydrogen atom has atomic weight 1 Dalton (Da = "atomic mass unit," equivalent to 1.660×10^{-27} kg, or 1.660×10^{-24} g). Helium has an atomic weight of 4 Da, while nitrogen and methane molecules are 28 and 16 Da, respectively. Hydrogen and helium are readily lost from a planetary atmosphere, while nitrogen and methane, as on Pluto, escape more slowly.

Chapter 15

1. Scott Sheppard webpage (https://home.dtm.ciw.edu/users/sheppard/). The satellite count was current as of May 2016.

2. See Marcialis, "The First 50 Years of Pluto-Charon Research"; Milton L. Humason, "Photographs of Planets with the 200-inch Telescope," in *Planets and Satellites*, edited by Gerard P. Kuiper and Barbara M. Middlehurst, 572 plus 15 plates (Chicago: University Chicago Press, 1961.

3. Samuel P. Langley, "On the Allegheny System of Electric Time Signals," *Journal of the Society of Telegraph Engineers* 1 (1873): 433–41. Langley also wrote, "The advantages of this uniform and wide distribution of exact time in facilitating the transportation of the country, and in enhancing the safety of life and of merchandise in transit between the Western and the Atlantic cities, seem to be sufficiently evident."

4. The Flagstaff Station is located only twelve miles from Lowell Observatory, and the collegial relationship of the two neighboring institutions is attested by the fact that both Hall and Hoag later became Lowell directors. In recent years, as additional instrumentation and capabilities have been added, the Navy Precision Optical Interferometer (NPOI), operated by Lowell under contract with the Flagstaff Station and the Naval Research Laboratory in Washington, has been set up at the Lowell Observatory's dark-sky site on Anderson Mesa, twenty-three miles from Flagstaff. The NPOI uses an array of up to six mirrors spaced tens to hundreds of meters apart to measure relative star positions far more accurately than with conventional telescopes. Positional astronomy certainly has come a long way from the days of Flamsteed and Le Monnier, or even Piazzi and Bessel.

5. The biographical information about James W. Christy is based on an interview of Christy and his wife, Charlene, at their Flagstaff home by William Sheehan and Kevin Schindler on June 21, 2016.

6. James W. Christy, "The Discovery of Pluto's Moon, Charon, in 1978," in *Pluto and Charon*, edited by S. Alan Stern and David J. Tholen, xvii–xxi (Tucson: University of Arizona Press, 1997).

7. James W. Christy and Robert S. Harrington, "The Satellite of Pluto," *Astrophysical Journal* 83 (1978): 1005–8.

8. Cruikshank, Pilcher, and Morrison, "Pluto: Evidence for Methane Frost."

9. According to classical scholar Michael Armstrong of William Smith and Hobart Colleges (personal communication to William Sheehan, June 22, 2016), "In 'Charon,' the initial *ch* is a chi; it would be pronounced (we think) originally like a *k* with an immediate puff of breath behind it. That is what they tell me, but it's not something I can speak or hear or even grasp. More traditionally, the chi would probably sound like breathed *ch* in Loch. It certainly shouldn't sound like an English *sh* (sheep) or *ch* (church). Thus, if one wishes to be strictly observant of the classical source, Shar-on is not acceptable.

   "Also, the alpha is short and the *o* is an omega. So it should be 'KHAH-roan.' But I doubt that anyone in English would bother about the short *a* and the long *o*."

10. Leif E. Andersson, "Eclipse Phenomena of Pluto and Its Satellite," *Bulletin of the American Astronomical Society* 10 (1979): 586.

11. See John Westfall and William Sheehan, *Celestial Shadows: Eclipses, Transits and Occultations* (New York: Springer, 2015), 233.

12. From this, resolving the Pluto-Charon system visually would seem to be an utter impossibility. However, when Pluto is near the perihelion of its orbit and Charon is at its maximum separation from Pluto, and one knows exactly where to look, Charon can just be seen in very large telescopes. The first person to do so was David Tholen, using the 2.2-m (88-inch) telescope at Mauna Kea, and then by Jean Lecacheux, with the 1-m Cassegrain at Pic du Midi. In 1992, coauthor Sheehan and another amateur astronomer attempted to replicate Lecacheux's feat. Pluto was just past perihelion, which it had reached in 1989. Though Pluto and Charon were not resolvable, a distortion was visualized that was consistent with Charon's position as determined afterward from an ephemeris.

13. David J. Tholen, personal communication to Dale P. Cruikshank, n.d.

14. Marc W. Buie, David J. Tholen, and Keith Horne, "Albedo Maps of Pluto and Charon: Initial Mutual Event Results," *Icarus* 97 (1992): 211–27.

15. Robert L. Marcialis, George H. Rieke, and Larry A. Lebofsky, "The Surface Composition of Charon—Tentative Identification of Water Ice," Science 237 (1987): 1349–51.

16. Marc W. Buie et al., "Water Frost on Charon," *Nature* 329 (October 8, 1987): 522–23.

17. Water rapidly frozen at very low temperature assumes the noncrystalline state, but as it is warmed, crystals begin to form. The first crystalline state begins at

T~100° K, and is cubic in form. When the rising temperature reaches T~160° K, the crystalline form changes to hexagonal, which is the phase we are most familiar with; snowflakes have a hexagonal symmetry because of this effect. Once the crystalline form is reached, lowering the temperature does not cause the ice to change back to the amorphous phase.

18. NH_4OH is ammonium hydroxide, while the notation $NH_3 \cdot nH_2O$ means that one ammonia molecule is "complexed" with one or more water molecules in an arrangement with a different kind of chemical bond than NH_4OH.

19. Jason C. Cook, "Near-Infrared Spectroscopy of Charon: Possible Evidence for Cryovolcanism on Kuiper Belt Objects," *Astrophysical Journal* 663 (2007): 1406–19.

Chapter 16

1. Buie et al., "Water Frost on Charon"; Marcialis, Rieke, and Lebofsky, "The Surface Composition of Charon."

2. Charles Darwin to J. D. Hooker, February 1, 1871.

3. Robert Shapiro, *Origins: A Skeptic's Guide to the Creation of Life on Earth* (New York: Bantam Books, 1986), 110.

4. At first the two tried to collaborate, but they soon diverged—sharply—over questions about the Moon, and became irreconcilable foes. We wrote about the conflict between Urey and Kuiper in chapter 9.

5. Stanley L. Miller, "A Production of Amino Acids Under Possible Primitive Earth Conditions," *Science* 117 (1953): 528–29.

6. James D. Watson and Francis H. C. Crick, "Molecular Structure of Nucleic Acids," *Nature* 171 (April 25, 1953): 737–38.

7. Konstantin Batygin, Gregory Laughlin, and Alessandro Morbidelli, "Born of Chaos," *Scientific American* 314 (May 2016): 28–37.

8. William Martin et al., "Hydrothermal Vents and the Origin of Life," *Nature Reviews Microbiology* 6 (2008): 805–14.

9. S. Pizzarello, G. W. Cooper, and G. J. Flynn, "The Nature and Distribution of the Organic Material in Carbonaceous Chondrites and Interplanetary Dust Particles," in *Meteorites and the Early Solar System II*, edited by Dante S. Lauretta and Harry Y. McSween Jr., 625–51 (Tucson: University of Arizona Press, 2006).

10. Keay Davidson, *Carl Sagan, A Life* (New York: Wiley, 1999), 63.

11. Shapiro, *Origins*, 20.

12. Quoted in David Morrison, "Carl Sagan, 1934–1996," *Biographical Memoirs of the National Academy of Sciences* (2014), 2, http://www.nasonline.org/publications/biographical-memoirs/memoir-pdfs/sagan-carl.pdf.

13. Carl Sagan, Bishun N. Khare, and John S. Lewis, "Organic Matter in the Saturn System," in *Saturn*, edited by Tom Gehrels and Mildred S. Matthews, 788–807 (Tucson: University of Arizona Press, 1984). This paper carries a footnote to the description of *tholin*: "This term, invented by Sagan and Khare is not universally accepted in the literature at this time."

14. Adam P. Johnson et al., "The Miller Volcanic Spark Discharge Experiment," *Science* 322 (2008): 404.

15. Christopher K. Materese et al., "Ice Chemistry on Outer Solar System Bodies: Electron Radiolysis of N_2-, CH_4- and CO-Containing Ices," *Astrophysical Journal* 812 (2015): 150–58.

Chapter 17

1. John Gillespie Magee Jr., "High Flight" (unpublished, 1941). Just weeks after completing this poem, on which rests his chief claim to posthumous fame, Magee died in a mid-air collision over Lincolnshire, England, at the age of only nineteen.

2. There have been many editions of Willy Ley's classic *Rockets, Missiles, and Space Travel* (New York: Viking).

3. William Sheehan, talk at Lowell Observatory on the eve of the New Horizons flyby, July 13, 2015 (unpublished notes).

4. The main advantage of numerical integration methods is that for computers they were very easy to carry out, in contrast to the difficult procedures for determining perturbations. However, they have some disadvantages as well, including the fact that to compute the position of one planet at a particular time, one must compute the positions for all the other planets, including all the intermediate time-steps for each one. In the days when computers were much slower than now, this could require a lot of computing time.

5. Donald P. Hearth, interview with Craig B. Waff, Boulder, Colorado, August 7, 1988; cited in Andrew J. Butrica, "Voyager: The Grand Tour of Big Science," *From Engineering Science to Big Science: The NACA and NASA Collier Trophy Research Project Winners*, edited by Pamela E. Mack, The NASA History Series (Washington, DC: National Aeronautics and Space Administration, 1998), 254.

6. Carl Sagan, *Pale Blue Dot* (New York: Random House, 1994).

7. Patrick Moore, *Guide to the Planets* (New York: Norton, 1955), 174–75.

8. On the other hand, astrologers and science-fiction writers have had an enduring fascination with Pluto, starting immediately after its discovery in 1930. For example, in the December 1931 issue of London-based *The Astrologers' Quarterly*, editor and author Charles Carter gave a list of a dozen previously

unexplained accidents demonstrating "that Pluto is capable, in combination with other planets, of causing severe physical hurt." The February 1940 issue of the pulp magazine *Fantastic Adventures* featured a full-cover image of an astronaut on Pluto and some local inhabitants, with the caption, "This world of cold and eternal twilight would most likely be inhabited by winged bat-people with heavy protecting fur. Details on page 97."

9.   David Jewitt and Jane Luu, "Discovery of the Candidate Kuiper Belt Object 1992 QB$_1$," *Nature* 362 (April 22, 1993): 730–32. The provisional designation reveals that it was the twenty-seventh object found in the second half of August that year. It has also received an asteroid number: 15760.

10.   Cohen, C. J. and Hubbard, E. C., "Libration of the Close Approaches of Pluto to Neptune," *Astrophysical Journal* 70 (1965): 10–13.

11.   Her seminal papers are Renu Malhotra, "The Origin of Pluto's Peculiar Orbit," *Nature* 365 (October 28, 1993): 819–21; "Orbital Resonances in the Solar Nebula: Strengths and Weakness," *Icarus* 106 (1993): 254–73.

12.   Renu Malhotra and J. G. Williams, "Pluto's Heliocentric Orbit," in *Pluto and Charon*, edited by S. Alan Stern and David J. Tholen, 148–49 (Tucson: University of Arizona Press, 1997).

13.   Many papers describe various aspects of the Nice model, but see Alessandro Morbidelli, Harold F. Levison, and Rodney Gomes, "The Dynamical Structure of the Kuiper Belt and Its Primordial Origin," in *The Solar System Beyond Neptune*, edited by M. A. Barucci et al., 275–92 (Tucson: University of Arizona Press, 2008).

14.   The origin of the Pluto-Charon pair is currently thought to have resulted from the collision of the proto-Pluto and a KBO of comparable size early in the development of the Solar System. The damaged Pluto healed as its own gravity pulled it back into the present spherical shape, while Charon formed from the vast cloud of debris liberated in the violent encounter. This scenario, supported by complex computer simulations, explains many of the properties of Pluto and Charon. The author of this work is Robin Canup of the Southwest Research Institute in Boulder, Colorado. Robin Canup, "A Giant Impact Origin of Pluto-Charon," *Science* 307 (2005): 546–50.

15.   In 2001, the National Research Council at the behest of the Space Studies Board and the National Academy of Sciences convened groups of planetary scientists to survey the state of the subject at that time and to formulate a slate of recommendations to NASA and the National Science Foundation (NSF). This exercise was termed the Decadal Survey, and while it was the first of its kind to address planetary science, it was closely patterned after several such surveys that had been conducted by the astronomy and astrophysics community. Six panels of six to fourteen people each studied the topics Mars, the inner planets,

the giant planets, large satellites, primitive bodies, and astrobiology. After several meetings that also incorporated input from the broader community through white papers reviewing and advocating for various specific subjects, facilities, or space missions, they made a prioritized list of recommendations for presentation to the Steering Group. The Steering Group evaluated each report, and in view of programmatic and budget guidelines, engineering readiness, and scientific importance, selected a set of recommendations for the use of NASA and the NSF. For NASA, space missions were the most expensive and prominent single items by far, and for the NSF observatory facilities weighed in significantly.

16. National Research Council, *New Frontiers in the Solar System: An Integrated Exploration Strategy* (Washington, DC: National Academies Press, 2003), 232.

17. Phase B in the NASA mission-approval process is for "design and definition" of the spacecraft and all other mission details. After favorable review and evaluation, a mission can be selected for phase C/D, allowing the team to proceed with fabrication, testing, and delivery of flight hardware, ending in integration with the spacecraft and launch vehicle.

18. The summary description of this mission was "a flyby mission of several Kuiper Belt objects, including Pluto-Charon, to discover their physical nature and understand their endowment of volatiles." National Research Council, *New Frontiers in the Solar System*, 5.

19. Michael J. Neufeld, "How We Got to Pluto," *Physics Today* 69 (2016): 41–47. A more detailed paper by the same author is Michael J. Neufeld, "First Mission to Pluto: Policy, Politics, Science, and Technology in the Origins of New Horizons, 1989–2003," *Historical Studies in the Natural Sciences* 44 (2014): 234–76. See also S. Alan Stern, "The New Horizons Pluto Kuiper Belt Mission: An Overview with Historical Context," *Space Science Reviews* 140 (2008): 3–21.

20. At least as early as 1980, Brian G. Marsden (1937–2010) set in motion his plan to demote Pluto to the status of a minor planet. See Patrick Moore, "The Golden Year of the Ninth Planet," *Journal of the British Astronomical Association* 90 (1980): 376–81. Marsden was for many years the head of the Minor Planet Center (MPC), the clearing house for asteroid and comet discoveries where the orbits are calculated and revised as new observations come in. Marsden took the imperious step of assigning to Pluto the minor planet number 134340, a completely nondescript number that itself is a snub to Pluto, its discoverer, and those who argued for keeping its planetary status. The MPC is co-located with the IAU Central Bureau for Astronomical Telegrams (CBAT), by which new discoveries are communicated to scientists. Marsden was also in charge of the CBAT.

21. In an official, though whimsical act of comic futility, the California State Assembly passed HR 36 in the closing days of the 2005–6 session condemning

"the International Astronomical Union's decision to strip Pluto of its planetary status for its tremendous impact on the people of California and the state's long term fiscal health."

Chapter 18

1. Hal A. Weaver et al., "Discovery of Two New Satellites of Pluto," *Nature* 439 (February 22, 2006): 943–45.
2. By convention, newly discovered objects receive their provisional designations according to the year, week, and sequential number in which they are found. Sometimes newly discovered objects are "lost" because an insufficient number of observations are available to discern the path of motion across the sky, and in the case of extremely faint objects, attempts to find them again are not always successful. This is the reason that the IAU requires that they be observed sufficiently well to have their orbits confirmed. Only then do they receive permanent designations and become eligible for naming.

Chapter 19

1. Four moons of Jupiter, nine of Saturn, five of Uranus, and one of Neptune.
2. William B. McKinnon et al., "Convection in a Volatile Nitrogen-Ice-Lich Layer Drives Pluto's Geological Vigour," *Nature* 534 (June 2, 2016): 82–85, https://doi.org/10.1038/nature18289.
3. Francis Nimmo et al., "Reorientation of Sputnik Planitia Implies a Subsurface Ocean on Pluto," *Nature* 549 (December 1, 2016): 94–96.
4. Tidal interactions between Io, Europa, Ganymede, and Jupiter cause the interiors of these satellites to be heated, which makes Io experience strong and continuing volcanic activity and creating below the ice crust of Europa a global ocean of liquid water. Liquid water may also exist inside Ganymede.
5. Robin Canup, "On a Giant Impact Origin of Charon, Nix, and Hydra," *Astronomical Journal* 141 (2011): 35.
6. Some planetary scientists assert that certain geological structures on Titan (e.g., Doom Mons), and that plumes seen on Triton by Voyager 2 in 1989 fall in a broad definition of cryovolcanic activity. Also, ejections of gas and ice particles from Saturn's moon Enceladus and the more recently recognized plumes emanating from the south polar region of Jupiter's moon Europa qualify as cryovolcanic phenomena.
7. Jeffrey M. Moore et al., "The Geology of Pluto and Charon Through the Eyes of New Horizons," *Science* 351, no. 6279 (2016): 1284–293, https://doi.org/10.1126/science.aad7055.

8. Precession is the slow change in orientation of a planet's rotational axis. Currently Earth's axis is tilted 23.4° from the plane of the orbit and the orientation of the axis in space changes in a cycle about 26,000 years long. This is called the precession of the equinoxes.

9. William M. Grundy et al., "Surface Compositions Across Pluto and Charon," *Science* 351, no. 6279 (2016): 1283 and supplemental material, https://doi.org/10.1126/science.aad9189.

10. Underlying these calculations is the realization that, as a matter of plain geometry, if a planet is tilted by more than 54° on its axis, the total amount of sunlight that a point on the north or south pole absorbs over a year is greater than the total amount of sunlight that a point on its equator absorbs over the same time. For such planets, the equatorial region will thus be colder on average than the poles. Uranus is another planet with a highly tilted axis, at 82.2°.

   In the case of Earth, the obliquity would vary chaotically from 0° to 85° were it not for the presence of the Moon. See J. Laskar, F. Joutel, and P. Robutel, "Stabilization of the Earth's Obliquity by the Moon," *Nature* 361 (January 28, 1993): 615–17. The Moon, which has served as the great climate regulator for Earth, is thought to be an accident of accretion, formed by a blow—whether glancing or direct is currently debated—by a Mars-size planetesimal early in the history of the Solar System. For how Earth's obliquity and thus climate would have varied without a moon, see G. E. Williams and James Kasting, "Habitable Planets with High Obliquities," *Icarus* 129 (1997): 254–67. These authors used an energy-balance climate model to simulate Earth's climate at obliquities up to 90° to show that Earth's climate would become regularly severe in such circumstances, with large seasonal cycles and accompanying temperate extremes in mid- and high latitudes. Such conditions would surely have been damaging to many forms of life. Earth-like planets, with a moon and corresponding small obliquities, may be extremely rare in the cosmos, with obvious implications for the existence of life.

11. John Milton, in *Paradise Lost*, attributes the tip of Earth's axis as punishment, meted out by angels, for Adam and Eve's sin in the Garden of Eden (Book X, lines 668–706). In addition to pushing Earth over on its axis, the angels, according to Milton, also deformed the hitherto circular orbit of Earth, making it eccentric. The angels could have done far worse by tilting Earth's axis even more and eliminating the Moon.

12. At present, Mars's axial tilt, or obliquity—25.2° from the perpendicular—is very nearly the same as Earth's. The current agreement is, however, a coincidence. Both Earth and Mars bulge slightly at the equator because of the centrifugal force of their rotations. The gravitational pull of the Sun on these equatorial bulges causes the axial tilts of both Earth and Mars to vary over time. But Earth's axial tilt

is largely stabilized by the presence of the Moon, and so ranges through only 4°. Mars, which lacks a large and massive satellite, wobbles in much more extreme fashion—at the current epoch, its axial tilt ranges between extremes of 15° and 35° over a period of 120,000 years, with the present value lying close to the mean.

13. Richard P. Binzel et al., "Climate Zones on Pluto and Charon," *Icarus* 287 (2017): 30–36.

14. G. Randall Gladstone et al., "The Atmosphere of Pluto as Observed by New Horizons," *Science* 351, no. 6279 (2016), https://doi.org/10.1126/science.aad8866.

15. Measurements made with the Atacama Large Millimeter/submillimeter Array (ALMA) in Chile; E. Lellouch et al., "(134340) Pluto," *IAU Circular* 9273 (2015): 1.

16. See chapter 18.

17. Frances Bagenal et al., "Pluto's Interaction with its Space Environment: Solar Wind, Energetic Particles, and Dust," *Science* 351, no. 3512 (2016): https://doi .org/10.1126/science.aad9045.

18. Nicolas Altobelli et al., "Flux and Composition of Interstellar Dust at Saturn from Cassini's Cosmic Dust Analyzer," *Science* 352, no. 6283 (2016), https://doi .org/10.1126/science.aac6397.

19. Organa crater is about 5 km (3 miles) in diameter, and appears relatively fresh because it is surrounded by a pattern of surface material ejected by the impact of the body that created it—splash marks.

20. G. Randall Gladstone et al., "The Atmosphere of Pluto as Observed by New Horizons."

21. William M. Grundy et al., "Formation of Charon's Red Polar Caps," *Nature* 539 (November 10, 2016): 65–68. Colored tholins can be made from methane alone, but the addition of nitrogen greatly expands the range of constituent chemicals; the effect of nitrogen on the color isn't clear from laboratory experiments conducted so far.

22. Robin Canup, "On a Giant Impact Origin of Charon, Nix, and Hydra."

23. Mark R. Showalter and Douglas P. Hamilton, "Resonant Interactions and Chaotic Rotation of Pluto's Small Moons," *Nature* 522 (May 25, 2015): 45–49.

24. Hal A. Weaver et al., "The Small Satellites of Pluto as Observed by New Horizons," Science 351, no. 6279 (2016), https://doi.org/10.1126/science.aae0030.

Chapter 20

1. Carl Sagan, *The Cosmic Connection* (New York: Doubleday, 1973), 69.

2. The Apollo 11 lunar landing craft that carried the first astronauts to the surface of the Moon bears a plaque with the statement, "Here men from the planet Earth first set foot upon the Moon, July 1969, A.D. We came in peace for all mankind."

3.   Kenneth E. Edgeworth, "The Evolution of Our Planetary System," *Journal of the British Astronomical Association* 53 (1943): 181–88.

4.   Kenneth E. Edgeworth, "The Origin and Evolution of the Solar System," *Monthly Notices of the Royal Astronomical Society* 109 (1949): 600–609. The same words appear in appendix 6 of Edgeworth's 1961 book, *The Earth, the Planets, and the Stars* (London: Chapman & Hall).

5.   Jan Oort, "The Structure of the Cloud of Comets Surrounding the Solar System and a Hypothesis Concerning Its Origin," *Bulletin of the Astronomical Institutes of the Netherlands* 11 (1950): 91–110. Later researchers have refined Oort's calculations and reduced the size of the cloud. Thus, according to Marsden, Sekanina, and Everhart, the average value for orbital semimajor axes for long period comets is about 22,000 AU, giving for the average extent of the Oort cloud twice this distance, or about 44,000 AU. See Brian G. Marsden, Zdenek Sekanina, and Edgar Everhart, "New Osculating Orbits for 110 Comets and Analysis of Original Orbits for 200 Comets," *Astrophysical Journal* 83 (1978): 64–71.

6.   Fred L. Whipple, "A Comet Model. I. The Acceleration of Comet Encke," *Astrophysical Journal* 111 (1950): 375–94.

7.   See, for instance, H. U. Keller et al., "Deep Impact Observations by OSIRIS Onboard the Rosetta Spacecraft," *Science* 310 (2005): 281–83.

8.   Edgeworth, *The Earth, the Planets, and the Stars*, 19.

9.   Chandrasekhar, of course, was a colleague at Yerkes. Oort had not only been a professor at the University of Leiden, where Kuiper had studied, he was in residence at Yerkes in the fall of 1947, where he gave a seminar on the structure of the Galaxy that contains many ideas that could be applied to the origin of the Solar System. Von Weizsäcker himself came to the University of Chicago as a visiting professor in the Committee on Social Thought.

10. Kenneth E. Edgeworth, "Origin of the Solar System," *Nature* 153 (March 11, 1944): 140–41.

11. Kuiper, "On the Origin of the Solar System," in *Astrophysics*.

12. John Davies, *Beyond Pluto* (Cambridge: Cambridge University Press, 2001).

13. Richard P. Binzel, preface to *The Solar System Beyond Neptune*, edited by Maria A. Barucci et al. (Tucson: University of Arizona Press, 2008), xix.

14. David Jewitt, foreword to *The Solar System Beyond Neptune*, edited by Maria A. Barucciet al. (Tucson: University of Arizona Press, 2008), xv–xvii.

15. Ibid.

16. On the putative "ninth planet," see Chadwick A. Trujillo, and Scott S. Sheppard, A Sedna-like Body with a Perihelion of 80 Astronomical Units," *Nature* 507 (February 28, 2014): 471. See also Konstantin Batygin and Michael E. Brown, "Evidence for a Distant Giant Planet in the Solar System," *Astrophysical Journal* 151, no. 2 (2016): 22.

17. The phase angle is the angle between the Sun and the observer, measured from the object. For most KBOs the maximum phase angle observable from Earth-based telescopes is about one degree, which is too small for determining the details of the surface microstructure of the bodies. Pluto's maximum phase angle seen from Earth is just under two degrees.

18. Radioactive elements in the early solar nebula contributing to the internal heating of accreted bodies include aluminum-26 (^{26}Al), which rapidly decays to a stable isotope, as well as potassium-40 (^{40}K), uranium-238 (^{238}U), uranium-235 (^{235}U), and thorium-232 (^{232}Th), all of which decay much more slowly and contribute to longer term heating.

19. Some of that dust filters into the Solar System from elsewhere in the Galaxy. These interstellar dust particles are found in some primitive meteorites.

Index

Note, Place names on Pluto with asterisk (*) are approved by the IAU at press time. Other names for features on Pluto and Charon are provisional.

About the Authors

Dale P. Cruikshank received his PhD from the University of Arizona in the Lunar and Planetary Laboratory in 1968 under Gerard Kuiper and the eminent geologist Spencer R. Titley. He is currently an astronomer and planetary scientist at NASA Ames Research Center and a coinvestigator on the Cassini-Huygens mission and NASA's New Horizons mission to Pluto and the Kuiper Belt. He is the editor of the University of Arizona Press book *Neptune and Triton* (1995) and has contributed articles in the journals *Icarus*, *Astronomy and Astrophysics*, *Comptes Rendus Physique*, and many others. In 2006 he received the Kuiper Prize of the Division for Planetary Sciences. He was awarded NASA's Exceptional Scientific Achievement Medal in 1994 and again in 2017, as well as NASA's Exceptional Service Medal in 2006. He is an elected fellow of the California Academy of Sciences, the American Geophysical Union, and the American Association for the Advancement of Science.

William Sheehan received his MD degree from the University of Minnesota in 1987, and completed his residency in psychiatry at the University of Minnesota. Professionally, he is a psychiatrist in private practice in Flagstaff, Arizona. He is also a well-known amateur astronomer, historian of science, and author of twenty books, including the following titles for the University of Arizona Press: *Planets and Perception: Telescopic Views and Interpretations* (1988), *Worlds in the Sky: Planetary Discovery from Earliest Times Through Voyager and Magellan* (1992), and *The Planet Mars: A History of Observation and Discovery* (1996). Among his other books are *The Immortal Fire Within: The Life and Work of Edward Emerson Barnard* (Cambridge University Press, 1995), *Galactic Encounters: Our Majestic and Evolving Star-System, from the Big Bang to Time's End* (Springer, 2015), and *Celestial Shadows: Eclipses, Transits and Occultations* (Springer, 2015). He is a regular contributor to *Sky and Telescope*, and serves as a member of the International Astronomical Union's Working Group for Planetary System Nomenclature (WGPSN). Asteroid 16037 has been named "Sheehan" in his honor. He won a Guggenheim fellowship in 2001, and in 2004 received the Gold Medal of the Oriental Astronomical Association for his outstanding work on Mars, notably for his research on and recognition of important contributions by Japanese students of the red planet.